PATHWAYS TO MODERN PHYSICAL CHEMISTRY

An Engineering Approach with Multidisciplinary Applications

PATHWAYS TO MODERN PHYSICAL CHEMISTRY

An Engineering Approach with Multidisciplinary Applications

Edited by

Rainer Wolf, PhD
Gennady E. Zaikov, DSc
A. K. Haghi, PhD

AAP | APPLE ACADEMIC PRESS

Apple Academic Press Inc.
3333 Mistwell Crescent
Oakville, ON L6L 0A2
Canada

Apple Academic Press Inc.
9 Spinnaker Way
Waretown, NJ 08758
USA

Exclusive worldwide distribution by CRC Press, a member of Taylor & Francis Group

No claim to original U.S. Government works

Printed in the United States of America on acid-free paper

International Standard Book Number-13: 978-1-77188-322-1 (Hardcover)

International Standard Book Number-13: 978-1-77188-323-8 (eBook)

Library and Archives Canada Cataloguing in Publication

Pathways to modern physical chemistry : an engineering approach with multidisciplinary applications / edited by Rainer Wolf, PhD, Gennady E. Zaikov, DSc, A.K. Haghi, PhD.

Includes bibliographical references and index.
Issued in print and electronic formats.
ISBN 978-1-77188-322-1 (hardcover).--ISBN 978-1-77188-323-8 (pdf)

1. Chemistry, Physical and theoretical. 2. Polymers. 3. Textile industry. 4. Materials. 5. Chemical engineering. I. Haghi, A. K author, editor II. Zaikov, G. E. (Gennadi™i Efremovich), 1935-, author, editor III. Wolf, Rainer, 1935-, editor

QD453.3.P38 2016 541 C2016-905116-1 C2016-905117-X

Library of Congress Cataloging-in-Publication Data

Names: Wolf, Rainer (Chemist), editor. | Zaikov, G. E. (Gennadiæi Efremovich), 1935- editor. | Haghi, A. K., editor.
Title: Pathways to modern physical chemistry : an engineering approach with multidisciplinary applications / editors, Rainer Wolf, PhD, Gennady E. Zaikov, DSc, A.K. Haghi, PhD.
Description: Toronto : Apple Academic Press, 2017. | Includes bibliographical references and index.
Identifiers: LCCN 2016034083 (print) | LCCN 2016036107 (ebook) | ISBN 9781771883221 (hardcover : alk. paper) | ISBN 9781771883238 (ebook)
Subjects: LCSH: Chemistry, Physical and theoretical. | Chemistry, Technical. | Chemical engineering.
Classification: LCC QD453.3 .P38 2017 (print) | LCC QD453.3 (ebook) | DDC 541--dc23
LC record available at https://lccn.loc.gov/2016034083

Apple Academic Press also publishes its books in a variety of electronic formats. Some content that appears in print may not be available in electronic format. For information about Apple Academic Press products, visit our website at **www.appleacademicpress.com** and the CRC Press website at **www.crcpress.com**

LIST OF CONTRIBUTORS

M. I. Artsis
N.M. Emanuel Institute of Biochemical Physics, Russian Academy of Sciences, 4 Kosygin str., Moscow 119334, Russia

A. A. Berlin
Semenov Institute of Chemical Physics RAS, Kosygin Str. 4, Moscow 119991, Russia

S. N. Bondarenko
Volzhskii Polytechnic Institute, Branch of Volzhskii State Technical University, ul. Engel'sa 42a, Volzhskii, Volgogradskaya oblast, Russia

A. D. Borkar
Department of Chemistry, Nabira Mahavidyalaya, Katol, Dist. Nagpur 441302 (M.S) India

A. V. Bychkova
Emanuel Institute of Biochemical Physics, Russian Academy of Sciences, Moscow, Russia.

T. I. Chalykh
Plekhanov Russian University of Economics

Rafael Askhatovich Cherkasov
Kazan Federal University, The Russian Federation, 420008, Kazan.

Nodar Chkhubianishvili
Georgian Technical University

Rustam Rifkhatovich Davletshin
Kazan Federal University, The Russian Federation, 420008, Kazan

Natalia Viktorovna Davletshina
Kazan Federal University, The Russian Federation, 420008, Kazan

R. Ja. Deberdeev
Kazan National Research Technological University, Kazan, Russia

R. Ya. Deberdeev
Kazan National Research Technological University, Kazan, Russia

T. R. Deberdeev
Kazan National Research Technological University, Kazan, Russia

Zibin Dimitry I.
Moscow University of Fine Chemical Technology, Vernadsky Prospekt, 86, Moscow, 119571 Russia

A. A Efremova
Kazan National Research Technological University, Kazan, Russia

Airat Rizvanovich Garifzyanov
Kazan Federal University, The Russian Federation, 420008, Kazan

R. M. Garipov
Kazan National Research Technological University, Kazan, Russia

L. Yu. Gubaidullin
Bashkir Branch of the USSR Academy of Sciences , Ufa, Russia

K. Z. Gumargalieva
N.N. Semenov Institute of Chemical Physics, Russian Academy of Sciences, 4 Kosygin str., Moscow 119991, Russia

A. K. Haghi
Department of Textile Engineering, Faculty of Engineering, University of Guilan, P.O. Box: 3756, Rasht, Iran

A. A. Idiyatova
Kazan National Research Technological University, Kazan, Russia

A. A. Ijagbuji
Institute of Technology, East Ukrainian National University, Severodonetsk, 93400, Ukraine.

V. I. Irzhak
Russian Academy of Sciences, Moscow, Russia

Max B. Jefferey
University of Melbourne, Parkville, Victoria, Australia

V. F. Kablov
Volzhskii Polytechnic Institute, Branch of Volzhskii State Technical University, ul. Engel'sa 42a, Volzhskii, Volgogradskaya oblast, Russia

I. G. Kalinina
N.N. Semenov Institute of Chemical Physics, Russian Academy of Sciences, 4 Kosygin str., Moscow 119991, Russia

S. G. Karpova
Russian Academy of Sciences, Moscow, Russia

N. A. Keybal
Volzhskii Polytechnic Institute, Branch of Volzhskii State Technical University, ul. Engel'sa 42a, Volzhskii, Volgogradskaya oblast, Russia

Yu. N. Khakimullin
Kazan National Research Technological University, Kazan, Russia

A. I. Khasanov
Kazan National Research Technological University, Kazan, Russia

I. A. Kirsh
Moscow State University of Food Production

V.I. Kodolov
Basic Research-Educational Center of Chemical Physics and Mesoscopy, UdSC, UrD, RAS

Komova Ninel N.
Moscow University of Fine Chemical Technology, Vernadsky Prospekt, 86, Moscow, 119571 Russia

G. A. Korablev
Izhevsk State Agricultural Academy

Sergey Alekseevich Koshkin
Kazan Federal University, The Russian Federation, 420008, Kazan

A. L. Kovarskii
Emanuel Institute of Biochemical Physics, Russian Academy of Sciences, Moscow, Russia.

Lali Kristesashvili
Georgian Technical University

V. V. Kurnosov
B Verkin Institute for Low Temperature Physics and Engineering

A. G. Liakumovich
Kazan National Research Technological University, Kazan, Russia

L. P. Lyusova
Russian Academy of Sciences, Moscow, Russia

Sh. Maghsoodlou
Department of Textile Engineering, Faculty of Engineering, University of Guilan, P.O. Box: 3756, Rasht, Iran

Williams W. McCarthy
University of Melbourne, Parkville, Victoria, Australia

V. S. Minkin
Russian Academy of Sciences, Moscow, Russia

Yu. V. Minkina
Russian Academy of Sciences, Moscow, Russia

L. F. Nasibullina
Russian Academy of Sciences, Moscow, Russia

Yu. A. Naumova
Russian Academy of Sciences, Moscow, Russia

O. V. Oshchepkov
Russian Academy of Sciences, Moscow, Russia

T. C. Philips
Institute of Technology, East Ukrainian National University, Severodonetsk, 93400, Ukraine.

D.A. Pomogova
Moscow State University of Food Production

A. A. Popov
Russian Academy of Sciences, Moscow, Russia

S. Poreskandar
Department of Textile Engineering, Faculty of Engineering, University of Guilan, P.O. Box: 3756, Rasht, Iran

E. V. Poshtarëva
Institute of Technology, East Ukrainian National University, Severodonetsk, 93400, Ukraine

P. G. Pronkin
Emanuel Institute of Biochemical Physics, Russian Academy of Sciences, Moscow, Russia.

F. Raeisi
Department of Textile Engineering, Faculty of Engineering, University of Guilan, P.O. Box: 3756, Rasht, Iran

A. N. Reisser
Moscow State University, Moscow, Russia

M. A. Rosenfeld
Emanuel Institute of Biochemical Physics, Russian Academy of Sciences, Moscow, Russia

K. U. Rudenko
Volzhskii Polytechnic Institute, Branch of Volzhskii State Technical University, ul. Engel'sa 42a, Volzhskii, Volgogradskaya oblast, Russia

V. V. Schwarzkopf
Moscow State University, Moscow, Russia

Je. A. Sergeeva
Russian Academy of Sciences, Moscow, Russia

V. N. Serova
Kazan National Research Technological University, Kazan, Russia

O. N. Sorokina
Emanuel Institute of Biochemical Physics, Russian Academy of Sciences, Moscow, Russia.

O. V. Stoyanov
Kazan National Research Technological University, The Russian Federation, 420015, Kazan.

A. S. Tatikolov
Emanuel Institute of Biochemical Physics, Russian Academy of Sciences, Moscow, Russia.

N. V. Ulitin
N. V. Ulitin of Kazan National Research Technological University, Kazan

R. R. Valyaev
Russian Academy of Sciences, Moscow, Russia

A. I. Zagidullin
Kazan National Research Technological University, Kazan, Russia

G. E. Zaikov
Emanuel Institute of Biochemical Physics, Russian Academy of Sciences, ul. Kosygina 4, Moscow, 119991 Russia

I. I. Zakharov
Institute of Technology, East Ukrainian National University, Severodonetsk, 93400, Ukraine.

L. A. Zimina
N.M. Emanuel Institute of Biochemical Physics, Russian Academy of Sciences, 4 Kosygin str., Moscow 119334, Russia

LIST OF ABBREVIATIONS

ABDM	alkylbenzyl-dimethylammonium
ADSL	asymmetric-digital-signal line
AFM	atomic force microscope
AIBN	azo-bis-izobutironitrile
APT	atomically precise technologies
BP	benzoyl peroxide
BSA	bovine serum albumine
CA	cellulose acetate
CCD	charge-coupled device
CD	cyclic dimer
CNT	carbon nanotubes
CRT	cathode ray tube
CVD	chemical vapor deposition
DA	demulsifying agent
DFT	density functional theory
DLS	dynamic light scattering
DMF	dimethyl formamide
DMF	dimethylformamide
DMSO	dimethyl sulphoxide
DSC	differential scanning calorimetry
DX	Diamet X
EA	endic anhydride
EDLC	electric double-layer capacitor
EEET	electronic excitation energy transfer
EMI	electromagnetic induction
ESR	erythrocyte sedimentation rate
FG	fibrinogen
FTIR	Fourier transform infrared
GDT	gas discharge tube
GNF	graphite nanofibers
$H_2N_2O_2$	hyponitrous acid
H_2O_2	hydrogen peroxide
HDPE	high-density polyethylene
HNO_2	nitrous acid

HNO_3	Nitric acid
HRTEM	high-resolution transmission electron microscopy
HSA	human serum albumine
IBM	research laboratory in Zurich
IR	infrared
ISO	International Organization for Standardization
ITO	indium-tin-oxide
MA	maleic anhydride
MES	magnetic electrospinning
MMD	molecular mass distribution
MnO_2	manganese dioxide
MNP	magnetic nanoparticle
MNS	magnetic nanosystems
MWNTs	multi-walled nanotubes
N_2O	nitrogen protoxide
NMP	1-methyl-2-pyrrolidone
NMR	nuclear magnetic resonance
NST	nanoscience and technology
OEA	oligoesteracrylates
PA	phthalic anhydride
PA	polyamide
PA6	polyamide-6
PAA	polyacrylic acid
PANI	polyaniline
PANI-co-PNPANI	poly(aniline-co-N-phenylaniline)
PCL	polycapro lactone
PE	polyethylene
PEO	polyethylene oxide
PEO	polyethylene oxide
PET	polyethylene terephthalate
PETP	polyethyleneterephthalate
PMMA	polymethyl methacrylate
PMSQ	multifunctional polymethylsilsesquioxane
PP	polypropylene
PS	polystyrene
PSO	polysulfide oligomers
PTFE	polytetrafluoroethylene
PU	polyurethane
PVA	polyvinyl alcohol
PVP	polyvinyl pyrrolidone

SAN	styrene–acrylonitrile copolymer
SEM	scanning electron microscope
SPM	scanning probe microscope
STM	scanning tunneling microscope
SWNTs	single-walled nanotubes
TC	technical committee
TCP	trichloropropane
TGA	thermal gravimetric analysis
THF	tetrahydrofuran
TIS	turbo ion spray
TS	technical specification
TS	transition state
TSC	thermally stimulated current
TSD	thermostimulated depolarization
UPE	unsaturated polyester
VA	vinyl acetate
VC	vinyl chloride
VOC	volatile organic compound
VRH	variable range hopping

LIST OF SYMBOLS

°	Degree
α	Alpha
θ	Theta
%	Percent
γ	Gamma
σ	Sigma
τ	Tau
β	Beta
$\sum$	Summation
δ	Delta
ε	Epsilon
μ	Micro
Γc	Gamma(c)
Π	3.14
Φ	Phi
ω	Omega
η	Eta
∞	Infinity
$>$	Greater-than
!	used for logical negation
$<$	Less-than
$=$	Equal
Δ	Delta
$\approx$	approximately equals to
∂	Partial derivative
λ	Lambda
ψ	Psi
π	3.14
$\sqrt{}$	Square root
ρ	Rho
$\equiv$	Identical to equivalent
f	*Frequency*
$\sim$	Approximately
υ	Nu

$\int$	Integral
*	Denotation
′	Modifier letter prime
″	Modifier letter double prime
$\tilde{A}$	Capital A with tilde

PREFACE

This book focuses on the recent trends in the chemistry and physics of materials for micro- and nanotechnologies. This valuable volume covers the occurrence, synthesis, isolation, production, properties, applications, and modification, as well as the relevant analysis techniques to reveal the structures and properties of different materials.

This volume is intended as a reference for basic and practical knowledge about the synthesis, characterization, and application of materials in micro-and nanoscale for students, engineers, and researchers. This book also centers on the production of selected types of nanofibers and their applications, including the nanocomposites.

A combined theoretical and technical approach is utilized to illuminate the concept of advanced materials for diverse applications.

PART I

Polymer Science

CHAPTER 1

EFFECT OF A PRIMARY AROMATIC AMINE ON PROPERTIES AND STRUCTURE OF HDPE

R. JA. DEBERDEEV, V. V. KURNOSOV, JE. A. SERGEEVA, and O. V. STOYANOV

CONTENTS

ABSTRACT

Low-pressure (high-density) polyethylene (HDPE) produced by gas-phase methods has a number of valuable technical properties, including its usage in protective coatings. Previously,[1] it has been shown that for good properties of coatings on steel on the basis of HDPE, material modifications are necessary. From a practical point of view, polymer modification immediately in the melt in the process of coating formation by mechanical mixing of components is of particular interest. During the work, the action of additives with effects on the properties of HDPE on steel surface coatings and also of their structure and properties was posed.

1.1 INTRODUCTION AND EXPERIMENTS

Powdered nonstabilized HDPE produced by gas-phase method in Kazan Public Joint Stock Company "Organicheskiy Sintez" was used with particle diameter $<315 \mu m$; their characteristics are shown in Table 1-1.

TABLE 1-1 Characteristics of HDPE

Density, kg/m3	**959**
Molecular weight, 10^3	
– average M_w	267
– average M_n	16.3
CH_3-groups/100 carbon atoms	0.6
Content of >C=C< /1000 carbon atoms	0.9
Distribution of >C=C<bonds, %	
– vinylic –CH=CH_2	93
– vinylidenic –C=CH_2	3
– trans vinylenic >C=C<	4
Melting temperature, °C	138
Melt flow index MFI g/10 min	0.55

As base modification, 4,4′-diamino-3,3′-dichlorodiphenylmethane produced at the Beresneky chemical factory (technical name Diamet X [DX] was used, the characteristics of which are shown in Table 1-2 in the following.

TABLE 1-2 Characteristics of 4,4′-diamino-3,3′-dichlorodiphenylmethane

Elementary composition	$C_{13}H_{12}N_2Cl_2$
Molecular weight	267.15
Aggregate state	crystals
Melting temperature, °C	105
Release form	powder

Compositions for coatings were manufactured by mechanical dry blending of components. Samples were prepared by spraying of polymer powder compositions on polyfluoroethylene plastics (for formation of free film) and steel substrate (standard of Russia GOST 18178-72) for subsequent melting in a heating cabinet, and also by various methods of powder technology.

For defectless separation of coatings from metals, anodic dissolution of metallic substrate or cathodic peeling in 0.1 N NaCI solution[2] was used.

An estimate of HDPE structural parameters in the coating was prepared by the following methods: X-ray analysis, infrared spectroscopy (IR), and differential scanning calorimetry (DSC).[3–5]

The structural organization of coatings was studied by optical and electron microscopy using "Docuval", "JEM-100", "Tesla BS-500" microscopes on the butt-end and longitudinal microscopic sections made on ultramicrotome "Tesla BS-490A". These methods were previously described in detail.[6,7]

Molecular characteristics were studied by gel permeation chromatography method on GPC-200 "Waters" chromatograph in *o*-dichlorobenzene at 125°C[8] and by IR.[9–11]

Oxidation of HDPE was studied by DTA, TGA, and also by measuring the kinetics of oxygen adsorption at 200°C and 250 mm Hg initial pressure as described in ref. [12].

Methods of estimation of elasticity modulus, coefficient of volumetric thermal expansion, water resistance, tolerance to cathodic peeling, adhesion strength, and other physico-mechanical characteristics have been described.[1,13,14]

1.2 RESULTS

During the search for additives, mainly substances with some antioxidation propertieswere probed. By changing the concentration of such additives as well as temperature and time conditions of coating formation, we can run the process of HDPE oxidation and act upon the structure of borderline layers and coating as a whole. Moreover, the appearance of functional groups, which are able to react in such substances, defines the possibility of their involvement in the interface polymer—substrate. The result of our research was the discovery that primary aromatic amines, namely DX, generate an efficient modification effect on HDPE. It was found that there are rather wide concentration intervals and temperature/time conditions of coating formation (180–240°C) in which their properties turn out to be more appropriate. Adhesion strength during peeling increases five times compared to those of test HDPE coatings (Figure 1-1), tolerance to cathodic peeling (Figure 1-2) and water resistance (Figure 1-3) increases more than 10 times. A decrease of residual (internal) stress (Figure 1-4) was observed, and an increase of resilience of coating was obtained (Figure 1-5). The impact strength of coating after modification remains constant. DX concentration dependence of HDPE elastic modulus in the coating is described by a curve with a peak in the area of 1%. Coefficient of volumetric thermal expansion dependence on DX concentration is contrary in its character to elastic modulus (Figure 1-6).

Oxygen absorption kinetics with various DX contents shows that the modifier inhibits the thermo-oxidation of HDPE because it belongs to aromatic amines.[15] With increased DX concentration, an increase of thermo-oxidation induction period and a decrease of oxidation rate after the induction period can be observed (Figure 1-7).

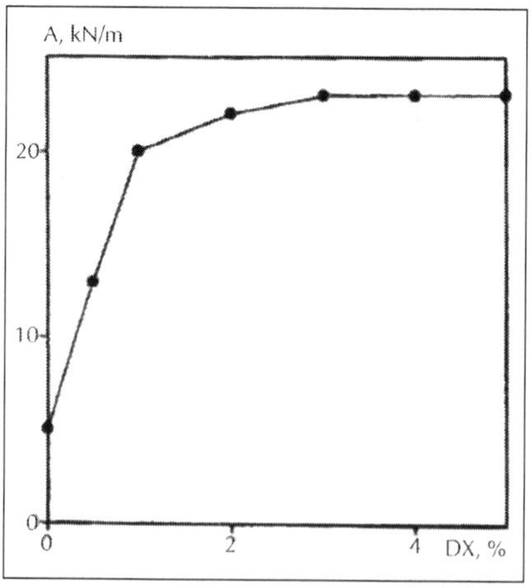

FIGURE 1-1 DX content dependence of adhesion strength A of HDPE coatings formed at 220°C in 20 min.

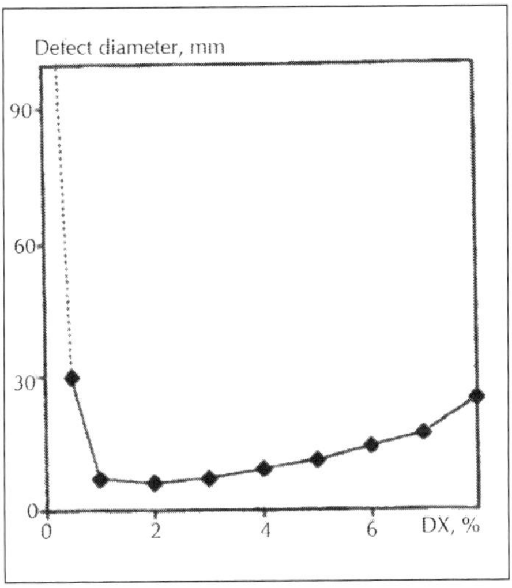

FIGURE 1-2 DX content dependence of defect diameter during cathodic peeling for coatings formed at 220°C in 20 min.

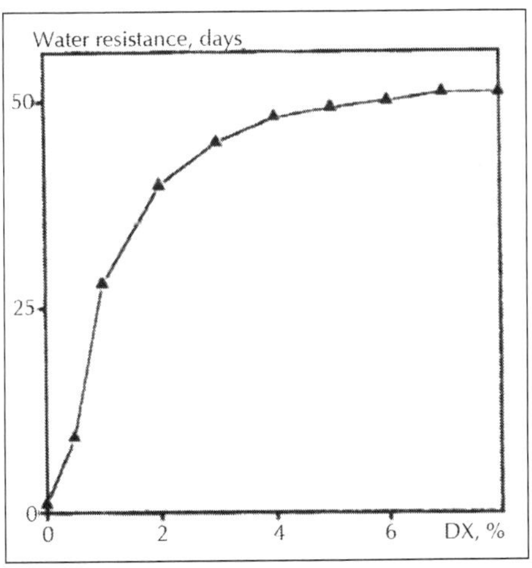

FIGURE 1-3 DX content dependence of water resistance of coatings formed at 220°C during 20 min.

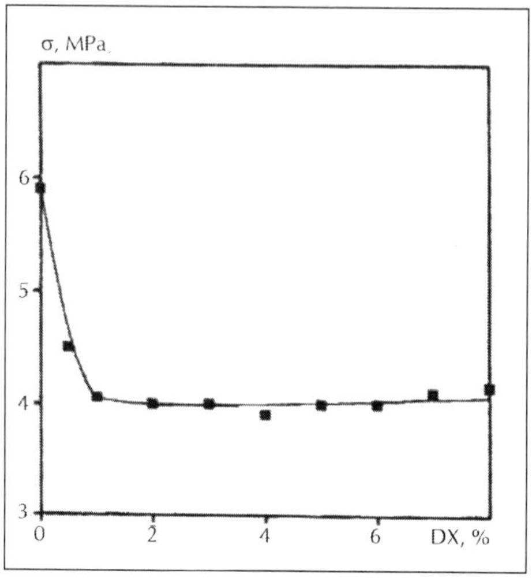

FIGURE 1-4 DX content dependence of residual (internal) stress 6 for coatings formed at 220°C during 20 min.

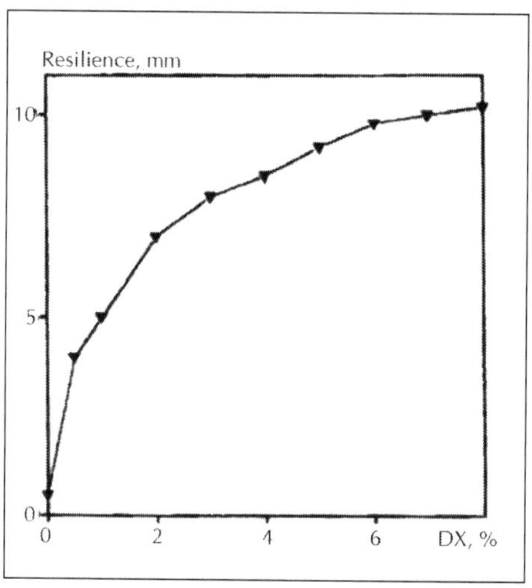

FIGURE 1-5 DX content dependence of resilience, according to Eriksen, of coatings formed at 220°C during 20 min.

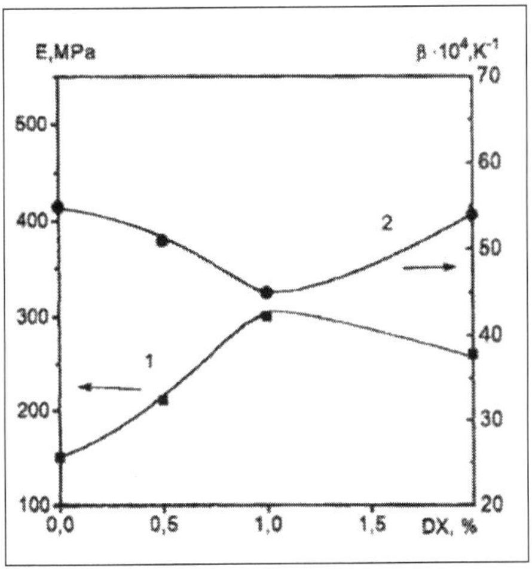

FIGURE 1-6 DX content dependence of elasticity modulus E (1) and coefficient of volumetric thermal expansion B (2).

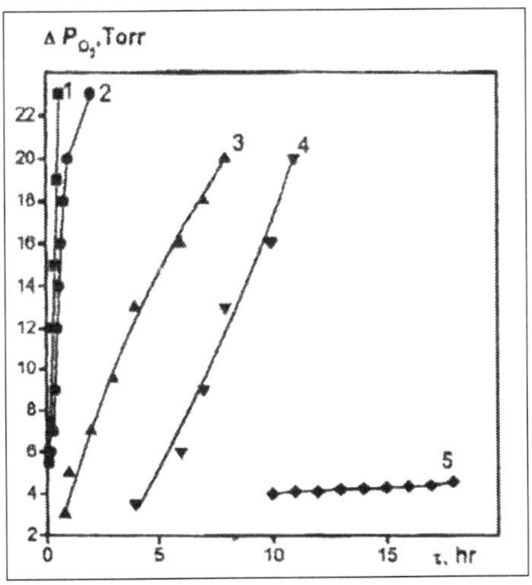

FIGURE 1-7 Kinetics of oxygen absorption (200°C) of initial HDPE (1) and with 0.2% (2), 1% (3), 2% (4), 6% (5) DX content. ΔPo_2 change of oxygen pressure in the oxidation process; τ–oxidation time.

According to DTA and TGA data, there has also been an increase in the polymer thermostability (Table 1-3): the relative heat of oxidation becomes lower, the beginning of thermo-oxidation and weight loss is displaced to higher temperatures.

The estimate of molecular mass distribution (MMD) by GPC shows in Table 1-4 that after the addition of DX to HDPE, the narrowness (as a result of thermo-oxidative destruction) of MMD and molecular weights M_w, M_n, and M_z decrease.

TABLE 1-3 Some characteristics of HDPE thermo-oxidation

Concentration DX (%)	Relative heat of oxidation	Onset temperatures (°C)	
		oxidation	mass loss
0	7,4	178	270
0,2	3,5	181	337
1	1,8	188	337
2	1,4	199	339
3	1,4	211	339

TABLE 1-4 Characteristics of molecular mass distribution of various HDPE samples

Sample	$M_w \cdot 10^3$	$M_n \cdot 10^3$	$M_z \cdot 10^3$	M_w/M_n	M_z/M_w
Initial HDPE	267	16,3	1852	16,4	6,9
HDPE coating	108	12,7	600	8,5	5,6
HDPE coating with 1% of DX	122	21,4	756	9,9	6,2

Chemical processes in the polymer during the stage of coating formation are manifested by some changes in chemical structural characteristics like the quantity of unsaturated bonds and oxygen-containing groups, branching of macromolecules. As can be observed from Figure 1-8, increased DX concentration leads to a loss of a number of carbonyl groups in the process of thermo-oxidation during coating formation; this is consistent with the information about antioxidative properties of the modifier mentioned earlier. Also a decrease of double bond concentration was found in the coatings (Figure 1-8). Thus, while it was established that DX inhibits HDPE thermo-oxidation, the changes of double bonds quantity cannot be explained by the disappearance of these groups during thermo-oxidation. The probable explanation of such dependence is an assumption of the existence of chemical interaction with double bonds of the modifier.

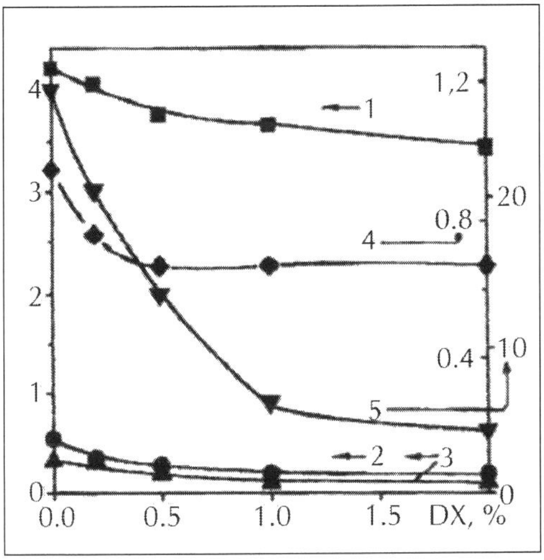

FIGURE 1-8 DX content dependence of > C = C < bonds/1000 carbon: vinylic (1), transvinylene (2), vinylidenic (3) groups; CH_3-groups/100 carbon atoms (4) and an index of absorption of carbonyl groups (5, cm^{-1}).

As shown by investigations carried out by optical and electron microscopy, the structural organization of coating is in excellent agreement with a six-layer model described recently.[8] However, the presence of the modifier and chemical changes have to account for special features in the supermolecular structures of modificated coatings compared with non-modificated samples. Thus, in layers in the vicinity of the walls after addition of DX, the appearance of transcrystalline layers and decline of the weak buffer layer thickness were observed (Figure 1-9). The structure of HDPE in the center layer of the coating which is specific for the polyethylene lamellar structure also shows some changes: The rise of modifier concentration leads to the appearance of smaller structural elements (Figure 1-10). It is necessary to note that the structural changes in the center layer take place without changes of the degree of crystallinity estimated by various methods, density, and other structural parameters (Table 1-5). In superficial layers during the stage of coating formation, the autocatalytic oxidation of polymer is mainly concentrated[16] and the amorphous structure is typical for it. Addition of DX leads to partial improvement of this structure.

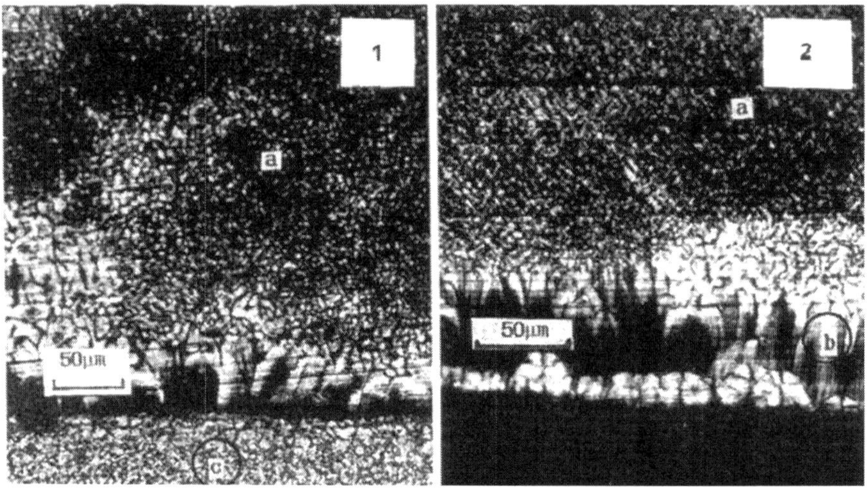

FIGURE 1-9 Morphology of the butt-end microscopic sections of HDPE coatings according to optical microscopy data in polarized light data. (1) nonmodified HDPE, (2) HDPE with 1% DX. a–center layer, b–transcristalline layer, c–buffer layer.

TABLE 1-5 Some structural characteristics of HDPE in coating

DX,%	Degree of crystallnity, %			Effective size of crystallites, nm		Wide period, nm	Melting tempera ture °C	Density, kg/m^3
	X-ray	DSC	IR	<110>	<200>			
0	66	54	60	9,35	7,58	33	132	958
0,2	64	56	60	9,35	7,62	33	132	958
0,5	62	58	61	9,40	7,62	33	134	958
1,0	62	61	62	9,33	7,62	33	132	960
1,5	-	60	63	-	-	-	132	960
3,0	-	61	63	-	-	-	131	960

1.3 DISCUSSION

In the superficial coating layer, the intensity of oxidative crosslinking reactions and the accumulation of oxygen containing groups decrease greatly due to the ability of DX to inhibit thermo-oxidation. As a result, the elasticity of layers and the rate of the relaxation process increase and this contributes to the decrease of residual stress in coating (Figure 1-4).

Formation of fine-crystalline structure in a greater number in the center layer (Figure 1-10) determining mechanical properties of coatings after the addition of DX is responsible for the rise of elastic modulus of polymer in coating.

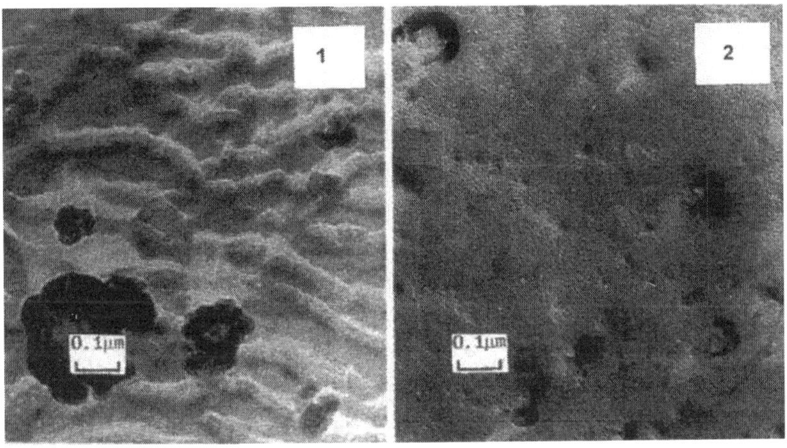

FIGURE 1-10 Electron micrographs of the center layer of HDPE coatings. (1) nonmodificated HDPE. (2) HDPE with 1% of DX.

(Figure 1-6). This effect is identical to that observed for the viscosity of suspensions: When the size of particles decreases at a constant volume concentration, viscosity becomes higher. A concurrent reduction of the coefficient of volumetric thermal expansion indicates that the structure formed during modification is more stable (Figure 1-6). Reduction of the number of polar oxygen-containing groups in this layer of coating, as in the superficial layer, leads to a decrease of residual stress (Figure 1-4). This has a positive influence on the adhesion strength (Figure 1-1) of coating.

Apparently, the formation of the transcrystalline layer (Figures 1-9 and 1-11) surpassing in mechanical strength the residual parts of polymer coating plays an essential role in increasing elastic modulus (Figure 1-6). The increasing cohesion strength takes place with a decrease of the transition (buffer) layer by the modification of HDPE with DX (Figure 1-11). The latter is defined by a reduction of quantity of low-molecular products during thermo-oxidative destruction, appearing into this region in the process of polymer crystallization.[17] Besides that, according to IR data for coatings, formed on an optically clear monocristalline substrate for IR radiation, the introduction of modifier leads to a rise of structural anisotropy in the buffer layer of coating (on the depth ~ 10μm).[18]

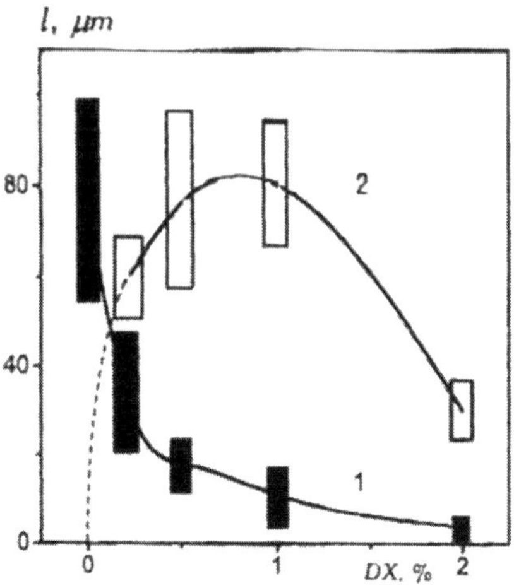

FIGURE 1-11 DX content dependence of transcrystalline layer distance from substrate (1) and its extention (2) 1 in the coating of HDPE.

Reinforcing bond layers of coatings are responsible for an increase of adhesion strength (Figure 1-1) because of the cohesive character of polymer fractures during adhesion strength measurement.

Hence, the modificated DX coatings have both increased adhesion strength because of improvement of layer structure bonding with the substrate and high water-tolerance because of better intensity of interface interaction of coating and substrate. The reason for a high intensity of interface interaction is probably the ability of primary amino groups to participate in donor-acceptor interaction with functional groups of the substrate, but it is necessary to carry out further investigations of this problem.

KEYWORDS

- **Adhesion Strength**
- **Cathodic Peeling**
- **Diamet X**
- **Low-Pressure (High-Density) Polyethylene**
- **Molecular Weight**
- **Thermo-Oxidation**

REFERENCES

1. Deberdeev, R. Ja.; Stoya-nov, O. V.; Huzahanov, R. M.; Kurnosov, V. V. LKM I ikh primenenie. 2 (1989) 52.
2. Shmakova, O. P.; Stepin, S. N.; Deberdeev, R. Ja.; Svetlakov, N. V. Dep. Oniitechim. Cherkassy (1983) 84xp-D83.
3. Martynov, M. A.; Vyle-gjanina, K. A. Rentgenografia polimerov. Khimiya, Leningrad, (1972), p. 216.
4. Godovsky, Ju. K. Thermophysical Methods of Polymer Investigation, Khimiya, Moskva, (1976), p. 216.
5. Godovsky, K. Infrakrasnaya spectroscopia polimerov. Pod. red. I. Dehanta, Khimiya, Moskva, (1976), p. 472.
6. Deberdeev, R. Ja.; Prival-ko, V. P.; Shmakova, O. P.; Bezruk, L. I.; Moisya, Ye. G.; Stoyanov, O.V.; Baranov, V. G. App. Poly. Comp.. 38 (1988) 12.
7. Privalko, V. P.; Deber-deev, R. Ja.; Rymarenko, N. A.; Shmakova, O. P.; Lipatov, Ju. S. Dokl. Akad. Nauk. SSSR. 11 (1986) 44.
8. Belenkyi, B. G.; Vilen-chik, L. Z. Chromatographia polimerov. Khimiya, Moskva, (1978), p. 344.
9. Goldenberg, A. L. App. Poly. Comp. 12 (1960) 59.
10. Goldenberg, A. L. Zh. Prikl. spektrosk. 3 (1973) 510.
11. Shaidullin, R. R.; Vandicova. I. I.; Plamovatyi, A. Ch.; Vahoreit, A. Z.; Morozova, L. G. Dep. VINITI. Moskva (1988) 9144 B88.
12. Shlapnikov, Ju. A.; Miller, V. B.; Neiman, M. B.; Torsueva, E. S.; Gromova, B. A. Vysokomol. soed. 12A (1960) 1409.
13. Mirontsov, L. I.; Privalko, V. P.; Antonov, A. I.; Muslyk, A. F.; Sopina, I. M. Polymer Lab. Report. 19 (1983) 3.
14. Kariakina, M. I. Materials Lab. Report. Moskva, Khimiya, (1977), p. 240.
15. Emanuel, N. M.; Bucha-chenko, A. L. Khimicheskaja phizika starenia i stabilizacii polimerov. Moskva, Nauka, (1982), p. 360.
16. Sirota, F. G. Modifikatsia strukturi i svoistv poliolefinov. Leningrad, Khimiya, (1984), p. 152.
17. Basin, V. E. Adgezionnaja prochnost. Moskva, Khimiya, (1981), p. 208.
18. Krasovsky, A. N.; Kurnosov, V. V.; Stoyanov, O. V.; Polakov, D. A.; Baranov, V. G.;Deberdeev, R. Ja.; Aliner, L. B. Vysokomol. soed. 32A (1990) 2174.

CHAPTER 2

CRITICAL CONVERSION OF CROSSLINKED EPOXYAMINE POLYMERS

T. R. DEBERDEEV, V. I. IRZHAK, R. YA. DEBERDEEV, and O. V. STOYNOV

CONTENTS

ABSTRACT

The simplest and commonest method for the description of the correlation between the structure of a network polymer and the kinetics of a chemical reaction is based on the statistical approach. According to this approach, the structure is defined as a single-valued function of the depth of conversion and the initial composition of the system. This method of calculating the structural parameters has been described in most detail in monographs[1,2] and reviews[3,4]. However, as is well known[2–4] when functional groups change their reactivity (substitution effect) during the process (e.g., during polycondensation with unequal reactivities of primary and secondary amino groups), the polymer structure becomes dependent on the reaction route, that is, the aforementioned unambiguity of the structure conversion relationship is not satisfied. In this case, a rigorous kinetic calculation for all structural elements is required. One should mention that this fact is not yet universally recognized and probabilistic approaches are rather popular (see e.g.[5]).

2.1 INTRODUCTION

It is evident that the known formulas for the critical conversion value derived on the basis of probabilistic approaches cannot characterize systems with a substitution effect. An exact solution of the kinetic problem seems to be rather complicated even when effective computational means are used. However, this solution may be considerably simplified by using the concept of bond blocks.[3] Of note, the pioneering communication on this subject has been recommended for publication by N.S. Enikolopov.[6]

The concept of bond blocks involves a combination of the kinetic and statistical approaches. This combination allows one to propose a generalized method for the estimation of the gel point so that the known formulas appear as special cases under certain conditions. This work concerns the derivation of this equation.

2.2 CONCEPT OF BOND BLOCKS

For the sake of convenience, let us consider a particular system composed of a bifunctional epoxy oligomer and a pentafunctional amine. The presence of primary and secondary amino groups in the latter compound is a prerequisite for the substitution effect.

Just as the statistical calculation, the concept of bond blocks uses the functional fractions of reactive groups rather than the molar concentrations of the reagents: n_i stands for the functional fractions of the corresponding groups capable of reacting with the amine whose functional fraction is denoted as N_j:

$$n_i = \frac{N_i}{\sum_j N_j}$$

where N_i is the molar concentration of the corresponding groups at a given moment (at a given conversion); at the same time $N_i = i\ m_i$ where m_i is the molar concentration of the reagent containing i active functional groups. α_i are the probabilities of the reactions for the corresponding groups (depths of conversion). The balance condition is the following:

$$\sum_i N_i \alpha_i = A\alpha_A$$

where the subscript *i* refers to the epoxy reagents and the subscript A denotes the amine.

All the values of α_i and n_i are found by solving the corresponding sets of kinetic equations.

The network structure is determined by the concentration of junctions of various functionalities, and the junctions are represented by fragments of the curing amine agent. Such structural elements comprise bond blocks, whose concentration should in this case be calculated by the kinetic method, that is, by solving the corresponding kinetic equations. The number of oligomeric fragments attached to a junction characterizes its functionality.

In the case of the bifunctional epoxy oligomer-pentafunctional amine system (e.g., the amine may be represented by diethylenetriamine, or DETA), one should distinguish between primary and secondary amino groups from the very onset of the reaction. Using digits in parentheses to denote the numbers of primary (first digit) and secondary (second digit) hydrogen atoms, we can list the following network-comprising groups: A(41) is the initial amine, junction of zero functionality; A(40) is the amine with a reacted secondary amino group, functionality 1, tail fragment in the network; A(22) is the product of the reaction of one primary amino group, functionality 1, tail fragment in the network; A(21) is a junction of functionality 2, central fragment; A(03) is a junction of functionality 2, central fragment; A(20) is a trifunctional junction; A(02) is a trifunctional junction; A(01) is a tetrafunctional junction; and A(00) is a pentafunctional junction.

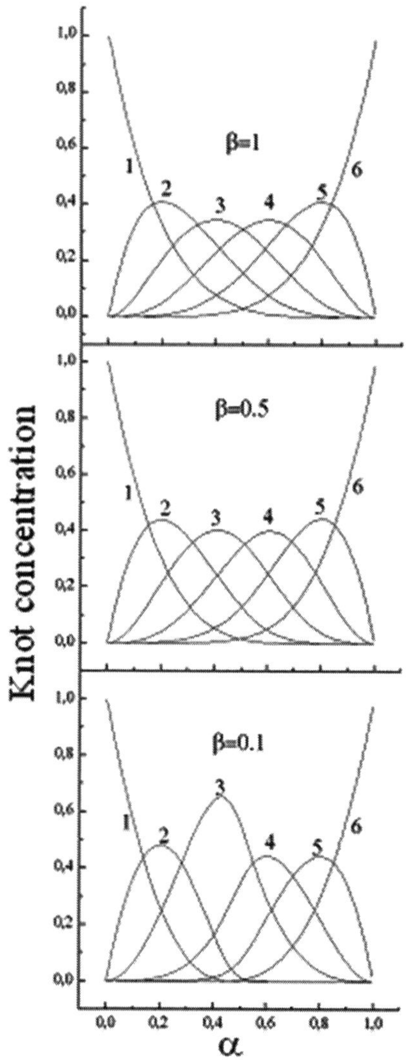

FIGURE 2-1 Concentration of (1) nonfunctional, (2) monofunctional, (3) bifunctional, (4) trifunctional, (5) tetrafunctional, and (6) pentafunctional network junctions vs. conversion at β = (a) 1, (b) 0.5, and (c) 0.1.

If the ratio of the kinetic constants of the secondary and primary amino groups is denoted as $\beta = k_2/k_1$, the set of equations for all the amino groups is written as

$$dB/dt = -B\{4A(41)+4A(40)+2A(22)+2A(21)+2A(20)+$$
$$+\beta\,[A(41)+2A(22)+A(21)+3A(03)+2A(02)+A(01)]\};$$
$$dA(41)/dt = -(4+\beta)BA(41);$$
$$dA(40)/dt = \{\beta A(41) - 4A(40)\}B;$$
$$dA(22)/dt = \{4A(41) - 2(1+\beta)A(22)\}B;$$
$$dA(21)/dt = \{4A(40)+2\beta A(22) - (2+\beta)A(21)\}B; \qquad (2\text{-}1)$$
$$dA(03)/dt = \{2A(22) - 3\beta A(03)\}B;$$
$$dA(02)/dt = \{3\beta A(03)+2A(21) - 2\beta A(02)\}B;$$
$$dA(20)/dt = \{\beta A(21) - 2A(20)\}B;$$
$$dA(01)/dt = \{2\beta A(02)+2A(20) - \beta A(01)\}B;$$
$$dA(00)/dt = \beta A(01)B$$

FIGURE 2-1 shows how the concentration of junctions with various functionalities changes in the course of the reaction. It is characteristic that the concentration-conversion curves corresponding to different p values appear similar: the maximum concentration of junctions with a given functionality corresponds to the same depth of conversion, irrespective of the relative reaction rate constant of the secondary amino group. At the same time, this maximum value itself changes depending on the kinetic parameters of the polycondensation reaction.

Since the reactivity of epoxy groups is assumed to remain unchanged in the course of the reaction, the calculation of the network structure may be performed within the statistical approach[3] using the calculated concentrations of the junctions.

According to the theory of branching processes,[1] the calculation algorithm is the following.

Let function u(x) characterize the probability of the generation of an oligomer chain in the next generation by the same chain in the preceding generation. According to the theory,

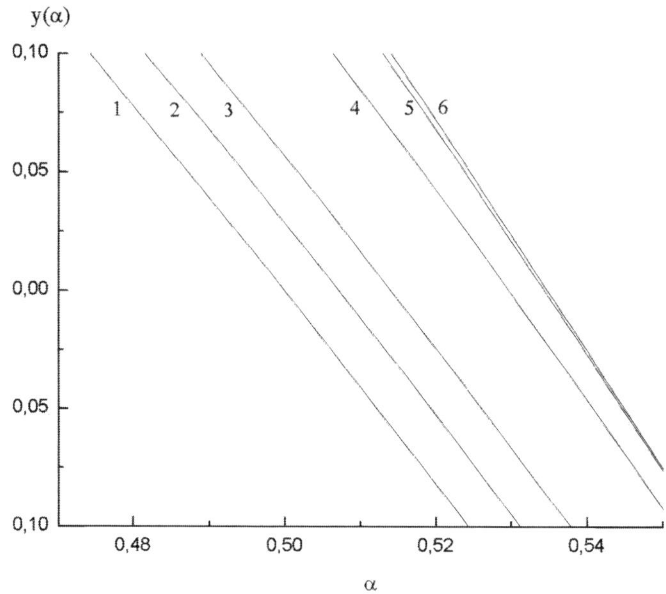

FIGURE 2-2 Graphical solution of equation for the critical conversion at β = (1) 1, (2) 0.7, (3) 0.5, (4) 0.2, (5) 0.1, and (6) 0.01.

$$u(x) = x\left(1-\alpha+\alpha\sum n_{ai}u^{i-1}(x)\right) \qquad (2\text{-}2)$$

where x is an arbitrary variable, α is the depth of conversion with respect to epoxy groups, and n_{ai} is the proportion of i-functional junctions.

Differentiation of Eq. (2-2) leads to the expression for the gelation condition:

$$\frac{\partial u(x)}{\partial x} = \left(1-\alpha+\alpha\sum_i n_{ai}u^{i-1}(x)\right) + x\alpha\sum_i (i-1)n_{ai}u^{i-2}(x) \qquad (2.3)$$

At x = 1, we have u(1) = 1 and

$$u'(1) = \frac{1}{1-\alpha\sum(i-1)n_{ai}} \qquad (2.4)$$

Hence, the critical conversion with respect to epoxy groups is determined as the solution to the equation:

$$\alpha\sum_i (i-1)n_{ai} = 1 \tag{2-5}$$

Formally, Eq. (2-5) coincides with the critical condition $\alpha_c(\bar{f}_w - 1) = 1$ for f-functional polycondensation, with the only difference that the weight-average functionality $\bar{f}_w$ changes during the process in our case; that is, this parameter is a function of the depth of conversion a rather than a characteristic of the initial system.

Figure 2-2 shows the graphical solutions of the equation $y(\alpha) = 1 - \alpha\sum_i (i-1)n_{ai}$ for various β. The α_c value is determined by the condition y(α) = 0.

The values of the critical conversion α_c are presented as functions of the kinetic constant β in Table 2-1.

TABLE 2-1 Dependence of the size of critical conversion (gel-point) from the reactionary ability of a secondary amine (system: diepoxyde – DETA, equifunctional ratio)

β	1	0.7	0.5	0.3	0.1	0.01
α_c	0.5	0.507	0.514	0.524	0.535	0.535

As is seen, the solution depends on β, a conclusion that is not derived in any way from the statistical theory.[1] However, it is not difficult to show that we arrive at the known formula when the requirements of the statistical approach are met for the dependence of n_{ai} on the conversion.

Indeed, according to the statistical approach,

$$A_i = A\frac{f!}{i!(f-i)!}\alpha_A^i(1-\alpha_A)^{f-i} \tag{2-6}$$

where A_i is the concentration of the i-functional junction and f is its maximum possible functionality (for DETA, $f = 5$).

Since $n_{ai} = \dfrac{iA_i}{\sum_i iA_i}$, the substitution of formula (2-6) into Eq. (2-5) yields the following result: $\alpha(f-1)\alpha_A=1$. This expression coincides with the known formula for copolycondensation of bifunctional and f-functional reagents.

Note that the critical conversion at low (but not zero) values of the p constant approaches a constant value of $\alpha_c = 0.535$. This is likely to be related to the fact that the polycondensation process becomes stagewise. At the first stage, the lower the p value, the higher the conversion of primary amino groups. The depth of conversion reaches 40%. The second stage involves the accumulation of polyfunctional junctions in amounts providing the network formation. Evidently, the value of the additional conversion must be quite definite in this case.

As follows from Figure 2-3, the ratio β of the constants exerts a marked effect on the gel point if the reagents are present in nonequivalent amounts.

2.3 CALCULATION OF GEL POINT FOR TERNARY SYSTEMS

2.3.1 MONOFUNCTIONAL MODIFYING AGENT

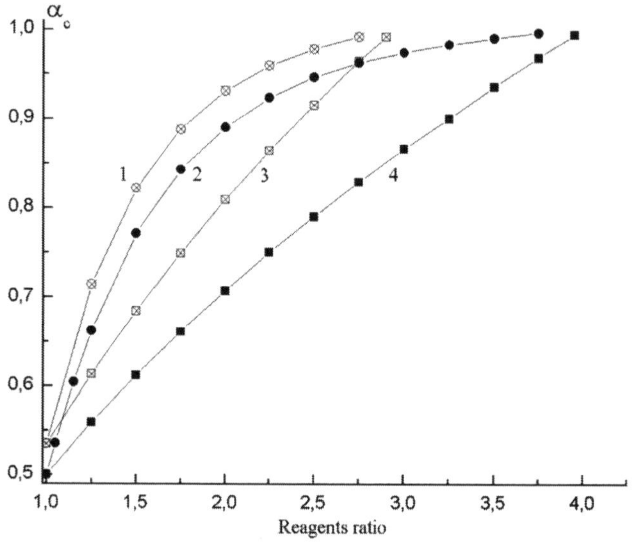

FIGURE 2-3 Critical conversion at various ratios between the (A) amine and (B) epoxy oligomer reagents and β = (1, 3) 0.1 and (2, 4) 1.0. (1, 2) A/B ratio and (3, 4) B/A ratio.

Let us initially consider a monofunctional epoxy compound as the modifying agent for the system under study.[7,8] In this case, the kinetic scheme is similar to the earlier set of equations (2-1). First, however, one should take into account the second epoxy component (B1) and, accordingly, additional kinetic constants; second, junctions should be distinguished not only by the types of their amino groups but also by the types of the added epoxy oligomers. Indeed, the functionality of the junction in this case will depend not on the total number of reacted hydrogen atoms in the amine but only on the number of these atoms that have reacted with a bifunctional reagent.Let the first and second subscripts denote the numbers of the added bi- and monofunctional (tail) oligomers, respectively. Then the $A_{01}(40)$ junction contains one tail and thus has zero functionality. In addition to $A_{00}(41)$, junctions $A_{01}(22)$, $A_{02}(21)$, $A_{02}(03)$, $A_{03}(20)$, $A_{03}(02)$, $A_{04}(01)$, and $A_{05}(00)$ also have zero functionality. Monofunctional junctions are $A_{10}(40)$, $A_{10}(22)$, $A_{11}(21)$, $A_{11}(03)$, $A_{12}(20)$, $A_{12}(02)$, $A_{13}(01)$, and $A_{14}(00)$; bifunctional junctions are $A_{20}(21)$, $A_{20}(03)$, A21(20), A21(02), A22(01), and A23(00); and trifunctional junctions are $A_{30}(20)$, $A_{30}(02)$, $A_{31}(01)$, and $A_{32}(00)$. Junctions $A_{40}(01)$ and $A_{41}(00)$ are tetrafunctional, and junction $A_{50}(00)$ is pentafunctional.

In this case, the condition of gelation is assumed to have the following form:

$$\alpha\sum_{i}(i-1)n_{ai}=1 \qquad \text{(2-5a)}$$

where, just as earlier, the subscript i stands for the functionality of the junction but with allowance for the presence of a monofunctional "chain terminator;" that is, the functionality is equal only to the number of the added bifunctional oligomers.

The strong dependence of the critical conversion on the concentration of the monofunctional modifying agent (see Table 2-2) is due to a decrease in the functionality of junctions, as is shown in Figure 2-4. In contrast to the data presented in Figure 2-1, the curves depicting the changes in the concentrations of junctions of different functionalities are shifted to the right along the abscissa axis; the higher the content of the modifying agent, the more pronounced is the shift. The final state of the system is independent of the kinetic constants in the preceding case; however, the presence of a monofunctional modifying agent, as follows from Figure 2-3, exerts a strong effect on the structure of the formed network so that the gel point cannot be achieved at all at a certain concentration of the modifier.

Similar results are also obtained when cyclocarbonate is used as the modifying agent.[7–9]

The specific kinetic feature of cyclocarbonate is that its functional group reacts with a primary amino group at a ratio of one group per two hydrogen atoms. The molecular structure of DETA initially contains two primary and one secondary amino groups. Therefore, depending on the ratio between the kinetic constants, various reactions of amine hydrogen atoms are possible from the very beginning: reaction of primary group with the carbonate, reaction of primary group with the epoxy group generating a secondary hydrogen atom, and finally reaction of secondary atom with the epoxy group. The kinetic calculation makes it possible to estimate the concentrations of amino groups of various structures:

$$A(41), A(40), A(22), A(21), A(20), A(03), A(02), A(01), A(00), A_1(21), A_1(20), A_1(02), A_1(01), A_1(00), A_2(01), A_2(00).$$

As earlier, the digits in parentheses denote the numbers of primary and secondary hydrogen atoms in the amine fragment. The subscript at A stands for the number of carbonate groups that have reacted with the amine.

Evidently, the first of the aforementioned groups have reacted neither with the epoxy nor with the carbonate group, A(40) and A(22) have reacted with one epoxy group, A(21) and A(03) have reacted with two epoxy groups, A(20) and A(02) have reacted with three groups, A(01) has reacted with four groups, and, finally, A(00) has reacted with five epoxy groups. $A_1(21)$ and $A_2(01)$ contain no fragments of the epoxy oligomer; $A_1(20)$, $A_1(02)$, and $A_2(00)$ have reacted with one epoxy group; $A_1(01)$ has reacted with two epoxy groups; and $A_1(00)$ has reacted with three epoxy groups. The functionality of a junction is equal to the number of the added bifunctional epoxy chains.

TABLE 2-2 Dependence of critical conversion on the content n_1 of the monofunctional component for various relative rate constants of the reactions of the modifying agent with primary (k_1) and secondary (β_2) amino groups

n_1	$k_1 = 1$	$k_1 = 5$	$k_1 = 1$	$k_1 = 0.5$	$k_1 = 0.5$
	$\beta_1 = \beta_2 = 1$	$\beta_1 = \beta_2 = 1$	$\beta_1 = \beta_2 = 0.5$	$\beta_1 = \beta_2 = 1$	$\beta_1 = \beta_2 = 0.5$
0	0.500	0.500	0.514	0.500	0.514
0.1	0.527	0.527	-	0.527	0.542
0.2	0.559	-	0.572	-	0.574
0.3	0.597	-	-	0.597	0.613
0.4	0.645	0.645	0.655	-	0.661
0.5	0.707	-	-	0.707	0.722
0.6	0.790	-	0.794	-	0.804
0.65	0.845	0.845	-	-	0.857
0.7	0.913	-	0.912	0.913	0.923
0.74	0.981	0.981	0.980	-	0.987

Note: β_1 is the relative rate constant of the reaction between the secondary amino group and the epoxy group of the basic oligomer.

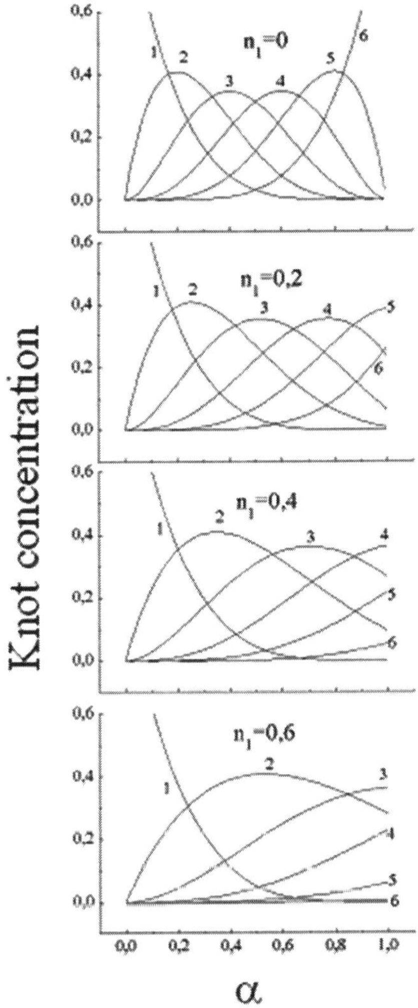

FIGURE 2-4 Concentration of (1) nonfunctional, (2) monofunctional, (3) bifunctional, (4) trifunctional, (5) tetrafunctional, and (6) pentafunctional network junctions vs. conversion at $\beta = 1$ and different n_1.

The critical conversion values calculated through Eq. (2-5) are presented in Figure 2-5.

Even if the reactivity of the cyclocarbonate is fairly high, its ultimate conversion is far from unity. Actually, in spite of having a relatively low reactivity, the epoxy component may react with a hydrogen atom even at early stages of the process and thus disturb the initial equifunctional ratio between the amine and the cyclocarbonate. Evidently, the lower the reactivity of the cyclocarbonate, the more pronounced is this factor. Since under the conditions of incomplete cyclocarbonate conversion, the reaction of the epoxy fragment is complete- the critical conversion but slightly depends on the ratio between the rate constants of the competing reactions.

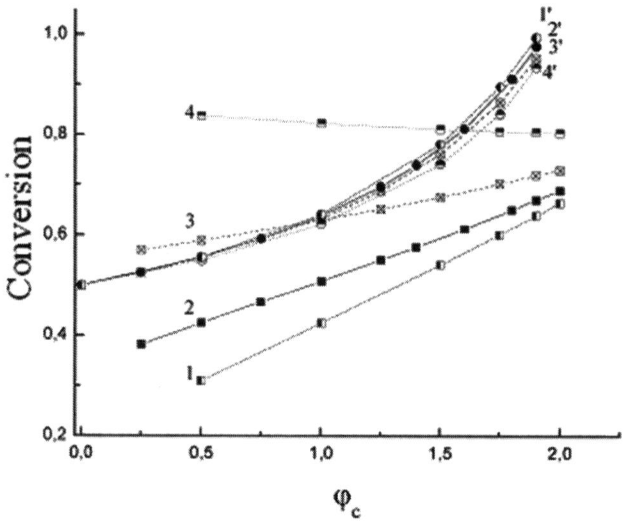

FIGURE 2-5 (1-4) Ultimate conversion of cyclocarbonate groups and (1'-4') critical conversion with respect to epoxy groups vs. relative concentration φ_c of cyclocarbonate groups at the equifunctional total ratio. Ratio of rate constants of the cyclocarbonate and epoxy groups $k_c/k_b = 0.5$ (1- 1'); 1 (2- 2'); 2 (3- 3'); 5 (4- 4').

The relative reactivity of the secondary amino group has a minor effect on the critical conversion value (Figures 2-6 and 2-7) but a notable effect on the ultimate depth of conversion with respect to the cyclocarbonate. As the rate constant of the second addition decreases, the probability of the reaction between the epoxy component and the primary amino group increases and the initial equifunctional cyclocarbonate: amine ratio is disturbed to an increasing extent and at earlier stages.

2.3.2 BIFUNCTIONAL MODIFYING AGENT

In contrast to a monofunctional modifying agent, a bifunctional modifier has no effect on the critical conversion because neither the components terminate network chains nor the functionality of junctions anymore depends on which of the components is attached to an amine fragment. In this case, the critical conversion is determined according to the following equation:

$$\left(\alpha_b n_b + \alpha_2 n_2\right)\sum_i (i-1)n_{ai} = 1 \tag{2-5b}$$

where the subscript b refers to the basic oligomer, the subscript 2 denotes the modifier, and $\left(\alpha_b n_b + \alpha_2 n_2\right) \equiv \alpha_{eff}$. The difference in the reaction rate constants of epoxy groups pertaining to different oligomers, as well as their relative concentrations, has an effect on a_{eff} rather than on the gel point.

Obviously, this conclusion is valid only when we deal with a modifier similar to a bifunctional epoxy oligomer, that is, when a hydrogen atom of the amine is consumed in the reaction of the functional group. When a reagent similar to bifunctional cyclocarbonate is used as the modifier, the situation is changed. The reason is not the termination of network chains, as in the case of monofunctional agents, but the change in the limiting functionality of a junction. Actually, since each carbonate group reacts with two hydrogen atoms, the presence of one carbonate fragment attached to the amine reduces the limiting functionality of the junction by unity, whereas the presence of two fragments reduces it by two units. Therefore, as is shown in Figure 2-7, the critical conversion increases with increasing concentration of a bifunctional modifying agent but this increase is less pronounced than in the case of a monofunctional agent.

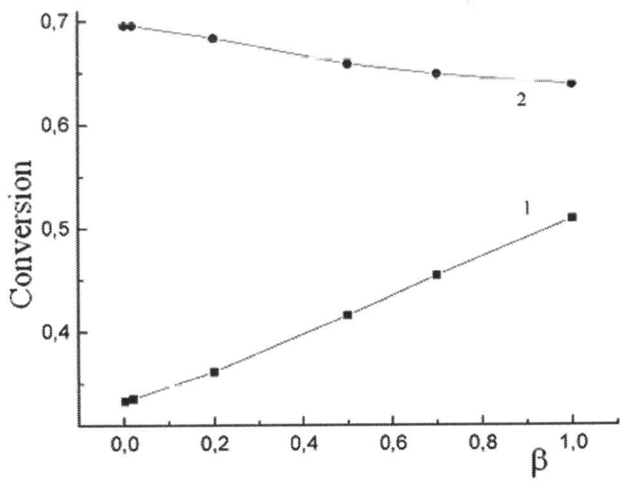

FIGURE 2-6 (1) Ultimate conversion of cyclocarbonate groups and (2) critical conversion with respect to epoxy groups vs. β value.

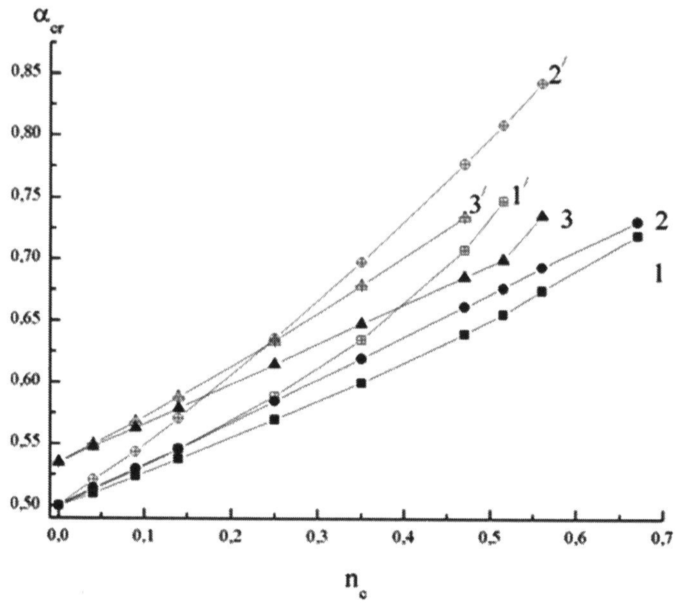

FIGURE 2-7 Critical conversion vs. concentration of (1-3) bifunctional and (1-3′) monofunctional cyclocarbonate at various constants: (1, 1′) k = 1, β= 1, (2, 2′) k = 5, β = 1, and (3, 3′) k =1, β = 0.1.

The earlier examples demonstrate that Eq. (5) offers a virtually universal approach for the estimation of the critical conversion value. The only exception concerns systems in which all reagents change their reactivity during the process. For such systems, a rigorous solution of the kinetic problem is required similar to the one obtained for three-dimensional polymerization (see e.g., 10–12).

The proposed approach to the problem of estimating the critical conversion value also makes it possible to characterize the structure of a network polymer: the content of the gel fraction and the concentration of elastically active chains.

According to the theory of branching processes, the network structure is determined by the chain termination parameter v. This parameter is found by solving Eq. (2-2) with allowance for u(1) = v. Thus,

$$v = 1 - \alpha + \alpha \sum n_{ai} v^{i-1} \tag{2-7}$$

or, after some simplification,

$$n_{a5}v^3 + (n_{a5} + n_{a4})v^2 + (n_{a5} + n_{a4} + n_{a3})v + 1 - n_{a1} - \alpha^{-1} = 0 \qquad (2\text{-}8)$$

The solution of this cubic equation (for a pentafunctional curing agent, such as DETA) makes it possible not only to express V as a function of the depth of conversion but also to derive the condition for finding the gel point. Indeed, since V = 1 at this point,[1] Eq. (2-8) yields formula (2-5).

KEYWORDS

- **Amine**
- **Critical Conversion**
- **Epoxy Groups**
- **Kinetics**
- **Oligomer**
- **Polycondensation**

REFERENCES

1. Irzhak, V. I.; Rozenberg, B. A.; Enikolopyan, N. S. Polymer Networks: Synthesis, Structure, Properties (Nauka, Moscow, 1979) [in Russian].
2. Kuchanov, S. I. Methods of Kinetic Calculations in Polymer Chemistry (Khimiya, Moscow, 1978) [in Russian].
3. Irzhak, V. I. Usp. Khim. Vol.66, №6,598 (1997).
4. Kuchanov, S. I. Adv. Polym. Sci. Vol.152, 157 (1999).
5. Tobita, H. J. Polym. Sci., Part B: Polym. Phys. Vol.36, 2423 (1998).
6. Tai, M. L.; Irzhak, V. I. Dokl. Akad. Nauk SSSR. Vol.259, 856 (1981).
7. Garipov, R. M.; Sof'ina, S. Yu.; Mochalova, E. N. et al., Plast. Massy, No. 7, 21 (2003).
8. Garipov, R. M.; Deberdeev, T. R.; Chernov, I. A. et al., Plast. Massy, No. 8, 6 (2003).
9. Garipov, R. M.; Sysoev, V. A.; Mikheev, V. V. et al., Dokl. Akad. Nauk SSSR. Vol.392, №4, 649 (2003).
10. Korolev, G. V.; Irzhak, T. F.; Irzhak, V. I. Vysokomol. Soedin. Ser. A Vol.43, №6, 970 (2001).
11. Korolev, G. V.; Irzhak, T. F.; Irzhak, V. I. Khim. Fiz. Vol.21, №1, 58 (2002).
12. Korolev, G. V.; Irzhak, T. F.; Irzhak, V. I. Vysokomol. Soedin. Ser. A Vol.44, №1, 5 (2002).

CHAPTER 3

DEFORMATION ELECTROMAGNETIC ANISOTROPY OF VARIOUS PHYSICAL STATES OF HIGHLY CROSS-LINKED POLYMERS

N. V. ULITIN, T. R. DEBERDEEV, R. YA. DEBERDEEV,
L. F. NASIBULLINA, and A. A. BERLIN

CONTENTS

ABSTRACT

Corona electrets based on a vinyl chloride-vinyl acetate copolymer and its compositions with talc are studied. A high value and stability of charge in polymeric corona electrets are due to simultaneous external and dipolar polarization. An increase in the filler concentration first increases and then decreases the parameters of the electret state of the material. The introduction of talc changes the type and depth of energy traps of injected charge carriers in the copolymer, as well as the mobility of dipolar groups of its macromolecules.

3.1 INTRODUCTION

In recent decades, a great deal of attention has been paid to the influence of various physical fields on the structure and properties of polymer materials and composites based on them. This, in particular, refers to such phenomena as electrets polarization of polymers and the electrets effect.[1–3] The versatility of application of electrets (electric analogues of permanent magnets) makes the search for new materials for this purpose and the design of electrets with persistent properties a highly topical issue. The most promising materials are composites manufactured by the bulk coupling of two or more chemically different components with a distinct interface between them. However, the general principles of electrets formation remain unknown for even many relatively simple two-component systems. As a result, it is of interest to study the specific features of polarization of various composites.

To make electrets, the corona poling of polymers is frequently used; its advantages are the simplicity of equipment and a fairly high rate of the process. During such an electrifying process, the injection of charge carriers (electrons, ions) into a dielectric and their retention by energy traps, which can be impurities, structural anomalies, and so on, take place.[1,2] Corona poling for the preparation of electrets with high and stable performance characteristics is mainly used in the case of nonpolar polymers. For ethylene-vinyl acetate copolymers of various brands, it is shown[4] that the stability of corona electrets obtained from these copolymers decreases with an increase in polymer polarity. This is explained by the fact that, during depolarization, charge transfer to the polymer surface and its relaxation take place, which are determined by the bulk conductivity of the material γ_v; the greater the number of polar groups in a polymer, the lower the conductivity.

However, the initial values of the surface charge density σ_{eff} of corona electrets based on polar polymers are higher as compared to nonpolar polymers. As the polarity of a copolymer increases, the initial σ_{eff} value first increases and then somewhat decreases. This behavior has been associated[4] with the fact that charge carriers entering a polymer polarize neighboring dipoles (polar groups) occurring in the surface layex,[1] and the dipoles become oriented in the field of injected charges. The oriented dipoles trap injected charges by attracting and retaining them via Coulomb attraction forces. The larger the number of polar (acetate) groups in the polymer, the greater the number of traps and the higher the initial effective surface charge density are. Charge carriers are released more readily from surface traps than from traps in the bulk: the proportion of bulk traps is much higher in nonpolar polymers. Detrapping is also facilitated by an increased concentration of charge carriers in the surface layer, which increases the conductivity of the polymer.[4] This mechanism is supported by the results obtained by Boev *et al.*,[5] who showed that most of the charge carriers injected into polar polymers occurred in the surface layer. There are virtually no charges to be observed in the bulk of such polymers.

Earlier,[6] a process was patented for the manufacturing of a coating on the basis of a corona electret obtained from a random vinyl chloride (VC)-vinyl acetate (VA) copolymer, which exhibited high and stable values of charge. It was interesting to study the manifestation of the electrets effect in this copolymer.

3.2 EXPERIMENTAL PROCEDURE

The subjects of study were a VC-VA copolymer of the brand A-15-0 (TU [Technical Specification] 6-01-1181-87) with a chlorine mass fraction of 52.5% and talc of the PMK-27 brand with a density of 2.75 g/cm^3 a particle size of 5.5 pm, and a specific surface area of 8 m^2/g.

The polymer was blended with a filler on a laboratory roller micromill at 135 ± 5°C for 3 min. Specimens in the form of plates 0.8 to 1.2 mm in thickness were prepared by molding according to GOST (State Standard) 12019-66 at 170 ± 5°C. The specimens were poled in a corona discharge using an electrode containing 196 sharp needles uniformly distributed over a square of 49 cm^2 in area. The distance between a specimen and the electrode was 20 mm, the poling voltage was 35 kV, and the poling time was 90s. Before poling, the specimens were held for 10 min in a thermal cabinet at 90°C.

The surface potential of electrets V_e was measured at regular intervals of 24 h by means of a vibrating electrode (noninvasive induction method) for a long period of time (1 year). The time between poling a plate and the first measurement of its surface potential was 1 h. Electrets were stored at an ambient temperature in the nonshorted state. The effective surface charge density σ_{eff} was calculated by the equation in the following:[2]

$$\sigma_{eff} = U_c \varepsilon_0 \varepsilon / \delta$$

where U_c is the compensating voltage in volts, ε_0 is the electric constant equal to 8.854×10^{-12} F/m, ε is the dielectric permittivity of the composite, and δ is the thickness of the specimen in meters.

The charge relaxation time T was determined as the time of a decrease in the surface potential by a factor of *e* (2.71). By the 360th day of storage, the spectra of thermally stimulated current (TSC) in the samples were recorded at a linear heating rate of 5 K/min using a special measurement cell with blocking aluminum electrodes and a Teflon spacer. The results of analysis of TSC curves were compared with DTA data (a Q-1500 differential thermal analyzer, a sample mass of 150 ± 50 mg, a heating rate of 5 K/min).

3.3 RESULTS AND DISCUSSION

The kinetics of variation in the magnitude of charge of a VC-VA copolymer corona electret during storage is described by a curve with a minimum (Figure 3-1; curve *1*). The initial decay of the surface potential of the corona electret is due to charge relaxation.

However, the dramatic decrease in the magnitude of charge in the period from the 80th to the 160th day is due to its screening by water molecules adsorbed on the surface from air. Under the given experimental conditions, this period was in summer when the humidity of the environment was maximal. As winter approached, the humidity decreased, and the considerable portion of water molecules was removed from the electret surface, thus recovering the surface potential. This phenomenon is characteristic of traditional electrets: a decrease in charge upon an increase of humidity in the environment and its recovery upon a decrease in the humidity.[7] At the same time, the charge relaxation rate increases under a high air humidity. This may be due to an increase in the surface conductivity of polymers and preferable relaxation of a homocharge in electret surface and near-surface layers.[8,9]

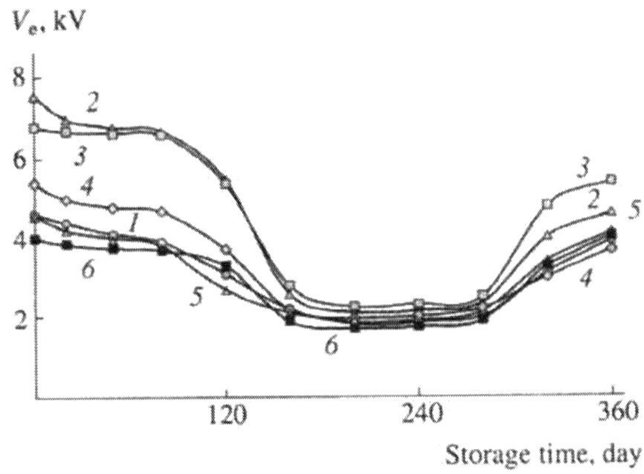

FIGURE 3-1 Surface potential of a corona electret obtained from (7) a VC-VA copolymer and *(2–6)* its compositions as a function of storage time. The talc content is (2) 2, *(3)* 4, *(4)* 6, (5) 8, and (6) 10 vol %.

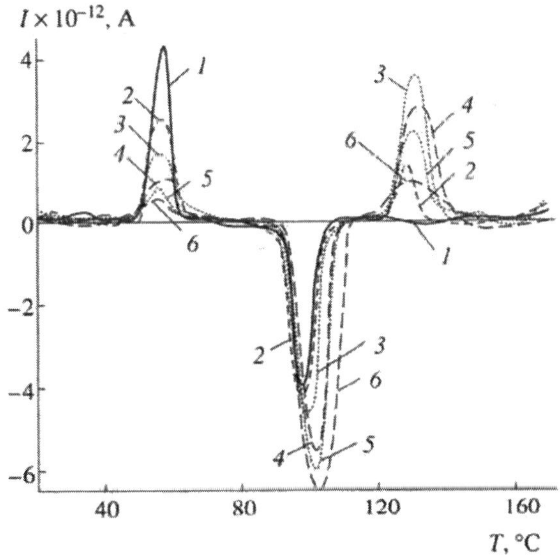

FIGURE 3-2 Thermostimulated depolarization current curves in (7) a VC-VA copolymer and *(2–6)* its composites. The talc content is (2) 2, *(3)* 4, *(4)* 6, (5) 8, and *(6)* 10 vol %.

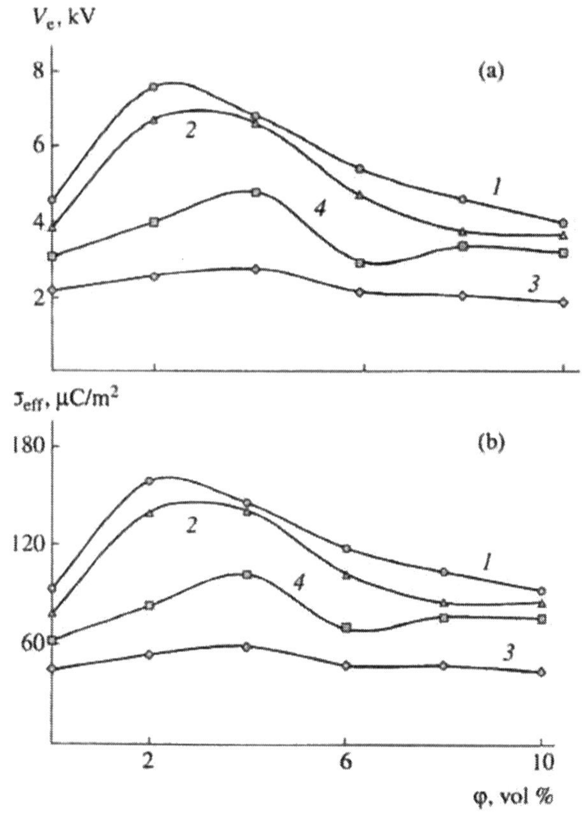

FIGURE 3-3 (a) Surface potential and (b) effective surface charge density of composite corona electrets as a function of the talc content. The storage time of electrets is (*1*) 1 h and (*2*) 80, (*3*) 160, and (*4*) 320 days.

An inspection of the TSC spectra of the copolymer (Figure 3-2, curve 7) reveals the presence of two trapping levels in the material, which may be attributed to two kinds of traps. First-level traps degrade at 60–65°C, which corresponds to the copolymer glass-transition temperature as determined by DTA. Probably, obstacles to rotation of dipolar groupings are mainly removed at this temperature, thus leading to the disappearance of dipolar-segmental polarization. The activation energy of charge relaxation corresponding to this peak is ~0.70 eV as calculated by the Garlic–Gibbson method.[2]

The presence of a negative peak indicates that the second level of trapping is destroyed at a temperature of 95–105°C, which corresponds to the DTA-determined temperature of transition of the copolymer to the visco-plastic state. At this temperature, intermolecular physical and hydrogen bonds weaken, leading to a release of the injected charge. The activation energy of the process is ~1.30 eV. Hence, it follows that the charge carriers are localized in deep traps, thus ensuring a stable and high value of the surface potential.

Based on their TSC spectra, it may be concluded that the high values and the stability of charge in VC-VA copolymer corona electrets are due to the external and dipolar types of polarization, with the polarization due to the injected charge determining the magnitude of the charge and the dipolar polarization being responsible for its stability. Preheating the copolymer to 90°C prior to corona charging transfers it to the rubbery state in which dipolar groupings are capable of orienting in a poling field or of "freezing" to form a heterocharge upon cooling.

When inhomogeneous materials are poled in a corona, charge carrier trapping at interfaces takes place. Charge buildup at the interfaces is due to the difference in conductivity between the contacting phases (Maxwell–Wagner effect). In view of this behavior, it appears quite interesting to study the electret properties of talc-filled VC-VA copolymer compositions.

Electrets based on the compositions of polar polymers with dielectric fillers have been studied previously.[10–13] However, despite a similarity in the nature of the polymers and fillers examined, the results obtained were controversial. For example, Kalogeras *et al*[10] found that PMAA thermoelectrets ranked in properties below PMAA composites containing 0.1 to 1.0-wt % silica gel. Thermostimulated depolarization (TSD) currents in composites were substantially higher than in the pure polymers. Naturally, the calculated activation energies of charge relaxation from energy traps in PMMA composites with silica are higher than those for PMAA thermoelectrets. In contrast, deterioration of electret characteristics by introducing dielectric fillers was observed in refs. [11-13]. Belyi *et al.*[11] noted that the charge density of metallopolymer electrets based on polyvinylbutyral decreased after getting filled with bank sand. S. Viridenok *et al*[12] observed a decrease in TSD peaks of polyamide corona electrets upon addition of grass fiber. A decrease in the effective surface charge density by introducing basalt fiber or glass fabric into polyvinylbutyral metallopolymer electrets was also reported in ref. [13]. The truth is that σ_{eff} insignificantly increased in the region of small amounts of a filler added as a result of the formation of more perfect supramolecular structures in polymer binders.

Experiments have shown that the presence of talc has an effect on the manifestation of electret properties by the VC-VA copolymer. The general course of the dependence of the surface potential (Figure 3-3a) and the effective surface density of electret charge (Figure 3-3b) on the filler content correlates with the earlier revealed pattern of the influence of various fillers on the electret properties of polyvinylbutyral[13]: an increase in the amount of filler first increases and then decreases the electret characteristics of composites.

The initial increase in V_e and σ_{eff} may be associated with a change of the type and depth of energy traps of injected carriers, since polymer filling always alters the supramolecular structure (size of structural units, their shape, and size distribution type) and packing density.[14] As a result of mechanochemical degradation of materials during intense mixing with finely divided compounds, free radicals[15] can be generated, which are also capable of acting as energy traps for charge carriers. In addition, fillers exert a considerable influence on the mobility of various kinetic units of a polymer and, hence, on its relaxation time spectrum.

The decrease in V_e and σ_{eff} with an increase in the filler content to 10 vol % may be due to the fact that talc particles do not contribute to the electric conductivity of samples upon poling and injected charge carriers mainly concentrate in surface traps characterized by a low trapping energy, without penetrating into the bulk of compositions.[3]

An inspection of TSD curves for filled electret compositions (Figure 3-2, curves *2–6*) showed a change in the intensity and position of TSD current peaks characteristic of the pure VC-VA copolymer. The decrease in the activation energy for traps of the first kind can be caused by a decrease in the flexibility of macromolecules in the presence of the dispersed filler and, as a consequence, an impediment to the rotation of dipolar groupings during the poling process. This is due to the adsorption of copolymer molecules on the solid filler surface, which form boundary copolymer layers with a reduced mobility of structural elements. The developed surface of dispersed talc is responsible for a rather high relative amount of macromolecules with reduced mobility in the composites examined.

The increase in the activation energy of the second-level trapping to ~1.55 eV (for a VC-VA copolymer composite with 10 vol % talc) is probably due to a decrease in the conductivity of the composition, which impedes the relaxation of external polarization of polymer compositions.

Furthermore, the inspection of the TSD spectra revealed the appearance of a new level of trapping of injected carriers, which is destroyed at 130–140°C. Most likely, this trapping level involves new type of traps that probably occur at the copolymer/filler interface and are presumably due to the Maxwell–Wagner effect. Moreover, corona poling can give rise to carboxyl, carbonyl, and other groups in macromolecules in the presence of fillers (primarily at the polymer/filler interface). These polar groups become oriented along a poling field even during their formation, and their disorientation (depolarization) can also take place only at high temperatures.

The activation energy of charge relaxation from this type of traps increases, depending on the filler content, from ~0.65 eV (for VC-VA copolymer compositions with 2-vol % talc) to ~1.40 eV (for VC-VA copolymer with 4-vol % talc) and decreases to ~0.6 eV as the filler content increases to 10 vol %.

The appearance of a new peak in TSC spectra for filled polymers was reported earlier, although it was observed only in the low-temperature region. For example, Sviridenok *et al*[12] revealed a low-temperature TSD peak (at ~55°C) in glass fiber-filled polyamide composites and attributed it to the charging of defects on the glass-fiber surface or at the polymer/filler interface. In the region of high temperatures (above the polymer-flow temperature), TSC of composite corona electrets have not been studied.

3.4 CONCLUSIONS

Corona electrets based on the VC-VA copolymer are materials that possess all the signatures of conventional electrets: they bear space electric charge (V_{e0} = 4.6 kV, σ_{eff} = 93 μC/m^2) retained for a long time (τ >360 days) and display a TSC spectrum upon heating. The existence of two levels of trapping of injected charge carriers corresponds to two kinds of traps. The temperature of degradation of first-level traps with an activation energy of ~0.70 eV corresponding to the copolymer glass transition temperature (60–65°C) indicates the disappearance of dipolar-segmental polarization. The relaxation of injected charge takes place at a temperature of 95–105°C, which coincides with the temperature of transition of the copolymer to the viscoplastic state and is responsible for the appearance of the negative TSC peak. The activation energy of this process is ~1.30 eV. The presence of talc alters the manifestation of the electret effect in the VC-VA copolymer: an increase in the amount of the filler first increases and then decreases the electret characteristics of the compositions. Their initial enhancement is due to a change in the depth and type of energy traps for injected carriers in the copolymer and to the Maxwell–Wagner effect. The decline in the electret characteristics of the VC-VA copolymer with an increase in the filler content to 10 vol % is due to an increase in the proportion of injected charge carriers concentrated in shallow surface traps. The study of the TSC spectra of filled electret compositions revealed a change in the intensity and position of peaks characteristic of the pure VC-VA copolymer, which is associated with the change in the flexibility of macromolecules in the presence of the filler, and the appearance of a new level of trapping injected carriers with an activation energy of 0.6–1.4 eV, whose destruction takes place at 130–140°C. This trapping level is due to the formation of the new type of traps that probably occur at the copolymer/filler interface and seems to be caused by the Maxwell–Wagner effect.

KEYWORDS

- **Copolymer**
- **Corona Electret**
- **Macromolecule**
- **Polarization-Depolarization**
- **Thermostimulated depolarization (TSD)**

REFERENCES

1. Electrets, Ed. by Sessler, G. M. (Springer, Berlin, 1980; Mir, Moscow, 1983).
2. Lushcheikin, G. A. Polymer Electrets (Khimiya, Moscow, 1984) [in Russian].
3. Irzhak, V. I.; Rozenberg, B. A.; Enikolopyan, N. S. Polymer Networks: Synthesis, Structure, Properties (Nauka, Moscow, 1979) [in Russian].
4. Kuchanov, S. I. Methods of Kinetic Calculations in Polymer Chemistry (Khimiya, Moscow, 1978) [in Russian].
5. Irzhak, V. I. Usp. Khim. Vol.66, №6, 598 (1997).
6. Kuchanov, S. I. Adv. Polym. Sci. Vol.152, 157 (1999).
7. Tobita, H. J. Polym. Sci., Part B: Polym. Phys. Vol.36, 2423 (1998).
8. Tai , M. L.; Irzhak,V. I. Dokl. Akad. Nauk SSSR. Vol.259, 856 (1981).
9. Garipov, R. M.; Sof'ina, S. Yu.; Mochalova, E.N. et al., Plast. Massy, No. 7, 21 (2003).
10. Garipov, R. M.; Deberdeev, T. R.; Chernov, I. A. et al., Plast. Massy, No. 8, 6 (2003).
11. Garipov, R. M.; Sysoev, V. A.; Mikheev, V. V. et al., Dokl. Akad. Nauk SSSR. Vol.392, №4, 649 (2003).
12. Korolev, G. V.; Irzhak, T. F.; Irzhak, V. I. Vysokomol. Soedin. Ser. A Vol.43, №6, 970 (2001).
13. Korolev, G. V.; Irzhak, T. F.; Irzhak, V. I. Khim. Fiz. Vol.21, №1, 58 (2002).
14. Korolev, G. V.; Irzhak, T. F.; Irzhak, V. I. Vysokomol. Soedin. Ser. A Vol.44, №1, 5 (2002).

CHAPTER 4

POLYSULFIDE OLIGOMER SOLIDIFICATION PROCESS

V. S. MINKIN, YU. N. KHAKIMULLIN, A. A. IDIYATOVA, YU. V. MINKINA, and R. YA. DEBERDEEV

CONTENTS

ABSTRACT

The features of vulcanization of commercial polysulfide oligomers by manganese dioxide have been investigated. The criteria of commercial manganese dioxide activity have been found based on the change of parameters and shape of EPR lines of vulcanizing agents. The effect of solidification rate on physico-mechanical properties of two types of hermetics—Y-30M and AM-0.5—has been evaluated. It has been shown that the activity of used manganese dioxides affects, in the first place, the rate of solidification and does not affect the strain-strength properties and the final hardness.

4.1 INTRODUCTION

At the present time, the study of the effect of manganese oxide (IV) structure on the solidification rate of commercial polysulfide oligomers (PSO) is of great interest. At PSO solidification on practice, there often arises problems connected with different activity of vulcanizing pastes what is related, in the first place, with activity of incorporated in them manganese dioxide (MnO_2).[1,2].

At the vulcanization of PSO, the chain increases and seldom crosslinking takes place. The crosslinking occurs at the oxidation of end thiol groups of linear chains and their long branches forming at oligomer synthesis. At the same time, decrease of molecular mobility is found and connected with it is decrease of period of time of nuclear spin-spin relaxation, T. The latter in all cases, as it was shown before,[3,4] decreases to certain minimum value in the same way as the change of the chain density of vulcanizates, and it is a reliable structural kinetic parameter at analyzes of curing processes of Thiokol and based on it compositions.

4.2 EXPERIMENTAL PROCEDURE

4.2.1 *MATERIALS*

The properties and structure parameters of investigated commercial MnO_2 are shown in the following (Table 4-1).

According to TY 38. 50309-93, Thiokol type 2 (SH-group content 1.9%) and Thiokol type 1 (SH-group content 2.8%) were used.

TABLE 4-1 Properties and structure parameters

Commercial type of Mn02	Content of MnO, %	Width of EPR line, 5H, (±0.2), 290°K	g-factor (±0.001)	Line shape
1	80	560	1.956	Complex
2	78.2	1890	1.953	Simple
3	76.8	Narrow component-620 Wide-2013.2	1.956	Complex
4	82.7	1556.3	1.954	Simple

4.2.2 TEST PROCEDURES.

The procedure to measure the period of time of the spin-spin relaxation and T_2 in the process of PSO vulcanization have been described in ref. [4]. The EPR spectra were written on radiospectrometer JES -I MEX with A = 3, 2 cm.

4.3 RESULTS AND DISCUSSION

The EPR spectra of commercial shipments of MnO_2 are shown in Figure 4-1. All the EPR spectra of MnO_2 and vulcanizing pastes based on it are the absorption signals related to Mn^{4+} ions. In this case, the values of g-factors, width and shape of resonance lines correspond to the literature data for $Mn0_2$ or Mn^{4+} in glasses or polymers.[5,6] The widening of EPR lines is connected with the covalent bond of Mn^{4+} ions with oxygen atoms. It should be noted that its electron state ($3d^3$) is analogous to the state of Cr^{3+} and V^{2+}.[5,6]

The maximal value of the resonance line width occurs for sample 2 and 4 (table 4-1). The complex shape of EPR lines is clearly expressed for samples 1 and 3. It is interesting to note that the parameters of MnO_2 (and vulcanizing pastes based on it) spectra show the same tendencies of resonance line width change. However, in this case, the line width value for vulcanizing pastes is significantly lower due to existence of plasticizers (dibutylphthalate) in the composition. As the quantitative criteria of MnO_2 activity, we have used the width and shape of EPR spectra resonance lines because such parameters correlate with mobility and distribution character of manganese atoms (Mn^{4+}).

The data in Figure 4-2 show the kinetic curves of vulcanization of commercial Thiokols of type 1 at its solidification by MnO_2 (samples 1, 2).

By comparison of the data in table 4-1 and Figures 4-1 and 4-2 , it is seen that the found differences in mobility and localization of basic compound ions of oxidizer are really reflected in the PSO vulcanization kinetics. The activity of sample 1 is almost twice of that of sample 2. The effective constant of vulcanization rate for sample 1 (к,- 10^3, min'^{1}) is 3.7, and for sample 2 is 7.8.

The effect of oxidizer activity is also shown in commercial types of Thiokol differing on molecular mass (Figure 4-3).

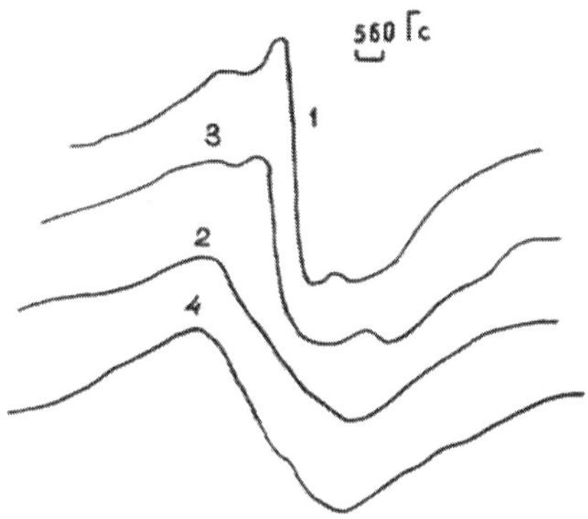

FIGURE 4-1 EPR-spectra of commercial types of MnO_2. Numbers on spectra correspond to the numbers of types shown in table 4-1.

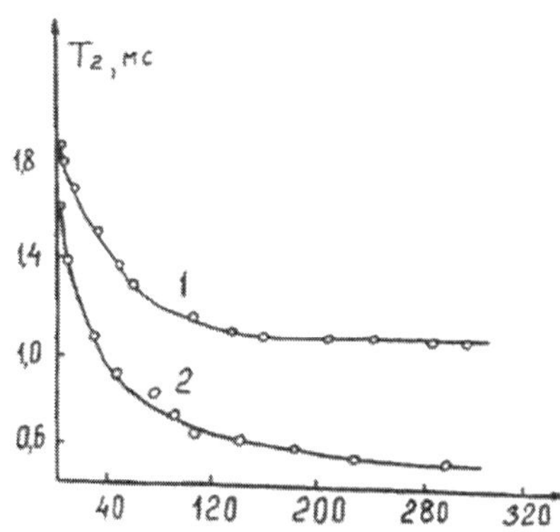

FIGURE 4-2 Kinetics of vulcanization of commercial liquid Thiokol type 1 by MnO_2 of different activity (samples 1 and 2, see table 4-1) in the oxidizer.

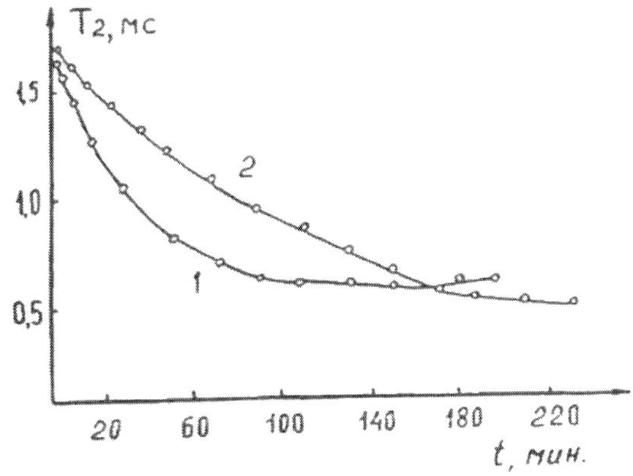

FIGURE 4-3 Kinetics of vulcanization of commercial liquid Thiokols type 1 (curve 1) and type 2 (curve 2) by commercial MnO_2 (sample 1, table 4-1).

The strong correlation between EPR width change and found activity of commercial vulcanizing agents has been noted for all investigated types of MnO_2. The vulcanizing agents showing the line width less or equal 600–700 Гс have enough high activity in PSO vulcanization reactions.

The oxidizers showing the line width above 1,OOOГс are characterized by low values of kinetic parameters.

The active types of $Mn0_2$ show complex shape of EPR lines caused by the existence of Mn^{4+} ions of different mobility (Figure 4-3). Indeed, the wide anisotropy line of Mn^{4+} ion absorption is superimposed by intense narrow component of the spectrum of the most mobile Mn^{4+} ions, the content of which in the original MnO_2 can be 20%–25% of the total Mn^{4+} content. Because of this, despite the big width of EPR lines of commercial types of $Mn0_2$, their activity is happened to be relatively high caused by existence in them of more mobile Mn^{4+} ions (the spectra narrow component) having enough high concentration. The ions are able to affect significantly the PSO oxidation rate on the beginning stage of liquid Thiokols vulcanization process. This is what we actually observe for samples 1 and 3.

The analysis of obtained structure parameters allows to evaluate even before vulcanization the commercial shipments of MnO_2 activity in PSO vulcanization reactions. It should be noted that different mobility, localization, and concentration of mobile Mn^{4+} ions in the vulcanizing agent (powder or paste) will lead to difference in PSO vulcanization rate. The latter, in its turn, will determine, basically, the 'life' of Thiokol compositions.

To confirm the made assumption, evaluation of solidification rate of two types of hermetics on their properties was conducted. The hermetics Y-30M (compositions 1,2) and AM-0.5 (compositions 3,4) differ by the nature of filler (carbon black П -803 and chalk correspondingly) and existence of epoxide-diene resin Э-40 in the AM-0.5 composition. To solidify compositions 1 and 3, MnO_2 (sample 1 in table 4-1) was used.

Another MnO_2 (sample 2, table 4-1) was used to solidify compositions 2 and 4. Hermetics solidification was conducted for 48 hrs at 70°C after their 'life' loss.

From the data in table 4-2, the conclusion can be made that the MnO_2 activity, in the first place, as it was assumed, affects the solidification rate (period of 'life', Shore A hardness after 24 hrs and 48 hrs) and practically does not affect the strain-stress properties and final hardness. Even at high solidification rate (the 'life' time about 10–12 min.), the structure defects leading to strength loss do not form.

It should be noted some adhesion improvement of hermetic AM-05 with high 'life' time to duralumin. That can be explained by the improvement of condition of contact forming on the border "hermetic-duralumin".

TABLE 4-2 Properties of Thiokol hermetics with different solidification rate

Hermetic type	"Life" time, min	Tensile strength, MPa	Elongation at break, %	Adhesion to duralumin, kN/m	Shore A hardness 24 h 48 h 336 h
Y-30M	10	2.87	275	-	48 51 56
Y-30M	420	2.51	275	-	31 38 54
AM-05	12	0.82	460	1.88	- -
AM-05	510	0.81	510	2.5	- -

4.4 CONCLUSIONS

The results of the investigation showed the ability to use the MnO_2 EPR-spectra to judge about their activity in the process of liquid Thiokols solidification. On some concrete examples, it was shown that the MnO_2 activity affects, first of all, the hermetic solidification rate.

KEYWORDS

- **EPR Spectra**
- **Hermetic**
- **Manganese Dioxide (MnO_2)**
- **Resonance Lines**
- **Solidification**
- **Vulcanization**

REFERENCES

1. Minkin, V. S.; Averko-Antonovich, L. A.; Kirpichnikov, P. A. // Vysokomol Soed., v.23, No.8, 1981, pp. 593–596.
2. Minkin, V. S., Kirpichnikov, P. A.; Liakumovich, A. G. Proc. Intern. Rubber Conf., Moscow, 1994, pp. 98–105.
3. Minkin V. S.; Averko-Antonovich, L. A.; Kirpichnikov, P. A. // Vysokomol Soed., v. 15, No.8, 1973, pp.24–26.
4. Averko-Antonovich, L. A.; Minkin. V. S.; Mukhutdinova, T. Z.; Yastrebov, V. N. // Vysokomol Soed., A, v. 16, No. 8, 1981, pp. 1709–1713.
5. A itshuler, C. A.; Kozyrev,B. M. "EPR of compounds of intermediate group elements" (Russ.), Moscow, Nauka, 1972, p. 262.
6. Kuska, H.; Rodgers, M. "EPR of complexes of transition metals" (Russ.), Moscow, Nauka, 1970, p. 82.

CHAPTER 5

SOLIDIFICATION OF POLYSULFIDE HERMETICS

YU. N. KHAKIMULLIN, R. R. VALYAEV, L. YU. GUBAIDULLIN, V. S. MINKIN, O. V. OSHCHEPKOV, A. G. LIAKUMOVICH, and R. YA. DEBERDEEV

CONTENTS

ABSTRACT

The features of the vulcanization process and properties of the polysulfide (Thiokol type) hermetics modified by unsaturated polyesters (UPE) have been investigated by using the ^{1}H NMR and ^{13}C NMR. The composition of UPE has been determined. It has been found that the rate of the interaction with the Thiokol SH-groups and the state of the UPE participation in the solidification of Thiokol hermetics depend on the double bond activity. The double bonds of the endic anhydride have the maximal activity. The UPE incorporation into the hermetic rises the chemical density of the vulcanizate chain network above the theoretical one, increases the strength, adhesion to glass, and duralumin what witnesses the UPE participation in formation of the tri-dimensional structure at the Thiokol hermetics solidification.

5.1 INTRODUCTION

Hermetics based on the liquid Thiokols are basically used in the aviation industry due to their oil-gasoline resistance, and in the building industry due to their atmosphere-water resistance and gas penetration resistance.[1,2] Taking into account that the hermetics are used mainly as coatings, the successful realization of the aforementioned advantages can be achieved only at conditions of providing high adhesion to the counterpart surfaces which would last for a long service time.

One of the most efficient ways to achieve high adhesion of the Thiokol hermetics to different substrates is their modification by the reactionable compounds. The modification of liquid polysulfides by unsaturated monomers or oligomers is very perspective.[2,3] Such modification allows not only to enhance adhesion but also to obtain the copolymer hermetics with new set of properties.

Indeed, the use of abietic acid, its derivatives, and oligoesteracrylates (OEA) allows to significantly increase adhesion to metals and glass.[4–8] The unsaturated compounds playing the role of temporary plasticizers actively participate in the vulcanization process.[9] As a rule, the reactionable compounds are incorporated into hermetic paste together with Thiokol, which will increase its viscosity with time and significantly reduce the storage time before its application. Because of this, in this work, we studied the possibility to use the unsaturated polyester (UPE) which does not interact with the -SH groups of Thiokol in conditions of joint storage and do interact in presence of oxidizing agents and participate in hermetic solidification. Simultaneously, we conducted the evaluation of the efficiency of unsaturated polyesters having different structure in the reactions with liquid polysulfides depending on the composition and nature (activity) of the available double bonds.

5.2 EXPERIMENTAL PROCEDURE

In this work, the liquid polysulfide (liquid Thiokol) containing 2.95 mass % of -SH groups and having viscosity 15.5 Pa.s was used. The polysulfide was obtained based on di-(β-chloroethyl)formal with 2 mol.% of the branching agent 1,2,3-trichloropropane (TCP). As the filler, the hydrophobic chalk in the amount of 80 mass parts. The liquid Thiokol solidification was achieved by using manganese dioxide in the form of a paste. The content ratio of the hermetic and the solidification pastes was 100:10 based on the mass, correspondingly. In the work, the following materials were used: pine rosin, oligo ester acrylates-ТГМ-3 tri(oxyethylene)-α,ω-dimethyl acrylate; МГФ-9 -α,ω-dimethyl acrylate-(bis-ethylene giycol phthalate), and UPE of different molecular mass and composition with the end carboxyl groups. The unsaturated polyesters are the resin-like materials with an acid number from 25 to 130 mg KOH/g and the unsaturation of 6–8 mass %. The UPEs were incorporated into the hermetic paste composition in the amount of 0.3 to 10.0 mass parts on 100 mass parts of Thiokol. The hermetic solidification was carried out both at normal conditions (20°C, 7 days) and under the accelerated regime (70°C, 24 hrs). The hermetic testing was performed under the standard procedures. The cross-link density was evaluated by using the Cluff-Cladding method.[10] The structure of the synthesized UPE was evaluated by using the ^{1}H and ^{13}C NMR-spectroscopy.[11] The spectra of polyesters were obtained by using "Gemini-200" NMR spectrometer made by "Varian" with the working frequencies for ^{1}H nucleus–200 MHz, and for ^{13}C–50 MHz. The spectra ^{13}C were obtained by the Furie method of transformation of the free induction signal decrease in the regime of the wide-band denouement of the spin-spin interaction with protons. The solvent was $CDCl_3$. The ^{1}H and ^{13}C signals in the NMR spectra were compared with ones in the spectra of the original components, with the spectra of different polyesters, with the calculation according to the additive schemes, and also by using the literature data.

5.3 RESULTS AND DISCUSSION

The data in Table 5-1 show that incorporation of the unsaturated compounds including rosin, consisting mainly of the abietic acid, and UPE leads to the hermetic life time increase.

TABLE 5-1 Properties of Thiokol-based hermetics modified by unsaturated compounds

Material	Content, mass, parts	Life Time, hrs	Tensile Strength, MPa	Elongation at Break,%	Adhesion to Glass, MPa
Control	0,0	0^{50}	0.93	28.5	0.51
Rosin	1.0	1^{20}	1.03	270	0.65
		1			
	2.0	1^{55}	1.30	245	0.78
	3.0	2^{20}	1.37	220	0.84
	5.0	3^{45}	1.32	205	0.90
ТГМ-3	5.0	2^{00}	0.81	350	0.77
МГФ-9	5.0	2^{45}	0.97	340	0.79
UPE with acid number:					
29.5	2.0	2^{45}	1.39	210	1.05
50.0	2.0	3^{10}	1.44	165	1.11
80.0	2.0	3^{55}	1.46	120	1.24
120.0	2.0	5^{20}	1.58	75	1.36

It is also seen that all the modifiers provide improvements in strength, adhesion to glass, reduction of the elongation at break of the hermetics. This witnesses that the additives participate in the solidification reactions with liquid polysulfide through their double bonds. The solidification mechanism, as it was shown before,[9] has a radical nature and is initiated by the inorganic peroxide (manganese dioxide) and the tert-amine (diphenyl guanidine). It should be noted, however, that by using rosin and the oligoester acrylates, the level of achieved properties in not high enough. The more strong effect on the adhesion and tensile properties is achieved by using the UPE. The increase of the carboxyl group in UPE leads to some increase of the parameters. This can be possibly related to the following two reasons:

1. By participation of the end carboxyl groups in formation of the chemical adhesion bonds on the border hermetic-substrate

2. By reduction of the molecular mass and viscosity of the polyester, acceleration of its diffusion to the contact border, and more active participation in the bond formation on the phases border[12]

Besides, as it was established before,[12] the use of UPE with the end carboxyl groups in the hermetic composition provides significant (in 2–3 times) increase of its storage time before application ("Life time") caused by retardation of the solidification reaction occurred between the end Thiokol SH- groups and the air oxygen. A noticeable change in hermetic properties is found at using different amount of UPE in the composition (Figure 5-l).

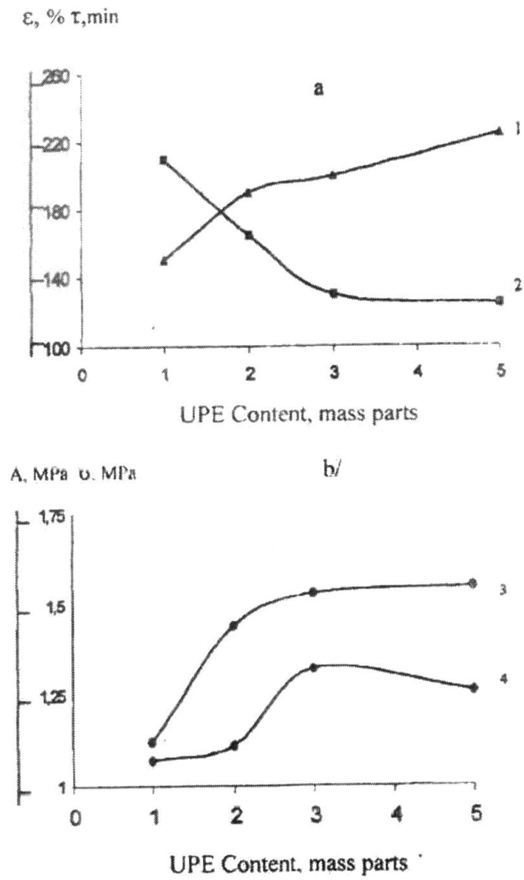

FIGURE 5-1 Dependence of hermetic life time, τ, (1) elongation at break, e, (2) tensile strength, c, (3) and adhesion to duralumin and glass, A, (4) on the unsaturated polyester, UPE, content.

The maximal value of adhesion was achieved at polyester content of 3 mass parts on 100 parts of liquid Thiokol. It should be noted that there is a correlation between adhesion and tensile properties of the hermetic with the UPE content increase. The character of the adhesion failure was cohesive. In the whole range of the UPE content change, the elongation at break decreases with the content increase.

Such character of the parameters change is connected with the active interaction of UPE along the double bonds with the polysulfide with the three-dimensional network formation. Indeed, the chemical density of the network chains (v_{chem}) increases with the polyester content increase (Figure 5-2).

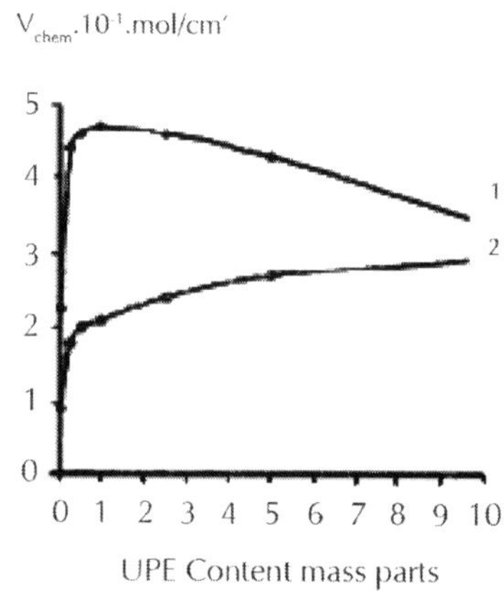

FIGURE 5-2 Dependence of the effective density (1) and the chemical density (2) of the crosslink network on the UPE content.

It should be noted that at UPE content more than 1 mass part, the v_{chem} starts to be in excess of the theoretically calculated value equal to $2.3 \cdot 10^4$ mol/cm^3 at TCP content in Thiokol 2.0 mol %.[13] Taking into account the fact that the UPE functionality along the double bonds is more then 3, it is possible to assume that its incorporation leads to an additional hermetic structure formation. At the same time, the increase of the effective network density, v_{eff} , takes place. The change of the v_{eff} with the UPE content increase has an extreme character. This can be explained by that incorporation of the spacial aromatic UPE structures in an amount more then 1–2 mass parts hinders the physical interaction between the Thiokol polar groups and filler.

The efficiency of the UPE in the polysulfide hermetics depends not only on their content but also on the nature of the double bond, which is determined by the structure of its comonomers containing the double bonds. To evaluate the effect of the double bond nature in the polyester on the hermetic properties, some polyesters were synthesized. The polyesters had in the main chain the maleic (MA), endic (EA) (for the chemical structure of the endic anhydride see below), and phthalic (PA) anhydrides both separately and in combination in different ratios.

The synthesized UPEs were investigated by using the methods of the NMR ^{1}H and ^{13}C spectroscopy.

The analysis of the obtained spectra witnesses the polycondensation reaction which occurs quantitatively with almost 100% conversion (Figures 5-3 and 5-4).

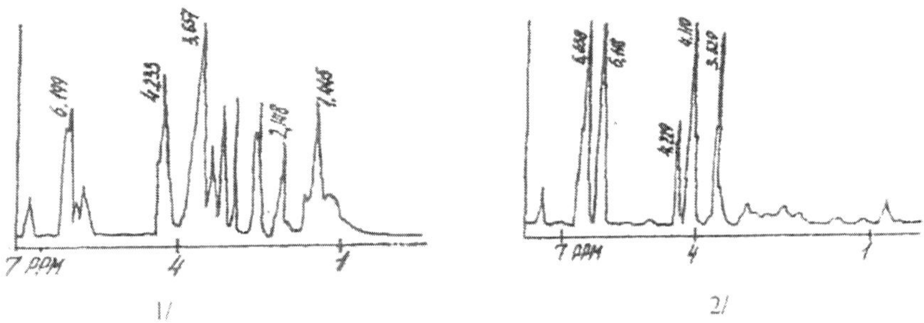

FIGURE 5-3 The ^{1}H NMR spectra of the UPE compositions: (1) EA:MA 0.9:0.1; (2) EA:MA 0.1:0.9.

The decrease of the EA content in polyester based on the combination of EA and MA leads to the sinibatic (in the same direction) decrease of the active double bonds. The comparison of the signal areas of the same groups (area C=C) allows to eliminate the errors caused by the differences in the relaxation times and the nuclear effect of Overhauser and quantitatively evaluate the polyester composition.[11,14] The decrease of the portion of EA in the UPE compound from 0.9 to 0.1 mass parts leads to decrease of the CH=CH groups by 30% which causes the polyester activity reduction in the hermetic solidification (Figures 5-3 and 5-4).

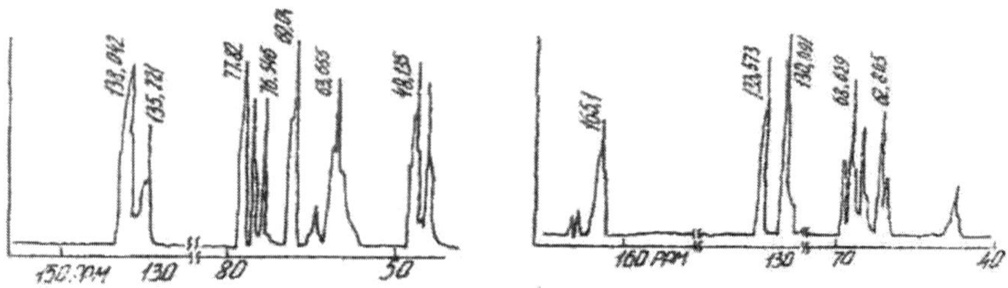

1/ 2/

FIGURE 5-4 The ^{13}C NMR spectra of the UPE compositions: (1) EA:MA 0.9:0.1; (2) EA:MA 0.1:0.9.

The data in Figure 5-5 show the kinetic curves of the hermetic composition solidification by different polyesters. The process kinetics was measured by the change of the time of the spin-spin relaxation, t_2, by the impulse NMR method. The analysis of kinetic curves witnesses that the composition containing higher amount of the endic anhydride in the UPE is solidifying faster. The substitution of MA by PA leads to retardation of the composition solidification process.

The effect of the UPE compositions on the solidification kinetics is very well in agreement with their structure, amount, and activity of the double bonds which are determined by the methods of NMR-high resolutions.

The final time of the spin-spin relaxation, t_2, well correlates with the effective network density, v_{eff}, formed at the hermetic compositions solidification: the network density increase should lead to a shorter time of the spin-spin relaxation, t_2. Such dependence allows to determine the effective network density of different compositions based on the found t, values. The data are shown in Table 5-2.

The use of UPE containing EA provides the increase of the hermetic network density. Decreasing of the EA content (sample 2, Table 5-2) leads to a decrease of the network density by 30%. This confirms the fact that in the relatively soft solidifying conditions, the MA double bonds are not effective enough while the EA double bonds interact with the Thiokol SH-end groups quantitatively, practically completely. In the row of the investigated anhydrides EA-MA-PA, the preference should be given to EA which provides the higher network density and better physical properties of the solidified hermetics. This is confirmed by the data in Figures 5-6 through 5-8.

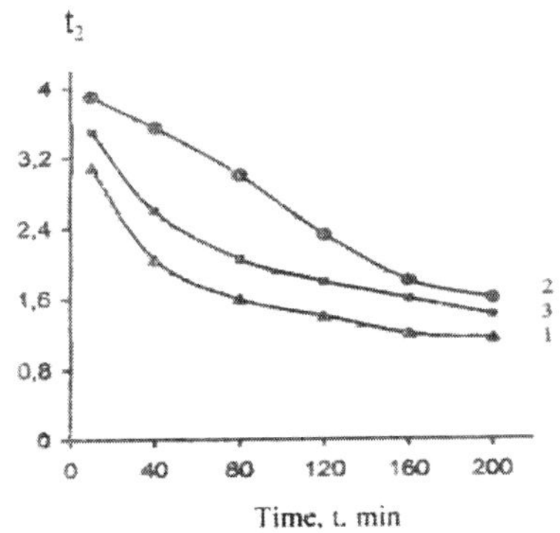

FIGURE 5-5 Kinetic curves of solidification of different compositions: (1) EA:MA (0.9:0.1); (2) EA:MA (0.1:0.9); (3) EA:PA (0.9:0.1).

TABLE 5-2 The final values of the spin-spin relaxation time, t_2, end the effective network density of different polyester compositions

No.	Polyester	t_2, ms	$v_{eff} \cdot 10^4$ mol/cm^3
1	EA:MA (0.9:0.1)	1.20	2.3
2	EA:MA (0.1:0.9)	1.55	1.6
3	EA:PA (0.9:0.1)	1.40	2.1

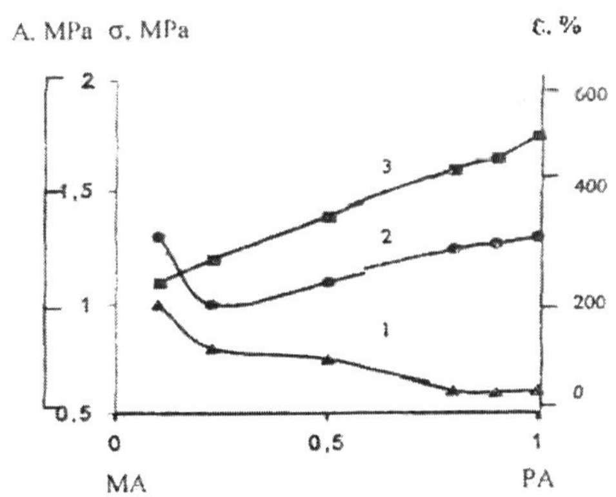

FIGURE 5-6 Dependence of adhesion, A, (1) tensile strength, σ, (2) and elongation at break, ε, (3) on the UPE composition (MA-PA).

As it is seen from Figure 5-6, the incorporation of double bonds in polyester by using as the copolymer MA leads to a significant increase of the hermetic strength, adhesion to glass, and to decrease of the elongation at break.

However, by using EA (Figure 5-7), the properties improvement is more remarkable.

The effect of the double bonds nature at incorporation into hermetic polyesters containing simultaneously maleic and the endic anhydride.

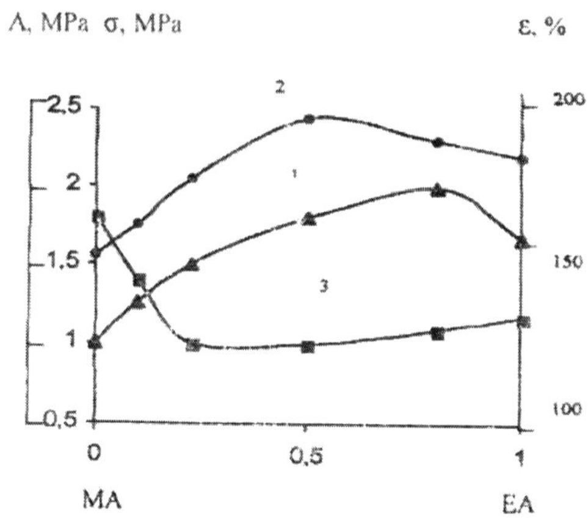

FIGURE 5-7 Dependence of adhesion, A, (1) tensile strength, σ, (2) and elongation at break, ε, (3) on the UPE composition (MA-EA).

The replacement of only 10% of MA by EA leads to increase of adhesion and tensile properties, and the elongation at break changes insignificantly.

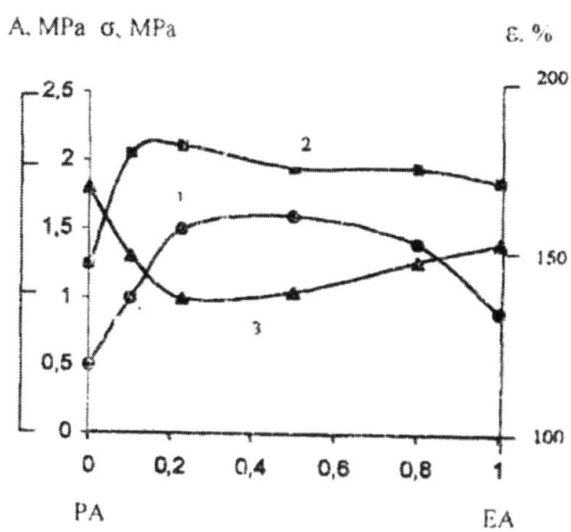

FIGURE 5-8 Dependence of adhesion, A, (1), tensile strength, σ, (2) and elongation at break, ε, (3) on the UPE composition (PA-EA).

We consider that the efficiency of the UPE as the modifier of the Thiokol hermetics is based, in the first place, on the difference in the electron density of the double bond. Therefore, the end double bond of oligoester acrylate is included in the isopropenyl group, and besides, it is conjugated with the carboxyl group which reduces somehow its activity in reactions of interaction both with the SH-groups and in the vulcanization reactions.

The MA incorporated in the polyester macromolecule is opened in the process of synthesis and forms fragments of maleic and fumaric acids, and, depending on the synthesis conditions, in the molecules some amount of the cis- and trans-double bonds will be present. The double bonds despite the conjugation show a high activity in the aforementioned reactions.

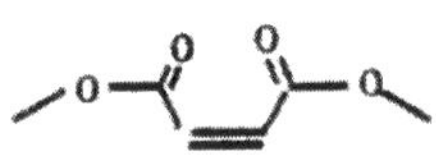

The maleic acid fragment, the cis-double bond.

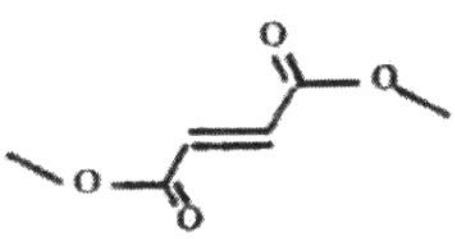

The fumaric acid fragment, the trans-double bond.

Finally, the nonsubstituted double bond incorporated into the polyester with EA molecule is located in the five-member ring and does not conjugate with any electron-active group.

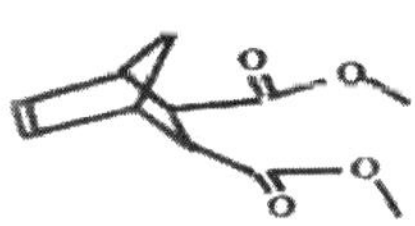

The molecule structure of the endic anhydride.

Besides, the ring undergoes certain stress which increases even more the activity of the double bond.

5.4 CONCLUSIONS

By using the ^{1}H and ^{13}C NMR methods, the composition of the unsaturated polyesters (UPE), which participate in the polysulfide (Thiokol type) hermetics solidification, was analyzed. The UPE plays a role of the additional structure-forming agent, and its activity depends on the nature of the double bonds. The incorporation of the UPE in the Thiokol hermetics in an amount of 1–3 mass parts leads to enhancing their adhesion to glass and duralumin and increasing of the tensile strength. The most advanced set of properties was achieved by using the UPE containing in its composition the endic anhydride.

KEYWORDS

- **Double Bond Activity**
- **Hermetic**
- **Maximal Activity**
- **Tri-Dimensional Structure**
- **Unsaturated Polyester**
- **Vulcanization Process.**

REFERENCES

1. Lucke, H. "Aliphatic Polysulfides". Publ. Huthing & Wepf. Verlag, Basel, 1994.
2. Averko-Antonovich, L. A.; Kirpichnikov, P. A.; Smyslova, R. A. "Polysulfide Oligomers and Hermetics Based on Them" (Russ.), L. Chimiya, 1983, 128 p.
3. Lee, T. C. P. Kautchuk i rezina (Russ.), 1995, No. 2, pp. 8-13.
4. US Patent No. 3,813,368, 1974.
5. Great Brit. Pat. No. 1,413,724, 1975.
6. Averko-Antonovich, L. A.; Kirpichnikov, P. A.; Romanova, G. V. Proc. KchTI (Russ.), 1969, v.40, Part 2, pp. 47–53.
7. Averko-Antonovich, L. A.; Kirpichnikov, P. A.; Romanova, G. V. Proc. KchTI (Russ.), 1969, v.40, Part 2, pp. 54–62.
8. Russ. Pat. (SSSR) No. 584,027, 1977
9. Minkin, V. S.; Averko-Antonovich, L. A.; Romanova G. V. Vysokomol. Soed., 1982, v.24B, No.7, pp. 806–809.
10. Cluff, E. F.; Cladding, E. K.; Parizer, R. J. Polym. Sci., 1960, v.45, No.8, pp. 341–345.
11. Deroum, E. "Modern Methods of NMR for Chemical Investigations" (Russ.). M. Mir, 1992, p.401.
12. Mukhutdinov, M. A.; Gubaidullin, Yu. N.; Yu, L. Liakumovich Kautchuk *i* rezina, 1998, pp. 33–35.
13. Mukhutdinova, T. Z.; Averko-Antonovich, L. A. Kautchuk i rezina, 1971, No. 12, pp. 10–13.
14. Bulai, A. Kh.; Slonim, I. Ya. Vysokomol. Soed., A, 1990, v.31, No. 12, pp. 952–956.

CHAPTER 6

THE IMPROVEMENT OF ADHESION PARAMETERS OF NEOPRENE-BASED ADHESIVE COMPOSITIONS

V. F. KABLOV, N. A. KEYBAL, S. N. BONDARENKO, and K. U. RUDENKO

Volzhskii Polytechnic Institute, Branch of Volzhskii State Technical University, ul. Engel'sa 42a, Volzhskii, Volgogradskaya oblast, Russia; E-mail: vpi@volpi.ru

CONTENTS

ABSTRACT

The possible mechanisms for an increase in the adhesion parameters of neoprene-based adhesive compositions modified with adhesion promoters on the basis of epoxy compounds and aniline derivatives are studied.

6.1 INTRODUCTION

Promising adhesion promoters for neoprene-based adhesive compositions are compounds containing amino and epoxy groups. As the amine-containing compounds, wastes of aniline production are of interest, because they are distinguished by their low volatility and stable enough structure whose disposal is an urgent environmental challenge. Still wastes of aniline production are characterized by the composition shown in Table 6-1.

TABLE 6-1 Composition of still wastes

Waste component	Content, mass parts
Aniline	15-80
Cyclohexylamine	0-10
Toluidine	2-4
Sodium hydroxide	1-3
Diphenylamine	3-20
Methaphenylene diamine	1-3
o, *p*-Aminophenol	1-6
High-molecular-weight tar compounds	6-45

According to our study, it was established that, upon interactions between bisphenol A-based ED-20 epoxy resin and still wastes, a product is formed that represents irregular, dark-brown, brittle granules that are well soluble in acetone, toluene, and gasoline and insoluble in water. Some properties of adhesion promoters prepared from this product are listed in Table 6-2.

TABLE 6-2 Properties of adhesion promoter

Parameters	Value
Melting point, °C	7
Mass fraction of volatile substances, %	0.11
Mass fraction of ash, %	0.20

6.2 EXPERIMENTAL PART

The data from an IR spectral study have confirmed the presence of epoxy, hydroxy, and amine groups, as well as residues of aromatic amines.

As is known, the chlorine atom in 3, 4 units of the neoprene macromolecule can easily be transferred to the allyl position (Figure 6-1), in which it possesses enhanced reactivity and is an active site of a polymer chain.[1]

The possible scheme of the modification of neoprene with developed adhesion promoters is confirmed by the infrared (IR) spectral data (Figure 6-2). As follows from the IR spectra, after the modification of neoprene with the developed adhesion promoters, new absorption bands appear in the 1450–1400 cm^{-1} range, which testifies to the transition from the amine to its salt.[2] Based on the scheme of the modification of neoprene and the analysis of published data, we constructed a model of a polymer with enhanced adhesive properties.[3] According to this model, the macromolecule of the adhesive should contain a set of specific functional groups (Figure 6-3).

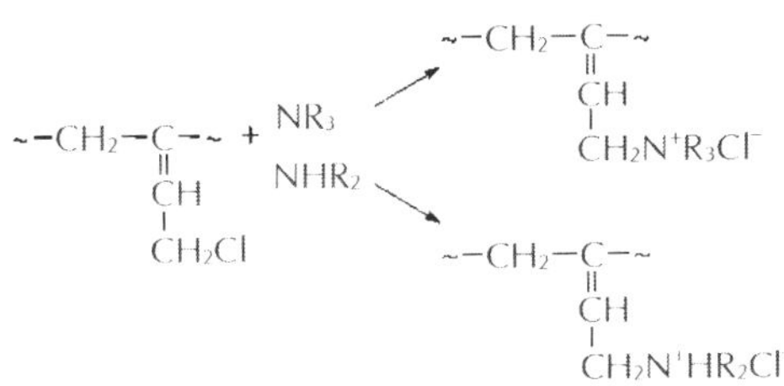

FIGURE 6-1 Scheme of neoprene modification.

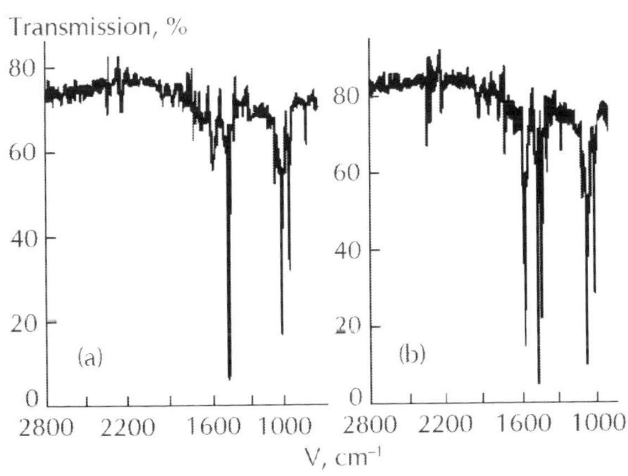

FIGURE 6-2 IR spectra of (a) initial and (b) modified neoprene.

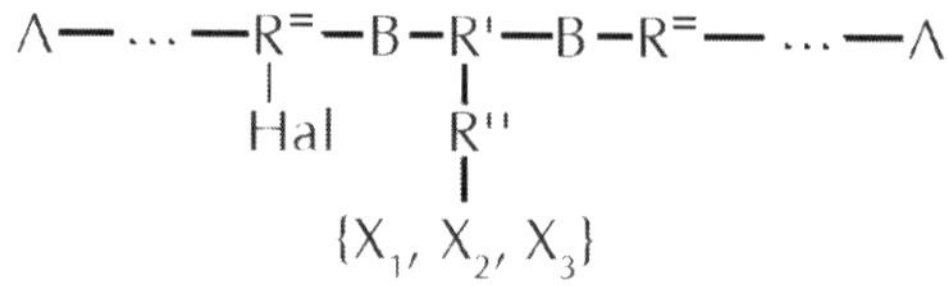

FIGURE 6-3 Specific functional groups of polymer with enhanced adhesive properties: A are the chain segments providing for the crystallizability of neoprene (units of 1,4-*trans*-isomer; R' is the methylene or ethylene group; R" is the alkylene group providing for the enhanced mobility of X groups; R= is the unsaturated (allyl) fragment (imparts the flexibility to macromolecule); B is the group responsible for the mobility of macromolecules; Hal is the halide atom; and X_1, X_2, and X_3 are functional groups containing nitrogen atoms, halide atoms, and hydroxyls.

To verify the validity of this model, we studied the adhesive bonding of vulcanizates based on polyisoprene (SIR) and ethylene-propylene (SREPT) rubbers using neoprene-based adhesive compositions modified with amine-containing adhesion promoters. Using local electron probe microanalysis, we studied the depth of the penetration of 88SA adhesive into vulcanizates during adhesive bonding. As a result, it was revealed that, after modification, the depth of the adhesive penetration into the SIR-based vulcanizate increases (according to chlorine profiling) from 16 to 36 μm (Figure 6-4).

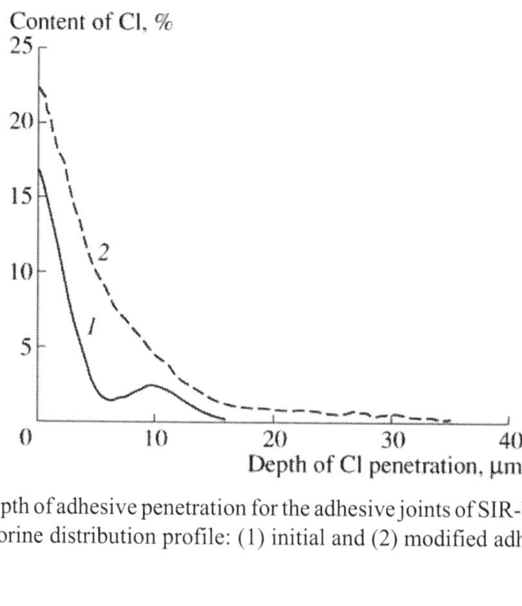

FIGURE 6-4 Depth of adhesive penetration for the adhesive joints of SIR-based vulcanizates by the data on chlorine distribution profile: (1) initial and (2) modified adhesive.

Analogous results were obtained for SREPT-based vulcanizates (Figure 6-5).

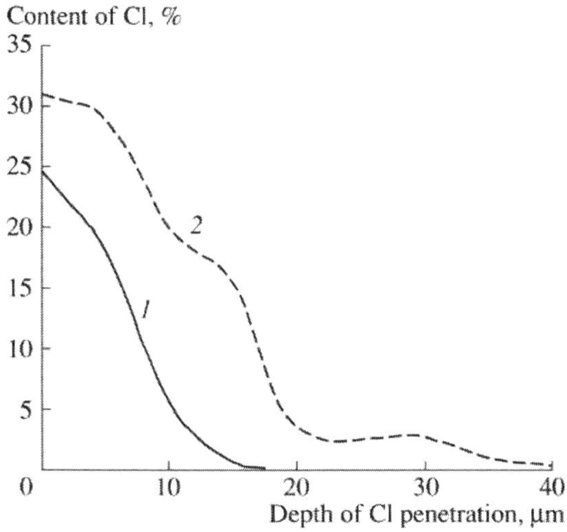

FIGURE 6-5 Depth of adhesive penetration for the adhesive joints of SREPT-based vulcanizates by the data on chlorine distribution profile: (1) initial and (2) modified adhesive.

As follows from the analysis of photomicrographs of the sections of adhesive-bonded samples of SIR-based vulcanizate taken with a scanning electron microscope (SEM), upon the modification of adhesive composition with amine-containing substances, the composition diffuses into the vulcanizate depth (Figure 6-6). The brighter color of the adhesive line with unmodified adhesive is explained by a complex contrast of image, in which the higher the chlorine concentration is, the brighter the chlorine-containing regions appear to be. The lighter-colored adhesive line with modified adhesive testifies to the diffusion of single segments of neoprene macromolecule into the vulcanizate bulk.

The surface of the adhesive film containing adhesion promoters is characterized by a more relief structure, which is supported by photomicrographs (Figure 6-7). Such a structure favors the better bonding of the adhesive film with the rough surface of vulcanizates, which increases the strength of the adhesive joint. This phenomenon is described by the mechanical theory of adhesion.

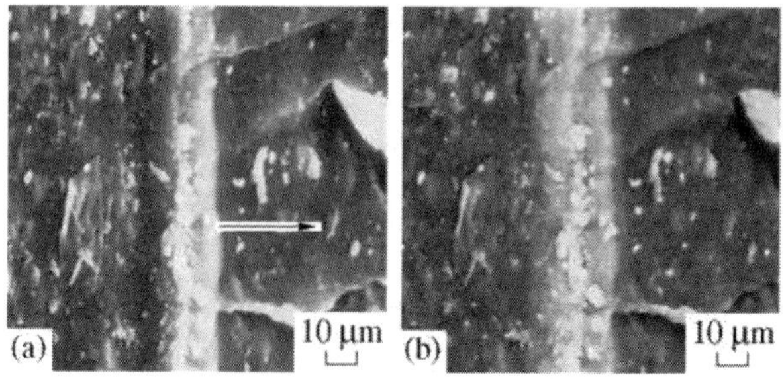

FIGURE 6-6 Photomicrographs of the sections of adhesive joints of SREPT-based vulcanizates bonded with: (a) 88SA adhesive and (b) modified 88SA adhesive.

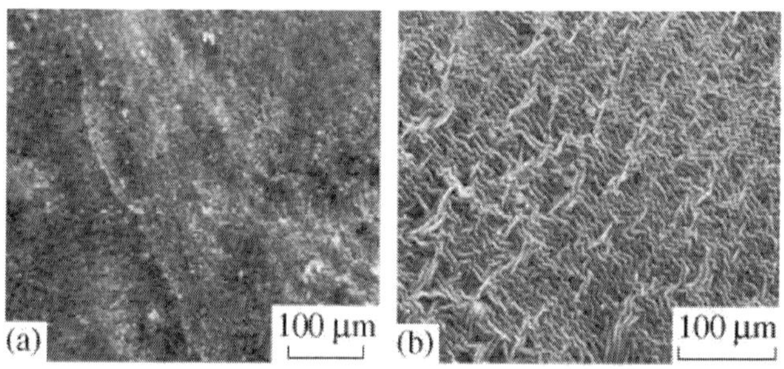

FIGURE 6-7 Photomicrographs of adhesive films based on neoprene: (a) initial and (b) modified films.

Based on the aforementioned, we performed comprehensive studies of the influence of the type and content of adhesion promoters on the strength of adhesive bonding of vulcanizates based on different rubbers to one another and to metal (Table 6-3).

TABLE 6-3 Comparative strength characteristics of initial and modified adhesive composition of 88SA grade

Parameters	88SA adhesive	Modified 88SA adhesive
Viscosity by VZ-246 method (0.6 mm), s	27.5	28.1
Shear strength, MPa, of adhesive joints:		
SIR-3 + SIR-3	1.02	1.39
SREPT-40 + SREPT-40	1.17	1.59
SRN-18 + SRN-18	0.95	1.38
Tensile strength, MPa, of adhesive joints:		
SIR-3 + Steel 3	1.20	1.91
SREPT-40 + Steel 3	1.38	1.93
SRN-18 + Steel 3	0.95	1.78

6.3 RESULTS AND DISCUSSION

According to the data obtained, the modified adhesive composition is superior in its adhesion characteristics to current commercial adhesive compositions of the 88 series. It is established that, upon the addition of developed adhesion promoters to neoprene-based compositions, the strength of adhesive bonding of vulcanizates based on different rubbers to metal increases by 35–45%, whereas the bond strength of vulcanized rubbers to one another rises by 40–80% (Table 6-3). These data refer to joints bonded with modified 88SA adhesive. For other adhesives of this series, for example, 88NT and 88NP adhesives, the shear and tensile strengths increase to even greater extents.

6.4 CONCLUSION

As a result of the studies performed, we revealed possible mechanisms of the improvement of adhesion parameters of neoprene-based adhesive compositions modified with adhesion promoters on the basis of epoxy compounds and aniline derivatives.

It was established that, upon the addition of developed adhesion promoters to neoprene-based adhesive compositions, chemical modification of neoprene macromolecules occurs, which leads to an increase in their flexibility and mobility and, hence, to deeper diffusion into the internal layers of bonded vulcanizates.

KEYWORDS

- **Adhesion**
- **Amine-Containing**
- **Aniline Derivatives**
- **Epoxy Compounds**
- **Neoprene.**

REFERENCES

1. Zakharov, N. D. Chloroprene Rubbers and Related Vulcanizates (Chemistry, Moscow, 1978) [in Russian].
2. Kazitsina, L. A.; Kupletskaya, N. B. Application of UV, IR, NMR, and Mass Spectroscopy in Organic Chemistry (Publishing house of the Moscow University, Moscow, 1979) [in Russian].
3. Vakula, V. L.; Pritykin, L. M. Physical Chemistry of Polymer Adhesion (Chemistry, Moscow, 1984) [in Russian].
4. Keibal, N. A.; Bondarenko, S. N.; Kablov, V. F.; Goryainov, I. Y. Study of Mechanisms for the Improvement of Adhesive Properties of Neoprene-Based Compositions by Their Modification with Amine-Containing Compounds // Polymer Science, Series D. Glues and Sealing Materials – 2008, Vol. 1, No. 3, pp. 151–153.

CHAPTER 7

ULTRASOUND EFFECT ON THE JOINT PROCESSING OF DIFFERENT CHEMICAL POLYMERS

I. A. KIRSH[1], T. I. CHALYKH[*], and D. A. POMOGOVA[1]

[1]*Moscow State University of Food Production*

**Plekhanov Russian University of Economics*

E-mail: irina-kirsh@yandex.ru, tichalykh@rambler.ru

CONTENTS

ABSTRACT

The results of the investigations of the ultrasound effect on the joint mixtures, which model polymer waste recycling, have been presented. The effect of processing cycles on the change in viscosity, molecular mass, chemical and physical properties of original polymers and their mixtures has been studied.

7.1 INTRODUCTION

In recent years, the problem of polymer waste recycling has drawn a significant attention. Most difficulties arise in the recycling of mixed waste, since polymers that make up the mixture are often thermodynamically incompatible. Separating waste polymeric film materials is not cost-effective, for example, in the case of multilayer films. In addition, after recycling, polymers can change their original properties, so there is a need for effective methods of modification in order for polymeric materials to be recycled for further use.[1,2]

The interest in various polymer modification methods has been continuously growing, including the use of ultrasonic vibration energy.[1,2] A considerable amount of work is devoted to the study of the effect of ultrasound on polymer properties and structure, which are described in detail in ref. [2–5]. Most of the research on the ultrasound effect have been performed using solutions of polymers where researchers studied the processes of destruction of polymers and molecular weight reduction that could pull together solubility parameters of thermodynamically incompatible polymers.[3,4]

In case of the ultrasonic effect on polymer melts, when the polymer is in a viscous state and has high viscosity as compared to the solutions, the literature has contradictory information. Some authors suggest that there is no destruction under the influence of ultrasound and there is only a change in the viscosity of polymers,[2] while others note a change in the properties of polymers due to the processes of destruction.[5] In this regard, we have considered it appropriate to investigate the influence of ultrasonic energy on a polymer melt to develop a new way of recycling of waste made up of polymers of different chemical nature.

7.2 THE EXPERIMENTAL TECHNIQUE

As a research subject, we have selected high-density polyethylene (PE) mark Kazpelen 15813-020 (GOST 16337-77), polypropylene (PP) mark Kaplen

PP-01003 (GOST 16337-77), polyamide-6 (PA) mark 210/310 (GOST 6-06-S9-83), polyethylene terephthalate (PET) mark G-80 (GOST R 51695-2000), and edges of a multilayer polyethylene-polyamide film mark FS5150 manufactured by the company Criovak, Volgograd, Russia.

We investigated the individual polymers as well as their model mixtures of varying composition. Initial samples were prepared using the extrusion method as strands. For further investigations, the samples were pressed into films (for study by IR spectroscopy). The films were obtained by pressing at temperatures: PE–130°C, PP–210°C, PA–240°C, PET–265°C.

The samples were prepared on a laboratory extruder with an ultrasonic vibration attachment mounted on the forming tool as described in ref. [1,6]. Molten polymer has been treated with the oscillation frequency of 22 kHz. Experimental samples based on the polymers and their mixtures PE-PA and PP-PET of various composition ratios have been obtained from the extruder with ultrasound. As a reference, polymers of the same brand marks have been used, which were obtained with the same apparatus but without exposure to ultrasound. The average time of the ultrasonic treatment of polymer melt was 3 seconds; processing of the polymer melt was conducted in cycles with the maximum number of cycles of 4. Each cycle comprised the processing of the polymer in the extruder to obtain strands, which were exposed to grinding in a knife-type crusher.

The evaluation of the polymer rheological properties has been carried out with a capillary viscometer (GOST 11645-86).

The average molecular weight of the polymers is calculated by formula (7-1):

$$\lg \eta = 3.4 \times \lg M + A \tag{7-1}$$

where η is the polymer viscosity, M is the average molecular weight of the polymer, and A is a constant.

Next, according to formula (7-2), the percentage of the polymers average molecular weight is calculated:

$$W_M = \frac{M_0 - M_i}{M_0} \times 100\% \tag{7-2}$$

where M_0 is the initial molecular weight of the polymer, M_i is the average molecular weight of the polymer with or without sonication.

Physical and mechanical properties of the polymers and their compositions have been determined according to the standard GOST 14236-81 (Polymer films. Tensile strength test method).

To assess structural changes in the polymers and polymeric compositions, a Fourier transform infrared spectroscopy (FTIR) spectroscopy method has been used. Since the infrared (IR) spectra of the studied polymers had changes in the areas of optical densities related to various oxygenated groups, such as carboxyl (1720sm^{-1}), carbonyl (1650, 1380, 1240sm^{-1}), hydroxyl, and ether groups (1165, 1090 cm^{-1}), to determine the effect of ultrasound on the structure of the polymers, we used a total coefficient for oxygen-containing group as an indicator, which can facilitate the comparative analysis on the selected areas of IR spectra. Table 7-1 contains the formulas for calculating the total coefficient for oxygen-containing groups of the starting polymers.

TABLE 7-1 Calculation of total coefficient for oxygen-containing groups of polymers

Polymer	Total coefficient for oxygen-containing groups
PE	$\sum K = \frac{D1720}{D1460} + \frac{D1650}{D1460}$
PP	$\sum K = \frac{D1720}{D1460} + \frac{D1650}{D1460} + \frac{D1165}{D1460}$
PA	$\sum K = \frac{D1720}{D1460} + \frac{D1380}{D1460} + \frac{D1165}{D1460}$
PET	$\sum K = \frac{D1720}{D870} + \frac{D1240}{D870} + \frac{D1090}{D870}$

7.3 RESULTS AND DISCUSSION

At the first stage, an average molecular weight of the polymers has been calculated. The initial molecular weights of the polymers are as follows: PE 80000, PP 70000, PA 35000, PET 45000. Figures 7-1 and 7-2 show the dependence of molecular weight reduction in the initial polymers on the time of sonication of PE, PA (Figure 7-1), and PP, PET (Figure 7-2).

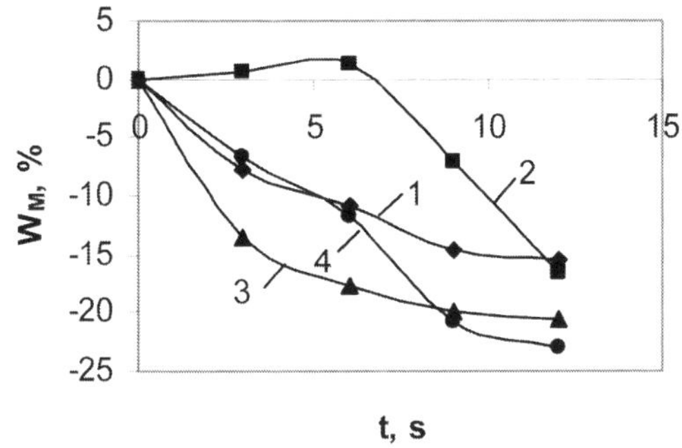

FIGURE 7-1 Changes in PE and PA molecular weight (in % of the initial value) caused by the different time of ultrasonic processing (1–PE with ultrasound, 2–PE reference, 3–PA with ultrasound, 4–PA reference).

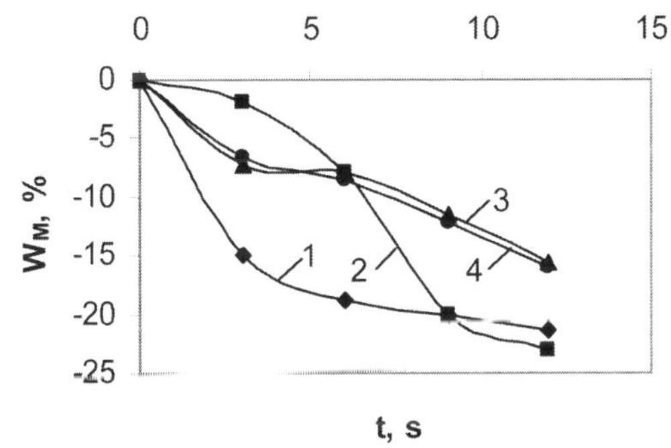

FIGURE 7-2 Changes in PP and PET molecular weight (in % of the initial value) caused by the different time of ultrasonic processing (1–PP with ultrasound, 2–PP reference, 3–PET with ultrasound, 4–PET reference).

As can be seen from the data obtained, as early as in the first cycle of sonication (for 3 and 6 seconds of exposure to ultrasound) there occurs a strong decrease in the average molecular weight of the polymers. Further, ultrasonic impact does not lead to any significant changes. Ultrasound has the greatest impact on PP and PA. It should be noted that the change in the average molecular weight of the PET depends little on the ultrasonic impact. In this case, PET in largely influenced by a multiple extrusion process, which is a determining factor in the change of the average molecular weight. These data suggest that the change in the average molecular weight of the polymers can be associated with the processes of destruction of polymers during the extrusion processing, intensified by the ultrasound. To confirm this assumption, the polymeric materials have been studied with IR spectroscopy (Fourier method). The obtained spectra of the polymers were processed, and the content and changes in the amount of oxygen-containing groups have been determined. As a result of the spectra processing, a total oxygen-containing group coefficient has been calculated (Table 7-2).

TABLE 7-2 Values of the total oxygen-containing group coefficient depending on the number of cycles of polymer processing

Polymer	Number of processing cycles	Polymer total oxygen-containing group coefficient	
		With ultrasound	Without ultrasound
PE	1	0.2370	0.0670
	2	0.2640	0.0900
	3	0.2690	0.1980
	4	0.2720	0.2850
PP	1	0.6313	0.2157
	2	0.6741	0.4180
	3	0.7292	0.7371
	4	0.7654	0.8100
PA	1	1.8700	1.0260
	2	1.9968	1.7070
	3	2.1725	1.8781
	4	2.1911	2.3071
PET	1	4.9800	4.7390
	2	5.1800	5.6361
	3	5.7400	6.1424
	4	5.9900	6.2900

As can be seen from the data, ultrasonic treatment leads to a significant change in the oxygen-containing groups in PE, PP, and PA, with the greatest increase in this indicator observed in the first and second processing cycle (3 and 6 seconds of sonication). Further, ultrasound processing does not cause any significant increase in oxygen-containing groups in polymers. This can be explained by the fact that the process of ultrasonic degradation is directly related to oxygen, which is distributed in the polymer system with stirring. Moreover, a parallel process can occur—a recombination of the formed radicals, because for all the studied polymers the values of oxygen-containing groups with 12 seconds sonication do not exceed the values for the oxygen-containing groups of samples obtained without the influence of ultrasound. Since it is known that the process of degradation of polymers is intensified with increasing temperature, rheological properties of polymers at different temperatures have been studied. For example, Figure 7-3 shows the dependence of the effective viscosity on the temperature of PP obtained with and without ultrasound.

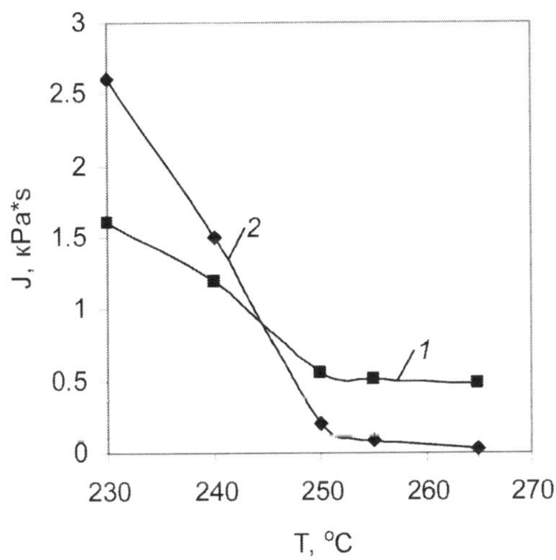

FIGURE 7-3 Dependence of the effective viscosity change on temperature for PP obtained with and without ultrasound (PP–1 with ultrasound, 2–PP reference).

An increase in temperature leads to a decrease in the effective viscosity of PP, although PP sonication leads to a less intense decrease of this parameter. This suggests that with increasing temperature, ultrasound may intensify the process of recombination of radicals, and this effect can be used to obtain compositions of thermodynamically incompatible polymers.

Physical and mechanical properties of the compositions based on PP and PET have been obtained and investigated in the ratio of 20:80, 15:85, 10: 90, 5:95, respectively (Table 7-3).

TABLE 7-3 Values of breaking stress and elongation at break as a function of the content of PP in PET after the first processing cycle

Amount of PP in PET%	Breaking stress, σ, MPa		Elongation at break, ε,%	
	With ultrasound	Without ultrasound	With ultrasound	Without ultrasound
0	46.0	50.0	375	300
5	42.0	48.0	200	45
10	36.0	47.0	125	8
15	28.0	39.0	85	5
20	20.0	22.0	25	3

The results show that even a small amount of PP in PET results in a decrease in breaking stress and increase in the elongation of the polymer compositions. Breaking stress in PET obtained with application of ultrasound at a content of 5% PP in PET is 20% lower than PET without ultrasound, but nonetheless these values are sufficiently high for a possible exploitation of the polymer compositions. The values of elongation at break of compositions treated with ultrasound, almost in the entire interval of component ratios, exceed the value of this parameter of the mixture obtained without the influence of ultrasound.

In the next step of the research, the physical properties of compositions based on PE and PA (Table 7-4) have been investigated.

TABLE 7-4 The values of breaking stress and elongation at break as a function of the content of the PE in PA after the first processing cycle

Amount of PE in PA%	Breaking stress, σ, MPa		Elongation at break, ε,%	
	With ultrasound	Without ultrasound	With ultrasound	Without ultrasound
0	8.8	12.1	420	600
10	18.3	10.8	320	114
20	18.9	8.2	205	53
30	13.4	5.3	88	30
40	10.8	2.1	65	19
50	6.3	1.7	52	9

It should be noted that the change in physical and mechanical properties of compositions with increasing amounts of PE in PA occurs to a lesser extent for the samples obtained with ultrasound. In this case, the parameters of strength properties of the sonicated composition is considerably higher than that of compositions obtained without ultrasound. This is probably due to the fact that sonication in PE and PA leads to an increase in oxygen-containing groups and allows for bringing closer the solubility parameters of polymers by changing the polarity of PE.

In this research, the recycling of multilayer PA-PE films mark FS5150 has been performed. The ratio of PE and PA in multilayer films was 80:20 mass. %, respectively. As a result, it was noted that in obtaining the composition of waste using sonication, the values of breaking stress were 24 MPa, elongation at break were 230%, which is high for film packaging materials.

7.4 CONCLUSIONS

As a result of the research, it was found that ultrasonic processing of polymer melts leads to a decrease in the average molecular weight of PP, PE, and PA by 15–20% compared to initial values. The maximum decrease in the average molecular weight of the polymers occurs in 9 seconds ultrasound processing, and further processing does not change this indicator significantly. The greatest change in the molecular weight after ultrasound processing has been observed in PP and PA. Ultrasonic treatment leads to a 1.5–2 times increase in the content of oxygen-containing groups in PE, PP, PA, and the largest increase in this indicator is observed for 3 and 6 seconds of ultrasonic treatment (1 and 2 processing cycle).

Ultrasound exposure intensity on polymer melts depends on the nature of the polymer. It was shown that ultrasonic treatment has little effect on the average molecular weight of PET and the amount of oxygen-containing groups.

It has been established that the application of ultrasound during the joint processing of thermodynamically incompatible polymers extends the range (quantitative ratio) of compositions based on polyamide and polyethylene, as well as on polypropylene and polyethyleneterephthalate.

KEYWORDS

- **Molecular Mass**
- **Physical and Mechanical Properties**
- **Polymers Ultrasound**
- **Viscosity**

REFERENCES

1. Ananyev, V. V.; Kirsh, I. A. Recycling polymer materials. M.: MSUAB, 2007, 126p. *(in Russian)*.
2. Ganiev, M. M. Improving the performance of polymer composites by ultrasonic. Kazan: KSTU. 2007, 81p. *(in Russian)*.
3. Myson, T. Chemistry and ultrasonic. M.: Mir 1993, 190p. *(in Russian)*.
4. Moor, D. The impact of ultrasonic on the polymer solutions. Urbana-Champaign. 2009, 85p.
5. Friedman, M. L.; Peshkovsky, S. L. Changing the polymers properties under the ultrasonic. Advance in Polymer Science. Berlin. 1993, p. 256.
6. Kirsh, I. A.; Chalykh, T. I. Comprehensive modification of secondary PET for packaging. // Foodstuffs commodity science, №11, 2014 // app.№2 – 2014, p.29–36. *(in Russian)*.

CHAPTER 8

ELECTRICAL TRANSPORT PROPERTIES OF POLY(ANILINE-CO-N-PHENYLANILINE) COPOLYMERS

A. D. BORKAR

Department of Chemistry, Nabira Mahavidyalaya, Katol Dist. Nagpur 441302 (M.S.) India; E-mail: arun.borkar@rediffmail.com

CONTENTS

ABSTRACT

Poly(aniline-co-N-phenylaniline)s copolymers have been synthesized by the chemically oxidative copolymerization of aniline and N-phenylaniline in aqueous hydrochloric acid medium under nitrogen atmosphere at 0–4°C. The molar feed ratio of monomers is varied to prepare copolymers of different composition. The copolymers are characterized by Fourier transform infrared spectroscopy (FTIR) and UV-visible spectroscopy. The solubility and spectroscopic analysis suggest that the product is a copolymer of aniline and N-phenylaniline. The electrical conductivity of the compressed pellets is measured by a method called two probe method. The electrical conductivity of copolymers is found to be less than polyaniline but processability has been improved significantly in solvents like 1-methyl-2-pyrrolidone (NMP), dimethyl sulphoxide (DMSO), and dimethyl formamide (DMF). The decrease in room temperature conductivity of the copolymers with increase of the N-phenylaniline is due to the introduction of phenyl groups in to the copolymer, which reduces the conjugation of the polymer chain and reduce the mobility of charge carrier along the main chain. From temperature dependence of electrical conductivity, charge localization length and hopping distance are calculated and the effect of substituent and dopant on crystallinity is discussed. The temperature dependence of dielectric data and conductivity suggests that copolymers are quasi 1D-disordered state composed of 3D-metallic crystalline regions in 1D-localized amorphous regions.

8.1 INTRODUCTION

In the recent years, conducting polymers have received considerable attention worldwide due to their novel electronic and electrical properties. These polymers have diverse applications ranging from energy storage, sensors, anticorrosive materials, electromagnetic interference shielding, electrostatic charge dissipation, organic light emitting diodes, plastic solar cells, and supporting material for catalysis.[1,2]] However, among other conducting polymers, polyaniline (PANI) has been extensively studied not only because its electronic conductivity can be tuned by adjusting the oxidation state and degree of doping of backbone but also due to its environmental stability as well as economic feasibility. Therefore, PANI is a promising futuristic material for various techno-commercial applications.

PANI is made up of a combination of fully reduced (B-NH-B-NH) and fully oxidized (B-N=Q=N-) repeating units, where B denotes a benzenoid and Q denotes a quinoid ring. Thus, different ratios of these fully reduced and fully oxidized units yield various forms of PANI, such as leucoemeraldine (100% reduced form), emeraldine base (50% oxidized form), and pernigraniline (fully oxidized form). However, all of these forms are electrically insulating in nature. Doping of emeraldine base with a protonic acid converts it into conducting form protonated emeraldine (emerdine salt).The main issue with PANI is processing difficulties due to its infusibility and relative insolubility in common organic solvents. It can be made processable/soluble either by polymerizing functionalized anilines[3] or by copolymerizing aniline with substituted monomers.[4] The copolymerization is a powerful method to improve processability of conducting polymers. In general, solubility of substituted PANIs in organic solvents is significantly higher than PANI. However, their thermal stability and electronic conductivities are substantially lower than doped PANI. In order to maintain balance between conductivity, stability, and processability, copolymerization has been performed.

In this paper, poly(aniline-co-N-phenylaniline) (PANI-co-PNPANI) copolymers have been synthesized by chemical peroxidation method. These copolymers are characterized by FTIR and UV-visible spectroscopy. Their electronic conductivities have been measured by two probe technique. Variable range hopping (VRH) model has been applied depending upon the nature of variation of conductivity with temperature.

8.2 EXPERIMENTAL PROCEDURE

8.2.1 SYNTHESIS OF HOMOPOLYMERS

First, PANI was chemically synthesized[5–7] using ammonium peroxodisulphate as an oxidant in an aqueous 1M HCl in nitrogen atmosphere at 0–4°C.

Second, Poly(N-phenylaniline) (PNPANI) was chemically synthesized[5–7] using ammonium peroxodisulphate as an oxidant in an aqueous 1M HCl and CH_3CN (1:1 ratio) in nitrogen atmosphere at 0–4°C.

8.2.2 SYNTHESIS OF COPOLYMERS

PANI-co-PNPANI copolymers were chemically copolymerized[8,9] from the monomers, aniline, and N-phenylaniline using ammonium peroxodisulphate as an oxidant in an aqueous 1M HCl and CH_3CN (1:1 ratio) in nitrogen atmosphere at 0–4°C. The molar feed ratio of starting aniline monomer was varied to result in copolymers of different compositions. The homopolymers and copolymers obtained from reaction medium were filtered and washed with distilled water and methanol several times to remove unreacted monomers and then dried in an air oven at 70°C for 8 hours.

8.2.3 CHARACTERIZATION

The solubility of the homopolymers and copolymers salt form was tested by dissolving each material in DMF. The mixture was kept for 24 hours at room temperature, after which the solution was filtered through sintered glass crucible G_4. The room temperature solubility was recorded. UV-Visible spectra of homopolymers and copolymers were recorded at room temperature in NMP, DMF, and DMSO in 190–700 nm range using UV-240 Shimadzu Automatic Recording Double Beam Spectrophotometer. FTIR spectra of homopolymers and copolymers were recorded on 550 Series II, Nicolet, using KBr pellet technique in the range of 400–4000 cm^{-1}. DC electrical conductivity of polymer samples was measured by the two probe technique in the temperature range of 298–398 K. Dry-powdered samples were made in to a pellet under hydraulic press IEBIG and placed between electrodes in a cell. Resistance was measured on a DC resistance bridge LCR Meter 926. The conductivity value was calculated from the measured resistance and sample dimensions.

8.3 RESULTS AND DISCUSSION

The solubility of homopolymers and copolymers in DMF are determined. PANI (0.0414 g/dl) and PNPANI (0.0732 g/dl) homopolymers show low and high solubility as compared to copolymers, which indicates the incorporation of the substituted monomer units in the copolymer which gives a solubility intermediate between the corresponding homopolymers. The substituent introduce flexibility into the rigid PANI backbone structure as a result copolymers shows higher solubility than PANI.

The empirical repeat unit for homopolymers and copolymers is shown in Figures 8-1 and 8-2 in the following.

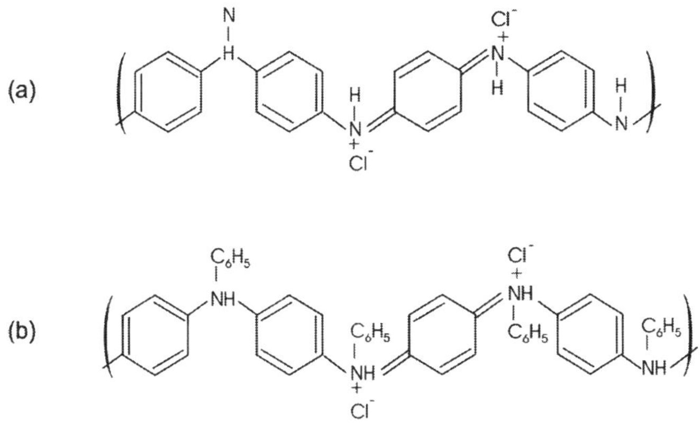

FIGURE 8-1 Repeat unit for homopolymers. (a) PANI (b) PNPANI.

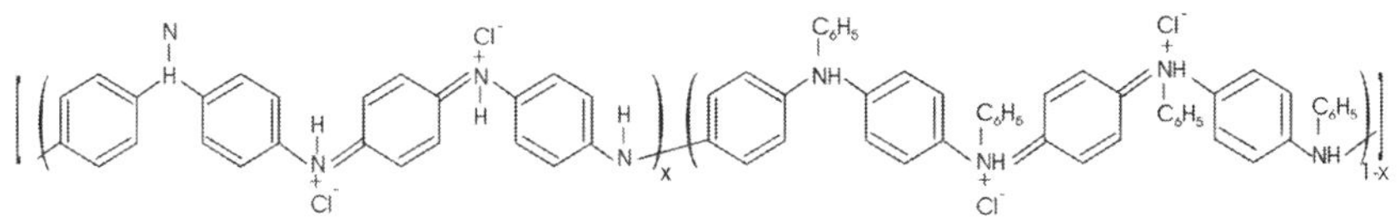

FIGURE 8-2 Repeat unit for copolymers.

The absorption bands of homopolymers and copolymers are recorded in NMP, DMF, and DMSO. The corresponding bands are presented in Table 8-1.

TABLE 8-1 UV-Vis. absorption bands of homopolymers and copolymers

Polymer/Copolymer	UV-Vis. absorption band (nm)					
	NMP		DMF		DMSO	
PANI	330	628	328	618	314	618
PNPANI	314	600	310	610	310	600
PANI-co-PNPANI (60:40)	292	560	298	560	294	580
PANI-co-PNPANI (30:70)	290	560	296	540	290	580

There are two absorption bands in the electronic spectra of homopolymers and copolymers. The band around 290–330 nm (4.278-3.760 eV) is assigned to $\pi - \pi^*$(bandgap) transition (which is related to the extent of conjugation between the adjacent rings in the polymer chain) and the band above 540 nm (exciton band) is due to interband charge transfer associated with excitation of benzenoid to quinoid moieties (formation of exciton). These bands change with solvents. The $\pi - \pi$ * band of copolymers shows hypsochromic shift with the increase in dielectric constant of the solvent.[10] The exciton band shows the bathochromic shift with an increase in dielectric constant of the solvent. The excitation leads to the formation of molecular exciton (positive charge on the adjacent benzenoid units and negative charge centred on quinoid unit).[11] This interchain charge transfer from HOMO to LUMO may lead to the formation of positive and negative polarons. A polymer in a solvent of high dielectric constant may exist in coil-like conformation (decrease in conjugation) and a less polar solvent provides thermodynamically more stable chain conformation and restricts the polymer to lower energy, high planarity state. Such shift may increase the conjugation of the system which yields a lower energy transition, red shift. The $\pi - \pi^*$ band in copolymers shifts to the lower wavelength as the percentage of N-phenylaniline in copolymer increase, which may be due to the addition of more phenyl groups which twist the torsion angle, which are expected to increase the average bandgap in the conjugated polymer chain. The continuous variation of the wavelength and intensity of the UV-vis bands may have resulted from the copolymerization effect of N-phenylaniline and aniline. In other words, the polymer formed by the oxidative polymerization of N-phenylaniline with aniline was the copolymer of two monomers rather than a mixture of two homopolymers.

The FTIR spectra of homopolymers and copolymers are shown in Figure 8-3 and spectral data are recorded in Table 8-2. The FTIR spectra of PANI show a broad band at 3431 cm^{-1} characteristics of N-H stretching. The band in the range 810-819 cm^{-1} assignable to 1,4 substituted aromatic ring indicating the bonding in polymers and copolymers were through 1,4 position. The bands of 1569 cm^{-1} and 1492 cm^{-1} are assigned to quinoid nitrogen (N=Q=N) and benzoid (N-B-N) ring stretching. In the spectra of copolymer, there appears an absorption band at 1310 cm^{-1} (C-N stretching) and 1500 cm^{-1} (benzenoid stretching) indicating the existence of phenyl group on a benzene ring. The relative intensity of 1600–1500 cm^{-1} is between those of PNPANI and PANI, and so the coexistence of N-phenylaniline and aniline units in the copolymer was further confirmed.

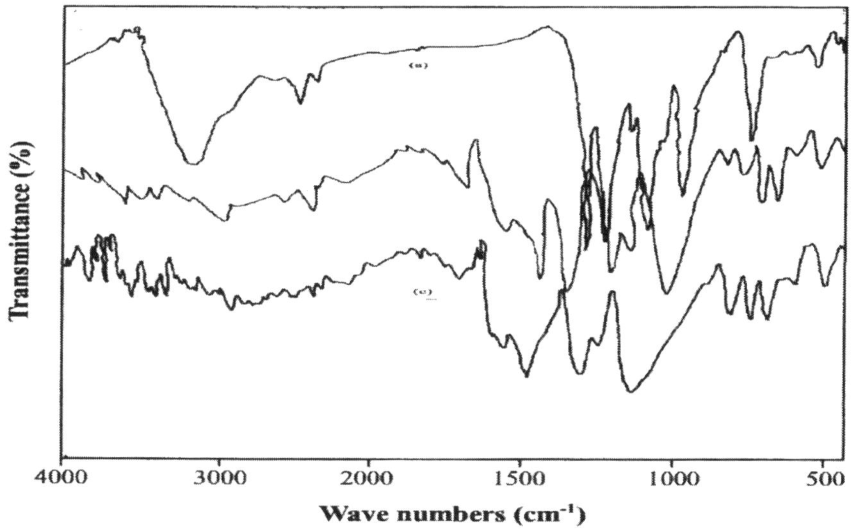

FIGURE 8-3 FTIR Spectra of (a) PANI (b) PNPANI (c) PANI-co-PNPANI (60:40).

TABLE 8-2 FTIR spectral data of homopolymers and copolymers

Polymer/Copolymer	Wavenumber (cm-1)						
PANI	819	1145	1246	1300	1492	1569	3431
PNPANI	810	1160	1250	1315	1510	1560	
PANI-co-PNPANI (60:40)	813	1141	1242	1310	1485	1556	

The temperature dependence of electrical conductivity data was fitted to an Arrhenius equation as follows:

$$\sigma(T) = \sigma_0 \exp{-Ea/2kT} \quad (8\text{-}1)$$

The measured values were plotted semi-logarithmically as a function of reciprocal of temperature (Figure 8-4[a]). The conductivity is found to increase with temperature. However, there are deviations at lower temperature. In the present studies, the temperature dependence of σ(T) is fitted to the Zeller equation:[12]

$$\sigma(T)\ \alpha\ \exp{-(T_0/T)^{1/2}} \quad (8\text{-}2)$$

where T_0 is the Mott characteristics temperature and is a measure of the hopping barrier.

Figure 8-4(b) describes the interchain conductivity where only the neighbor VRH of charge (which is quasi-1-dimensional) is considered.[13]

In Zeller equation, T_0 is related to delocalization length (α^{--1}), most probable hopping distance (R) and hopping energy (W) by the relations (8-3), (8-4) and (8-5):[14]

$$T_0 = 8\alpha/N\ (EF)\ Z\ K \tag{8-3}$$

$$R = (T_0/T)^{1/2}\ \alpha^{--1}/4 \tag{8-4}$$

$$W = ZKT_0/16 \tag{8-5}$$

where Z is the number of nearest neighboring chains (~4), K is the Boltzmann constant, and $N(E_f)$ is the density of states at Fermi energy for the sign of spin which is taken as 1.6 states per eV (2-ring unit suggested for PANI.[15]

The value of T_0 is determined graphically from log σ(T) vs. 100/$T^{1/2}$ (Figure 8-4[b]) and other parameters are computed from the data (Table 8-3). It is observed that T_0 for PNPANI is more than PANI (maximum localization length). Therefore, correspondingly the localization length and average charge hopping distance decrease on increasing the amount of N-Phenylaniline in copolymer chain. This increase in the electron localization is due to the presence of phenyl group in the copolymer chain. The temperature dependence of electrical conductivity fits equation (8-2) (Figure 8-4[b]) for homopolymers and copolymers, which suggests that the charge conduction is quasi-1D VRH between nearest neighboring chains.[16]

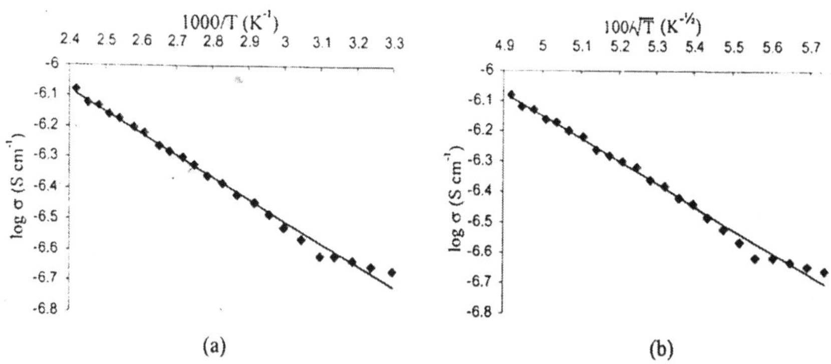

FIGURE 8-4 Temperature dependence of electrical conductivity of PANI-co-PNPANI (60:40). (a) Arrhenius equation and (b) Zeller equation.

TABLE 8-3 Transport properties of homopolymers and copolymers

Polymer/Copolymer	σ (S/cm)	T_0 (K)	α^{-1}(nm)	R(nm)	w(eV)	Ea(eV)
PANI	6.550 x 10^{-2}	3.315 x10^3	6.995	5.736	0.071	0.047
PNPAN	1.026 x 10^{-9}	7.260 x10^3	3.194	3.877	0.156	0.137
PANI-co-PNPANI(60:40)	2.191 x 10^{-7}	5.250 x10^3	4.416	4.558	0.101	0.133
PANI-co-PNPANI(30:70)	1.517 x 10^{-8}	5.804 x10^3	3.995	4.335	0.120	0.146

In case of HCl-doped PANI, Cl^- ions are small and the interchain separation is also small, resulting in appreciable coupling interaction between the chains. Thus, charge carriers could easily hop from one chain to other to give of 3D-VRH conduction. However, in the case of copolymers although the charge carriers hop between 3D-VRH but interchain hopping has been strongly inhibited leading to 1D-VRH. The reduction in interchain hopping may be attributed to the presence of Phenyl group in the copolymers that reduce the coupling interactions between the chains. Therefore, the net effect is the 1D-VRH conduction in case of copolymers.

The electrical conductivity of copolymer salts are measured and compared to that of homopolymers; the results are summarized in Table 8-3. The PANI and PNPANI show conductivity of order of 6.55 x 10^{-2} S cm^{-1} and 1.026 x 10^{-7} S cm^{-1}, respectively, while copolymer shows lower conductivity than PANI which indicates that the ionization potential, bandgap, and bandwidth are affected by the torsion angle between adjacent phenyl rings to relieve steric strain in the polymer chain. The conductivity of copolymer decreases with increasing the content of N-phenylaniline in copolymers is due to the introduction of phenyl groups which reduce the conjugation of the polymer chain and reduces the mobility of charge carrier along the main chain.

8.4 CONCLUSIONS

The chemical copolymerization of aniline with N-phenylaniline has been carried out. The ratio of the two monomers has influence on the copolymerization process due to different reactivity of monomers. Elemental analysis data infer that one anion is substituted for every two phenyl rings for complete protonation. Spectroscopic information indicates the presence of both the comonomers in the polymer chain. The temperature dependence of conductivity suggests that copolymers are quasi 1D-disordered state composed of 3D-metallic crystalline regions in 1D-localized amorphous regions. The substituent group decreases the conductivity of copolymers and the coplanarity of the polymer chain gets disrupted. It also reduces the mobility of the charge carriers along the main chain. The solubility, electrical conductivity, and thermal stability can be modified by varying comonomer composition and it is depending on the substituent group. A soluble polymer is more easily processable than insoluble and is thus more attractive to industry.

ACKNOWLEDGMENTS

The authors would like to thank the Director of RSIC Lucknow for recording the FTIR spectra.

KEYWORDS

- **Copolymers**
- **Polyaniline**
- **Poly(aniline-co-N-phenylaniline)**
- **Transport Properties**

REFERENCES

1. Kitani, A.; Kaya, M.; Sakshi, K. J. Electrochem. Soc. Vol. 133, 1995, 1069.
2. Aussawasathein, D.; Dong, J. H.; Dai, L. Synth. Met. Vol. 154, 2005, 37.
3. Dao, L. D.; Leclerc, M.; Guay, J.; Chevalier, J. W. Synth. Met. Vol. 29, 1989, 377.
4. Savitha, P.; Rao, P. S.; Sathyanarayana, D. N. Polym. Int. Vol. 54, 2005, 1243.
5. Gupta, M. C.; Umar, S. S. Macromolecules. Vol. 25, 1992, 138.
6. Gupta, M. C.; Borkar, A. D. Ind. J. Chem. Vol. 29A, 1990, 635.
7. Umare, S. S.; Huque, M. M.; Gupta, M. C.; Viswanath, M. G. Macromolecular Reports. Vol. 33A (Suppls.7&8), 1996, 381.
8. Borkar, A. D.; Gupta, M. C.; Umare, S. S. Polym. Plast. Technol. Eng. Vol. 40(2), 2001, 225.
9. Borkar, A. D.; Umare, S. S.; Gupta, M. C. Prog. Cryst. Grow. and Character. Mat. Vol. 32, 2002, 201.
10. Ghosh, S. G.; Kalapagam, V. Synth. Met. Vol. 1, 1989, 33.
11. Ginder, J. M.; Epstein, A. J. Phys.Rev. Vol. 41B, 1990, 10674.
12. Shante, V. K. S.; Verma, C. M.; Bloch, A. N. Phys. Rev. Vol.138, 1973, 4885.

13. Heeger, A. J.; Kivelson, S. A.; Schrieffer, J. R.; Su, W. P. Rev. Mod. Phys. Vol. 60, 1998, 781.
14. Pouget, J. P.; Jozefowicz, M. E; Epstein, A. J.; Tang, X. MacDiarmid, A. G. Macromolecules. Vol. 24, 1991, 779.
15. Jozefowicz, M. E.; Laversana, R.; Javadi, H. S.; Epstein, A. J.; Pouget, J. P.; Tang, X.; MacDiarmid, A. G. Phys. Rev. Vol. 39B, 1998, 12598.
16. Wang, Z. H.; Li, C.; Scherr, E. M.; MacDiarmid, A. G.; Epstein, A. J. Phys. Rev. Lett. Vol. 66, 1991, 1745.

PART II
Textile Engineering

CHAPTER 9

CARBON NANOTUBES: UPDATE AND NEW PATHWAYS

F. RAEISI, S. PORESKANDAR, SH. MAGHSOODLOU, and A. K. HAGHI

Department of Textile Engineering, Faculty of Engineering, University of Guilan, P.O. Box: 3756, Rasht, Iran

CONTENTS

ABSTRACT

Carbon nanotubes have enticed the vision of many scientists worldwide because of their small dimensions, strength, and the remarkable properties. These features make them a unique material with a whole range of applications in nanoscience and engineering, which are progressing so rapidly. In the following sections of this chapter, at first, an overview of the concepts of nanotechnology and its rate of growth in the last decade is provided, and then some complementary prescriptions of carbon nanotubes and their properties are explained. Finally, several interesting ap plications of carbon nanotubes based on some of the remarkable materials properties of nanotubes are described.

Abbreviation	Definition
ISO	International Standards Organization
TS	Technical Specification
NST	Nanoscience and Technology
STM	Scanning Tunneling Microscope
AFM	Atomic Force Microscope
IBM	Research Laboratory in Zurich
TC	Technical Committee
SPM	Scanning Probe Microscope
APT	Atomically Precise Technologies
HRTEM	High-Resolution Transmission Electron Microscopy
CNT	Carbon Nanotubes
MWNTs	Multi-Walled Nanotubes
SWNTs	Single-Walled Nanotubes
CVD	Chemical Vapor Deposition
UNC	University of North Carolina
GDT	Gas Discharge Tube
ADSL	Asymmetric-Digital-Signal Line
GNF	Graphite Nanofibers
EMI	Electromagnetic Induction
TGA	Thermal Gravimetric Analysis
CCD	Charge-Coupled Device
ITO	Indium-Tin-Oxide
EDLC	Electric Double-Layer Capacitor
CRT	Cathode Ray Tube

9.1 AN INTRODUCTION TO NANOTECHNOLOGY

The appearance of "nanoscience" and "nanotechnology" stimulated the burst of terms with "nano-" prefix. Historically, the term "nanotechnology" appeared before and it was connected with the appearance of possibilities to determine measurable values up to 10^{-9} of known parameters: 10^{-9}m-nm (nanometer), 10^{-9}s-ns (nanosecond), 10^{-9} degree (nanodegree, shift condition). Nanotechnology and molecular nanotechnology comprise a set of technologies connected with the transport of atoms and other chemical particles (ions, molecules) at distances contributing the interactions between them with the formation of nanostructures with different nature.[1]

The current dictionary definition of nanotechnology is "the design, characterization, production and application of materials, devices, and systems by controlling shape and size at the nanoscale". A slightly different nuance is provided by the same source as "the deliberate and controlled manipulation, precision placement, measurement, modeling, and production of matter at the nanoscale in order to create materials, devices, and systems with fundamentally new properties and functions." ISO also provides two meanings: (1) understanding and control of matter and processes at the nanoscale, typically, but not exclusively, below 100 nm in one or more dimensions where the onset of size-dependent phenomena usually enables novel applications; and (2) utilizing the properties of nanoscale materials that differ from the properties of individual atoms, molecules, and bulk matter, to create improved materials, devices, and systems that exploit these new properties. Another formulation encountered in reports is "the design, synthesis, characterization and application of materials, devices, and systems that have a functional organization in at least one dimension on the nanometer scale." The US Foresight Institute declares: "Nanotechnology is a group of emerging technologies in which the structure of matter is controlled at the nanometer scale to produce novel materials and devices that have useful and unique properties." The emphasis on control is particularly important: it is this that distinguishes nanotechnology from chemistry, with which it is often compared; in the latter, motion is essentially uncontrolled and random, within the constraint that it takes place on the potential energy surface of the atoms and molecules under consideration. In order to achieve the desired control, a special, nonrandom eutactic environment needs to be available. Reflecting the importance of control, a very succinct definition of nanotechnology is simply "engineering with atomic precision"; sometimes the phrase "atomically precise technologies" (APT) is used to denote nanotechnology; however, we should bear in mind the "fundamentally new or unique properties" and "novel" aspects that many nanotechnologists insist upon, wishing to exclude ancient or existing artifacts that happen to be small.[2]

In summary, nanotechnology has three aspects:[3]

1. A universal fabrication procedure
2. A particular way of conceiving, designing, and modeling materials, devices, and systems, including their fabrication which bears the same relation to classical engineering as "pointillism" does in classical painting
3. The creation of novelty

9.2 CRITICAL INSTRUMENTS IN NANOTECHNOLOGY

In the heady days of any new, emerging technology, definitions tend to abound and are first documented in reports and journal publications, then slowly get into books and are finally taken up by dictionaries, which do not prescribe, however, but merely record usage. Finally, the technology will attract the attention of the ISO, which may in due course publish TS prescribing in an unambiguous manner the terminology of the field, which is clearly an essential prerequisite for the formulation of manufacturing standards, the next step in the process. In this regard, nanotechnology is no different, except that nanotechnology seems to be arriving rather faster than the technologies with which we might be familiar from the past, such as steam engines, telephones, and digital computers. As a reflection of the rapidity of this arrival, the ISO has already in 2005 set up a technical committee devoted to nanotechnologies. Thus, unprecedented in the history of the ISO, we shall have technical specifications in advance of the emergence of a significant industrial sector.[3,4]Nanoscience and nanotechnology regard the understanding and control of matter at the nanoscale, which is a billionth of a meter. There is consensus in the scientific community that NST broadly involves: (i) research and technology development at the atomic, molecular, or macromolecular levels, in approximately the 1–100 nm range; (ii) creating and using structures, devices, and systems that have novel properties and functions because of their small and/or intermediate size; (iii) the ability to control or manipulate on the atomic scale.[5]

There is also consensus among scientists that NST came into being in 1981, when the STM was invented by Gerd K. Binnig and Heinrich Rohrer at the IBM Research Laboratory in Zurich.[6] In 1986, they were awarded the Nobel Prize for this discovery. The STM yields atomic-scale images of metal and semiconductor surfaces, something which had not been possible with the so-called Topografiner, invented by Russell Young in the late 1960s. The range of materials that can be imaged with a scanning device increased with the invention of the AFM by Gerd K. Binnig in 1986.[5,7]

These enabling instruments were invented at and with the support of the IBM Corporation, which was interested in scientific advances within the semiconductor industry. They soon realized that the STM and the AFM could be used in a vast array of scientific and technological fields, such as chemistry, biology, biotechnology, telecommunications, and many others.[5]

NST is an extremely interesting case in which the micro-mechanisms of science–technology interactions and the origins of entrepreneurship can be detected with great precision due to the novelty of the field and the relative wealth of the available documentation. This field can be characterized along three dimensions:[5,8]

- First of all, the rate of growth in the production of scientific results: Scientific fields that exhibit exponential growth or grow at significantly greater rates than average have completely different properties with respect to regimes that grow linearly.
- Second, the degree of diversity of directions of research: In some areas all research programs converge on a few areas, usually associated with crucial experiments based on a commonly held body of theory, while in other areas the agreement on general theories generates a proliferation of competing hypotheses and research programs, following a divergent dynamics. In NST, it is expected a proliferation pattern of research programs, driven by the specific combination of deeper understanding of the properties of matter at low levels of resolution and design objectives.
- And third, the importance and nature of complementarities in knowledge: While in big science the most important complementarities take place with large experimental facilities, in new emerging fields they are most likely to take the form of human capital and institutional complementarities. In particular, diversified knowledge bases are brought to the frontier of science, while both discovery and invention require a structured interdependence between institutions characterized by different goals (e.g., industry, academia, hospitals).

Based on these dimensions, a number of disciplines can be identified, including life sciences after the molecular biology revolution, computer science, materials science, and nanoscience. These broad disciplines share the following properties: they have been growing exponentially or much more than average for a long period, they follow a dynamic process of divergent research, and they are based on institutional and human capital complementarity.

9.3 THE RATE OF NANOSCIENCE AND NANOTECHNOLOGY GROWTH

In the case of nanoscience, it is clear that there has been impressive growth not only in individual fields such as carbon nanotubes, nanocoatings, or nanobiotechnology but also in the discipline as a whole. In less than 10 years, an army of almost 120,000 scientists worldwide has mobilized around the new discipline. Several thousand new institutions worldwide has entered the field. The scientific output of such collective action amounts to about 100,000 publications (see Figure 9-1).[8]

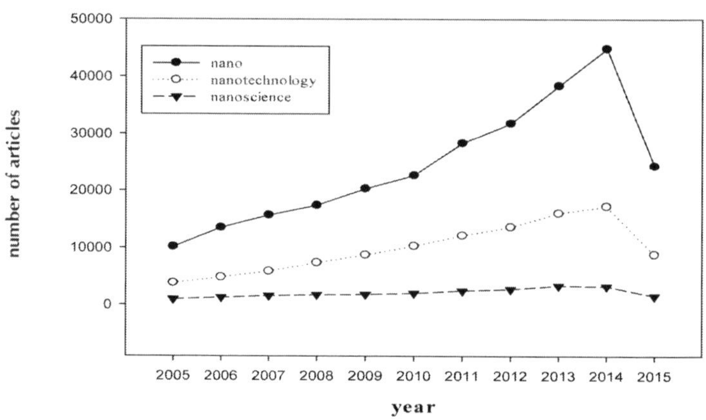

FIGURE 9-1 Rate of growth of nanoscience and nanotechnology according to the number of articles.

9.3.1 NANOTECHNOLOGY AS A PROCESS

We see nanotechnology as looking at things—measuring, describing, characterizing, and quantifying them, and ultimately reaching a deeper assessment of their place in the universe. It is also making things. The manufacturing aspect was evidently very much in the mind of the actual inventor of the term "nanotechnology," Norio Taniguchi from the University of Tokyo, who considered it as the inevitable consequence of steadily and exponentially improving engineering precision.[9] Clearly, the surface finish of a work piece achieved by grinding it cannot be less rough than atomic roughness, hence nanotechnology must be the endpoint of ultra-precision engineering.[3]

At the same time, improvements in metrology had reached the point where individual atoms at the surface of a piece of material could be imaged, hence visualized on a screen. The possibility had been, of course, already inherent in electron microscopy, which was invented in the 1930s, but numerous incremental technical improvements were needed before atomic resolution became attainable. Another development was the invention of the "Topografiner" by scientists at the US National Standards Institute.[10] This instrument produced a map of topography at the nanoscale by raster scanning a needle over the surface of the sample. A few years later, it was developed into the STM, and in turn the AFM, which is now seen as the epitome of nanometrology (collectively, these instruments are known as scanning probe microscopes, SPMs). Hence, a little more than 10 years after Feynman's lecture, advances in instrumentation already allowed one to view the hitherto invisible world of the nanoscale in a very graphic fashion. There is a strong appeal in having a small, desktop instrument such as the AFM able to probe matter at the atomic scale, which contrasts strongly with the bulk of traditional high-resolution instruments such as the electron microscope, which needs at least a room and perhaps a whole building to house it and its attendant services.

In parallel, people were also thinking about how atom-by-atom assembly might be possible. Erstwhile Caltech colleagues recall Richard Feynman's dismay when William McLellan constructed a minute electric motor by hand-assembling the parts in the manner of a watchmaker, thereby winning the prize Feynman had offered for the first person to create an electrical motor smaller than 1/64th of an inch. Although this is still how nanoscale artifacts are made, but perhaps for not much longer, Feynman's concept was about machines making progressively smaller machines ultimately small enough to manipulate atoms and assemble things at that scale. The most indefatigable subsequent champion of that concept was Eric Drexler, who developed the concept of the assembler, a tiny machine programmed to build objects atom-by-atom. It was an obvious corollary of the minute size of an assembler that in order to make anything of a size useful for humans, or in useful numbers, there would have to be a great many assemblers working in parallel. Hence, the first task of the assembler would be to build copies of itself, after which they would be set to perform more general assembly tasks.[3]

9.3.2 NANOTECHNOLOGY AS MATERIALS

The aforementioned section illustrates an early preoccupation with nanotechnology as a process—a way of making things. Before the semiconductor processing industry reduced the feature sizes of integrated circuit components to less than 100 nm,[3] however, there was no real industrial example of nanotechnology at work. On the other hand, while process—top–down and bottom–up, and we include metrology here—is clearly one way of thinking about nanotechnology, there is already a sizable industry involved in making very fine particles, which, because their size is less than 100 nm, might be called nanoparticles. Generalizing, a nano-object is something with at least one spatial dimension less than 100 nm; from this definition are derived those for nanoplates (one dimension less than 100 nm), nanofibers (two dimensions less than 100 nm), and nanoparticles (all three dimensions less than 100 nm); nanofibers are in turn subdivided into nanotubes (hollow fibers), nanorods (rigid fibers), and nanowires (conducting fibers) [3].

Although nanoparticles of many different kinds of materials have been made for hundreds of years, one nanomaterial stands out as being rightfully so named, because it was discovered and nanoscopically characterized in the nanotechnology era: graphene and its compactified forms, namely carbon nanotubes and fullerenes (nanoparticles).[3].

9.3.3 NANOTECHNOLOGY AS DEVICES AND SYSTEMS

One problem with associating nanotechnology exclusively with materials is that nanoparticles were deliberately made for various esthetic, technological, and medical applications at least 500 years ago, and one would therefore be compelled to say that nanotechnology began then. To avoid that problem, materials are generally grouped with other entities along an axis of increasing complexity, encompassing devices, and systems. A nanodevice, or nanomachine, is defined as a nanoscale automaton, or at least one containing nanosized components. Responsive or "smart" materials could of course also be classified as devices. A device might well be a system of components in a formal sense; it is not generally clear what meaning is intended by specifying "nanosystem," as distinct from a device. At any rate, materials may be considered as the most basic category, since devices are obviously made from materials, even though the functional equivalent of a particular device could be realized in different ways, using different materials.

More rigorously than ordering nanotechnology along an axis, these concrete concepts of materials, devices, and systems can be organized into a formal concept system or ontology.[3]

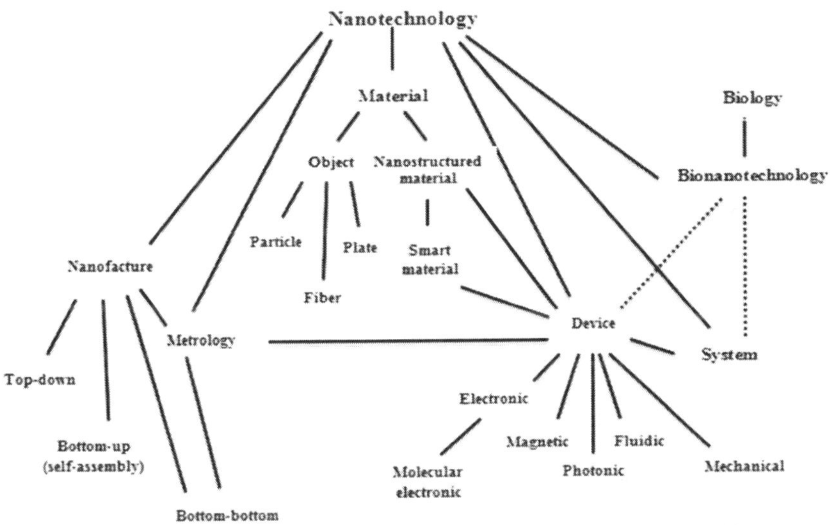

FIGURE 9-2 A concept system (ontology) for nanotechnology. Most of the terms would normally be prefixed by "nano" (e.g., nanometrology, nanodevice). A dashed line signifies that if the superordinate concept contributes, then the prefix must indicate that (e.g., bionanodevice, bionanosystem). Biology may also have some input to nanomanufacture (nanofacture), inspiring, especially, self-assembly processes.

9.4 NANOTECHNOLOGY AND NANOFIBERS

In this sub-chapter an attempt is made to classify nanofibers into one or more sub-category of nanotechnology.

To do so we briefly review some common sub-fields of nanotechnology itself. As far as "nanostructures" are concerned, one can view this as objects or structures whereby at least one of its dimensions is within nano-scale. A "nanoparticle" can be considered as a zero-dimensional nano-element, which is the simplest form of nanostructure. It follows that a "nanotube" or a "nanorod" is a one-dimensional nano-element from which slightly more complex nanostructure can be constructed of. Following this train of thought, a “nanoplatelet” or a “nanodisk” is a two-dimensional element which, along with its one-dimensional counterpart, is useful in the construction of nanodevices.

The difference between a nanostructure and a nanodevice can be viewed upon as the analogy between a building and a machine whether mechanical, electrical, or both. It goes without saying that as far as nanoscale is concerned, one should not pigeonhole these nano-elements for an element that is considered a structure can at times be used as a significant part of a device. For example, the use of carbon nanotube as the tip of an AFM would have it classified as a nanostructure. The same nanotube, however, can be used as a single-molecule circuit, or as part of a miniaturized electronic component, thereby appearing as a nanodevice. Hence the function, along with the structure, is essential in classifiying which nanotechnology sub-area it belongs to.

While nanostructures clearly define the solids' overall dimensions, the same cannot be said so for nanomaterials. In some instances, a nanomaterial refers to a nano-sized material while in other instances a nanomaterial is a bulk material with nanoscaled structures. Nanocrystals appear to be a misnomer. It is understood that a crystal is highly structured and that the repetitive unit is indeed small enough. Hence, a nanocrystal refers to the size of the entire crystal itself being nano-sized, but not of the repetitive unit.

Nanophotonics refers to the study, research, development, and/or applications of nanoscale object that emit light and its corresponding light. These objects are normally quantum gots. While the emission of photon is largest for bulk (3-dimensional), followed by quantum well (2-dimensional) and final quantum dot (0-dimensional), the ranking is reversed in terms of efficiency.

Although the term nanomagnetics is self explanatory, we wish to view it in terms of highly miniaturized magnetic data storage materials with very high memory. This can be attained by taking advantage of the electron spin for memory storage—hence the term “spin-electronics,” which has since been more popularly and more conveniently known as “spintronics".

In nanobioengineering, the novel properties at nanoscale are taken advantage of for bioengineering applications. The many naturally occurring nanofibrous and nanoporous structure in the human body further adds to the impetus for research and development in this sub-area. Closely related to this is molecular functionalization whereby the surface of an object is modified by attaching certain molecules to enable desired functions to be carried out—such as for sensing and/or filtering chemicals based on molecular affinity.

With the rapid growth of nanotechnology, nanomechanics is no longer the narrow field it used to be. This field can be broadly categorized into the molecular mechanics and the continuum mechanics approaches—which view objects as consisting of discrete many-body system and continuous media, respectively. While the former inherently includes the size-effect, it is a requirement for the latter to factor in the influence of increasing surface-to-volume ratio, molecular reorientation, and other novelties as the size shrinks.

As with many other fields, nanotechnology includes nanoprocessing—novel materials processing techniques by which nanoscale structures and devices are designed and constructed.

Depending upon the final size and shape, a nanostructure or nanodevice can be produced by the top-down or the bottom up approach. The former refers to the act of removal or cutting down a bulk to the desired size while the latter takes on the philosophy of using the fundamental building blocks—such as atoms and molecules —to build up nanostructures in the same manner as one would toward lego sets. It is obvious that the top-down and the bottom-up nanoprocessing methodologies are suitable for the larger and two smaller ends, respectively, in the spectrum of nanoscale construction. The effort of nanopatterning—or patterning at the nanoscale—would hence fall into nanoprocessing.

So where do all these descriptions point nanofibers to? It is obvious that nanofibers would geometrically fall into the category of 1-dimensional nanoscale elements that includes nanotubes and nanorods. However, the flexible nature of nanofibers would align it along with other highly flexible nanoelements such as globular molecules (assumed as 0-dimensional soft matter), as well as solid and liquid films of nanothickness (2-dimensional). A nanofiber is a nanomaterial in view of its diameter, and can be considered a nanostructured material, material if filled with nanoparticles to form composite nanofibers.

Where an application to bioengineering is concerned, such as the use of nanofibrous networks of tissue engineering scaffolds, these nanofibers play significant roles in nanobioengineering.[11] The study of the nanofiber mechanical properties as a result of manufacturing techniques, constituent materials, processing parameters, and other factors would fall into the category of nanomechanics. Indeed, while the primary classification of nanofibers is that of nanostructure or nanomaterial, other aspects of nanofibers such as its characteristics, modeling, application, and processing would enable nanofibers to penetrate into many subfields of nanotechnology. Finally, the processing techniques of nanofibers are diverse, and include both the top-down and the bottom-up approaches as we shall see in the next sub-chapter.[12]

9.5 NANOTECHNOLOGY AND NANOTUBES

In 1991, Japanese researcher Iijima studied the sediments formed at the cathode during the spray of graphite in an electric arc. His attention was attracted by the unusual structure of the sediment consisting of microscopic fibers and filaments. Measurements made with an electron microscope showed that the diameter of these filaments does not exceed a few nanometers and a length of one to several microns.[1]

Having managed to cut a thin tube along the longitudinal axis, the researchers found that it consists of one or more layers, each representing a hexagonal grid of graphite, which is based on hexagon with vertices located at the corners of the carbon atoms. In all cases, the distance between the layers is equal to 0.34 nm, which is the same as that between the layers in crystalline graphite. Typically, the upper ends of tubes are closed by multilayer hemispherical caps, each layer is composed of hexagons and pentagons, reminiscent of the structure of half a fullerene molecule. The extended structure consisting of rolled hexagonal grids with carbon atoms at the nodes are called nanotubes.[1]

Nanotubes are rolled into a cylinder (hollow tube) graphite plane, which is lined with regular hexagons with carbon atoms at the vertices of a diameter of several nanometers. Nanotubes can consist of one layer of atoms (single-wall nanotubes) and represent a number of "nested" one into another layer pipes (multi-wall nanotubes).[13]

9.5.1 *CARBON NANOTUBES*

Carbon is unique among the elements in its ability to assume a wide variety of different structures and forms. It is now a little more than ten years, since a new family of carbon cage structures, all based on a three-fold coordinated sp^2 network, was discovered, thereby inaugurating the science of fullerenes. Of these, C_{60} is the most abundant and perhaps the best known member. However, perhaps the most exciting among the recent additions to the fullerene family are carbon nanotubes, discovered soon after C_{60} was made available in gram quantities. Carbon nanotubes are hollow cylinders, consisting of a single sheet of graphite (graphene) wrapped into a cylinder. They are believed to have extraordinary structural, mechanical, and electrical properties, which are derived from the special properties of carbon bonds, their unique quasi-one-dimensional nature, and their cylindrical symmetry. For instance, the graphitic network upon which the nanotube structure is based is well known for its strength and elasticity, thereby providing mechanical strength, potentially unmatched by any material. Nanotubes can also be metallic or semiconducting, depending on their 'chirality'. This opens up the very interesting prospects of junctions and devices made entirely out of carbon. Because of these very unusual characteristics, and the potential compatibility of nanotubes with organic matter, their discovery has been greeted with a considerable amount of excitement within the scientific community. However, because they were originally synthesized in minute quantities, only relatively few experimental techniques were initially available for their study. Indeed, the original experimental work was only able to address the nanotube structure through HRTEM. Their discovery, however, has stimulated much theoretical work. In turn, these investigations have benefited tremendously from the substantial progress achieved in the past two to three decades in the development of theoretical methods, some of which now have a truly predictive power. Astonishing properties have been predicted, which in turn have stimulated further experiments, so that the progress has been very rapid. In the past two years, over 300 papers have been devoted wholly or in part to nanotubes.[13–16.]

Carbon nanotubes are strongly related to other forms of carbon, especially to crystalline 3D graphite, and to its constituent 2D layers (where an individual carbon layer in the honeycomb graphite lattice is called a graphene layer). In this chapter, several forms of carbon materials are reviewed, with particular reference to their relevance to carbon nanotubes. Their similarities and differences relative to carbon nanotubes with regard to structure and properties are emphasized.[13]

Since their discovery in 1991 (Iijima, 1991), the interest of the scientific community and of industry in carbon nanotubes has increased dramatically. Carbon nanotubes exhibit properties that include high thermal and electrical conductivity, mechanical resistance, low density, and tunable semiconductivity, which render them useful in a variety of industrial applications, such as components in electronics, energy-storage devices, solar cells, sensors, and as filler in polymeric composites in mechanical applications. Today, the annual global market in CNTs is estimated to be of the order of hundreds of tons. Carbon nanotubes are mainly used as components of electrodes in lithium batteries or as filler for electrical discharge in composites. Since the physical properties of CNTs largely depend upon their structure, researchers are endeavoring to produce different types of engineered CNTs with tailored physico-chemical features. Carbon nanotubes should therefore be considered not a single substance, but a family of different materials.[13]

Carbon nanotubes have attracted a great interest in their application in medicine. Like fullerenes, CNTs have been shown to rapidly cross the cell membranes and therefore they have been proposed as nanovectors. For this application, CNTs need to be modified to increase their compatibility in water and to bind other entities such as drugs or biomolecules. Furthermore, some authors have proposed using CNTs in the manufacture of alternative artificial hard tissues, tissue scaffold materials for bone formation, microcatheters, and as substrates for neuronal growth in nervous system disorders.

In 1991, Iijima of the NEC Laboratory in Japan reported the first observation of multi-walled carbon nanotubes in carbon-soot made by arc discharge.[17] About two years later, he made the observation of single walled nanotubes.[18] The past decade witnessed significant research efforts in efficient and high-yield nanotube growth methods. The success in nanotube growth has led to the wide availability of nanotube materials, and is a main catalyst behind the recent growth in basis physics studies and applications of nanotubes.[13]

The electrical and mechanical properties of carbon nanotubes have captured the attention of researchers worldwide. Understanding these properties and exploring their potential applications have been a main driving force in this area. Besides the unique and useful structural properties, a nanotube has high Young's modulus and tensile strength. A SWNT can behave as a well-defined metallic, semiconducting or semi-metallic wire depending on two key structural parameters, chirality and diameter.[15] Nanotubes are ideal systems for studying the physics in low dimensions. Theoretical and experimental work has focused on the relationship between nanotube atomic structures with electronic structures, electron-electron and electron-phonon interaction effects.[19] Extensive effort has been taken to investigate the mechanical properties of nanotubes, including Young's modulus, tensile strength, failure processes, and mechanisms. An intriguing fundamental question has been how mechanical deformation in a nanotube affects its electrical properties. In recent years, research work addressing these basic problems has generated significant excitement in the area of nanoscale science.[13]

Nanotubes can be utilized individually or as an ensemble to build functional device prototypes, as has been demonstrated by many research groups.

Ensembles of nanotubes have been used for field emission based flat-panel display, composite materials with improved mechanical properties and electromechanical actuators. Bulk quantities of nanotubes have also been suggested as high-capacity hydrogen storage media. Individual nanotubes have been used for field emission sources, tips for scanning probe microscopy and nano-tweezers. Nanotubes can also be used as the central elements of electronic devices, including field effect transistors, single-electron transistors, and rectifying diodes.[13]

The full potential of nanotubes for applications will not be realized until the growth of nanotubes can be well controlled. Real-world applications of nanotubes require either large quantities of bulk materials or device integration in scale-up fashions. For applications such as composites and hydrogen storage, it is desired to obtain high-quality nanotubes at the kilogram or the ton level using growth methods that are simple, efficient, and inexpensive. For devices such as nanotube based electronics, scale-up will unavoidably rely on some sort of self-assembly or controlled growth strategies on surfaces combined with microfabrication techniques. Significant work has been carried out in recent years to tackle these issues. Nevertheless, many challenges remain in the nanotube growth area. First, an efficient growth approach to structurally perfect nanotubes large scales is still lacking. Second, growing defect-free nanotubes continuously to macroscopic lengths has been difficult. Third, one needs to learn how to gain exquisite control over nanotube growth on surfaces and obtain large-scale ordered nanowire structures. Finally, there is the seemingly formidable task of controlling the chirality of SWNTs by any existing growth method.[13]

9.5.2 SYNTHESIS OF CARBON NANOTUBES

Carbon nanotubes are currently synthesized in carbon arcs[17,20–22] through laser vaporization,[23] catalytic combustion,[24] chemical vapor deposition,[25] and ion bombardment. The type of nanotube that is produced strongly depends upon the presence or absence of catalysts: MWNTs are most commonly produced via noncatalytic means, while SWNTs are usually the dominant products under catalytic growth conditions. However, oriented MWNTs have been grown by pyrolysis catalyzed by small co-clusters.[26]

9.5.3 PROPERTIES OF CARBON NANOTUBES

CNTs are highly desirable materials possessing unique structural, mechanical, thermal, and electrical properties which have been explained in this part in brief.

9.5.3.1 MECHANICAL PROPERTIES

It is noteworthy that the term resilient was first applied not to nanotubes, but in smaller fullerene cages, when Whetten et al. studied the high-energy collisions of C_{60}, C_{70}, and C_{84} bounces from a solid wall of H-terminated diamond.[27] They observed no fragmentation or any irreversible atomic rearrangement in the bouncing back cages, which was somewhat surprising and indicated the ability of fullerenes to sustain great elastic distortion. The very same property of resilience becomes more significant in the case of carbon nanotubes, since their elongated shape, with the aspect ratio close to a thousand, makes the mechanical properties especially interesting and important due to potential structural applications.

The utility of nanotubes as the strongest or stiffest elements in nanoscale devices or composite materials remains a powerful motivation for the research in this area. While the jury is still out on practical realization of these applications, an additional incentive comes from the fundamental materials physics. There is a certain duality in the nanotubes, molecular size and morphology, and at the same time possessing sufficient translational symmetry to perform as very small (nano-) crystals, with a well-defined primitive cell, surface, possibility of transport, and so on. Moreover, in many respects they can be studied as well defined engineering structures and many properties can be discussed in traditional terms of moduli, stiffness or compliance, geometric size and shape. The mesoscopic dimensions (a nanometer scale diameter) combined with the regular, almost translation-invariant morphology along their micrometer scale lengths (unlike other polymers, usually coiled), make nanotubes a unique and attractive object of study, including the study of mechanical properties and fracture in particular.

9.5.3.2 ELECTRONIC PROPERTIES

In this section, we give an introduction to the structure and electronic properties of the SWNTs. Shortly after the discovery of carbon nanotubes in the sort of fullerene synthesis, single-walled carbon nanotubes were synthesized in abundance using arc discharge methods with transition metal catalysts.[28–30] These tubes have quite small and uniform diameter, on the order of one nanometer. Crystalline ropes of SWNTs with each rope containing tens to hundreds of tubes of similar diameter closely packed have also been synthesized using a laser vaporization method[31] and other techniques, such as arc-discharge and CVD techniques. These developments have provided ample amounts of sufficiently characterized samples for the study of the fundamental properties of SWNTs. As illustrated in Figure 9-3, a SWNT is geometrically just a rolled-up graphene strip. Its structure can be specified or indexed by its circumferential periodicity.[32] In this way, a SWNT's geometry is completely specified by a pair of integers (n,m) denoting the relative position $c = na_1 + ma_2$ of the pair of atoms on a graphene strip which, when rolled into each other, forms a tube.[13]

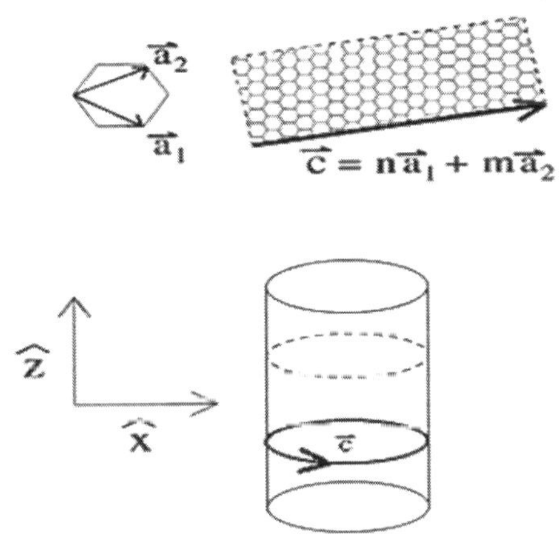

FIGURE 9-3 Geometric structure of an (n,m) single-walled carbon nanotube.

9.5.3.3 *THERMAL PROPERTIES*

The thermal properties of carbon nanotubes display a wide range of behaviors which are related both to their graphitic nature and their unique structure and size. The specific heat of individual nanotubes should be similar to that of two-dimensional graphene at high temperatures, with the effects of phonon quantization becoming apparent at lower temperatures. Intertube coupling in SWNT ropes, and interlayer coupling in MWNTs, should cause their low-temperature specific heat to resemble that of three-dimensional graphite. Experimental data on SWNTs show relatively weak intertube coupling, and are in good agreement with theoretical models. The specific heat of MWNTs has not been examined theoretically in detail. Experimental results on MWNTs show a temperature dependent specific heat, which is consistent with weak interlayer coupling, although different measurements show slightly different temperature dependencies. The thermal conductivity of both SWNTs and MWNTs should reflect the on-tube phonon structure, regardless of tube–tube coupling. Measurements of the thermal conductivity of bulk samples show graphite-like behavior for MWNTs, but quite different behavior for SWNTs, specifically a linear temperature dependence at low T which is consistent with one-dimensional phonons. The room-temperature thermal conductivity of highly aligned SWNT samples are over 200 W/m-K, and the thermal conductivity of individual nanotubes is likely to be higher still.

9.5.3.4 APPLICATIONS OF CARBON NANOTUBES

The discovery of fullerenes[34] provided exciting insights into carbon nanostructures and how architectures built from sp^2 carbon units based on simple geometrical principles can result in new symmetries and structures that have fascinating and useful properties. Carbon nanotubes represent the most striking example. About a decade after their discovery,[17] the new knowledge available in this field indicates that nanotubes may be used in a number of practical applications. There have been great improvements in synthesis techniques, which can now produce reasonably pure nanotubes in gram quantities. Studies of structure-topology-property relations in nanotubes have been strongly supported, and in some cases preceded, by theoretical modeling that has provided insights for experimentalists into new directions and has assisted the rapid expansion of this field.[15,35–37]

Quasi-one-dimensional carbon whiskers or nanotubes are perfectly straight tubules with diameters of nanometer size and properties close to that of an ideal graphite fiber. Carbon nanotubes were discovered accidentally by SumioIijima in 1991, while studying the surfaces of graphite electrodes used in an electric arc discharge. His observation and analysis of the nanotube structure started a new direction in carbon research, which complemented the excitement and activities prevalent in fullerene research. These tiny carbon tubes with incredible strength and fascinating electronic properties appear to be ready to overtake fullerenes in the race to the technological marketplace. It is the structure, topology, and size of nanotubes that make their properties exciting compared to the parent, planar graphite-related structures, such as are, for example, found in carbon fibers.[17]

The uniqueness of the nanotube arises from its structure and the inherent subtlety in the structure, which is the helicity in the arrangement of the carbon atoms in hexagonal arrays on their surface honeycomb lattices. The helicity local symmetry, along with the diameter, which determines the size of the repeating structural unit introduces significant changes in the electronic density of states, and hence provides a unique electronic character for the nanotubes. These novel electronic properties create a range of fascinating electronic device applications and this subject matter is discussed briefly elsewhere in this volume, and has been the subject of discussion in earlier reviews.[19] The other factor of importance in what determines the uniqueness in physical properties is topology, or the closed nature of individual nanotube shells; when individual layers are closed on to themselves, certain aspects of the anisotropic properties of graphite disappear, making the structure remarkably different from graphite. The combination of size, structure, and topology endows nanotubes with important mechanical, for example, high stability, strength and stiffness, combined with low density and elastic deformability and with special surface properties selectivity, surface chemistry, and the applications based on these properties form the central topic of this chapter. In addition to the helical lattice structure and closed topology, topological defects in nanotubes (five member Stone-Wales defects near the tube ends, aiding in their closure),[38] akin to those found in the fullerenes structures, result in local perturbations of their electronic structure;[39] for example, the ends or caps of the nanotubes are more metallic than the cylinders, due to the concentration of pentagonal defects.[39] These defects also enhance the reactivity of the tube ends, giving the possibility of opening the tubes,[40]functionalizing the tube ends,[41]and filling the tubes with foreign substances.[42–44]

The structure of nanotubes remains distinctly different from traditional carbon fibers that have been industrially used for several decades.[45] Most importantly, nanotubes, for the first time represent the ideal, most perfect and ordered, carbon fiber, the structure of which is entirely known at the atomic level. It is this predictability that mainly distinguishes nanotubes from other carbon fibers and puts them along with molecular fullerene species in a special category of prototype materials. Among the nanotubes, two varieties, which differ in the arrangement of their grapheme cylinders, share the limelight. MWNTs are collections of several concentric graphene cylinders and are larger structures compared to SWNTs which are individual cylinders of 1–2nm diameter (see Figure 9-4). The former can be considered as a mesoscale graphite system, whereas the latter is truly a single large molecule. However, SWNTs also show a strong tendency to bundle up into the ropes, consisting of aggregates of several tens of individual tubes organized into a one-dimensional triangular lattice. One point to note is that in most applications, although the individual nanotubes should have the most appealing properties, one has to deal with the behavior of the aggregates (MWNT or SWNT ropes), as produced in actual samples. The best presently available methods to produce ideal nanotubes are based on the electric arc[20,46] and laser ablation processes.[23] The material prepared by these techniques has to be purified using chemical and separation methods. None of these techniques are scalable to make the industrial quantities needed for many applications, for example, in composites, and this has been a bottleneck in nanotube R&D. In recent years, work has focused on developing chemical vapor deposition techniques using catalyst particles and hydrocarbon precursors to grow nanotubes;[25,26,47,48] such techniques have been used earlier to produce hollow nanofibers of carbon in large quantities.[45] The drawback of the catalytic CVD-based nanotube production is the inferior quality of the structures that contain gross defects like twists, tilt boundaries, and so on, particularly because the structures are created at much lower temperatures (600–1000°C) compared to the arc or laser processes (~2000°C).

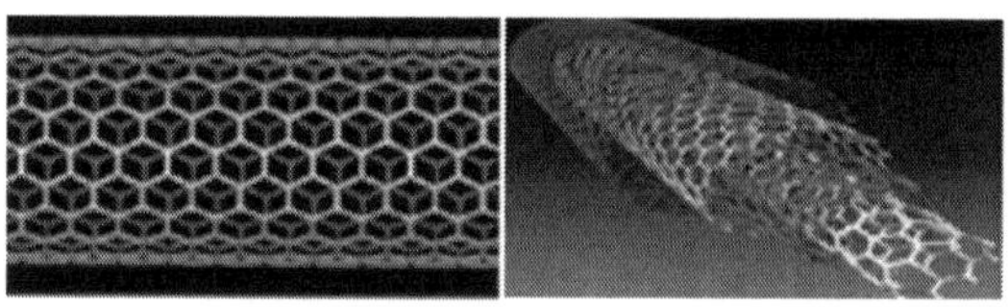

a b

FIGURE 9-4 Structure of single-walled (SWNT) (a) and multi-walled (MWNT) carbon nanotubes (b).

Since their discovery in 1991, several demonstrations have suggested potential applications of nanotubes. These include the use of nanotubes as electron field emitters for vacuum microelectronic devices, individual MWNTs and SWNTs attached to the end of an AFM tip for use as nanoprobe, MWNTs as efficiently supports in heterogeneous catalysis and as microelectrodes in electrochemical reactions, and SWNTs as good media for lithium and hydrogen storage. Some of these could become real marketable applications in the near future, but others need further modification and optimization. Areas where predicted or tested nanotube properties appear to be exceptionally promising are mechanical reinforcing and electronic device applications. The lack of availability of bulk amounts of well-defined samples and the lack of knowledge about organizing and manipulating objects such as nanotubes due to their sub-micron sizes have hindered progress in developing these applications. The last few years, however, have seen important breakthroughs that have resulted in the availability of nearly uniform bulk samples. There still remains a strong need for better control in purifying and manipulating nanotubes, especially through generalized approaches such as chemistry. Development of functional devices/structures based on nanotubes will surely have a significant impact on future technology needs. In the following sections, we describe the potential of materials, science-related applications of nanotubes, and the challenges that need to be overcome to reach these hefty goals.

In the following sections, we describe several interesting applications of carbon nanotubes based on some of the remarkable material properties of nanotubes. Electron field emission characteristics of nanotubes and applications based on this, nanotubes as energy storage media, the potential of nanotubes as fillers in high performance polymer and ceramic composites, nanotubes as novel probes and sensors, and the use of nanotubes for template-based synthesis of nanostructures are major topics that are discussed in the sections that follow.[13]

9.5.3.4.1 Potential application of CNTs in vacuum microelectronics

Field emission is an attractive source for electrons compared to thermionic emission. It is a quantum effect. When subject to a sufficiently high electric field, electrons near the Fermi level can overcome the energy barrier to escape to the vacuum level. The basic physics of electron emission is well developed. The emission current from a metal surface is determined by the Fowler–Nordheim equation:

$$I = aV^2 \exp\left(-b\Phi^{3/2}/\beta V\right)$$

where I, V, ϕ, β are the current, applied voltage, work function, and field enhancement factor, respectively.

Electron field emission materials have been investigated extensively for technological applications, such as at panel displays, electron guns in electron microscopes, microwave amplifiers.[49] For technological applications, electron emissive materials should have low threshold emission fields and should be stable at high current density. A current density of 1-10mA/cm^2 is required for displays[50] and >500mA/cm^2 for a microwave amplifier.[51] In order to minimize the electron emission threshold field, it is desirable to have emitters with a low work function and a large field enhancement factor. The work function is an intrinsic material property. The field enhancement factor depends mostly on the geometry of the emitter and can be approximated as: $\beta = 1/5r$ where r is the radius of the emitter tip. Processing techniques have been developed to fabricate emitters such as Spindt-type emitters, with a sub-micron tip radius.[49] However, the process is costly and the emitters have only limited lifetime. Failure is often caused by ion bombardment from the residual gas species that blunt the emission tips. Table 9-1 lists the threshold electrical field values for a 10mA/cm^2 current density for some typical materials.

Carbon nanotubes have the right combination of properties—nanometer-sized diameter, structural integrity, high electrical conductivity, and chemical stability—that make good electron emitters. Electron field emission from carbon nanotubes was first demonstrated in 1995,[52] and has since been studied intensively on various carbon nanotube materials. Compared to conventional emitters, carbon nanotubes exhibit a lower threshold electric field, as illustrated in Table 9-1. The current-carrying capability and emission stability of the various carbon nanotubes, however, vary considerably depending on the fabrication process and synthesis conditions.

The I-V characteristics of different types of carbon nanotubes have been reported, including individual nanotubes,[52,53] MWNTs embedded in epoxy matrices,[54,55]MWNT films[56,57] SWNTs[58–60] , and aligned MWNT films. Figure 9-5 shows typical emission I-V characteristics measured from a random SWNT film at different anode-cathode distances, and the Fowler- Nordheim plot of the same data is shown as the inset. Turn-on and threshold fields are often used to describe the electrical field required for emission. The former is not well-defined and typically refers to the field that is required to yield 1nA of total emission current, while the latter refers to the field required to yield a given current density, such as 10mA/cm^2. For random SWNT films, the threshold field for 10mA/cm^2 is in the range of 2–3V/μm. Random and aligned MWNTs [fabricated at the University of North Carolina (UNC) and AT&T Bell Labs] were found to have threshold fields slightly larger than that of the SWNT film and are typically in the range of 3–5V/μm for a 10mA/cm^2 current density (Figure 9-6). These values for the threshold field are all significantly better than those from conventional field emitters such as the Mo and Si tips which have a threshold electric field of 50–100V/μm (Table 9-1). It is interesting to note that the aligned MWNT films do not perform better than the random films. This is due to the electrical screening effect arising from closely packed nanotubes.[61] The low threshold field for electron emission observed in carbon nanotubes is a direct result of the large field enhancement factor rather than a reduced electron work function. The latter was found to be 4.8 eV for SWNTs, 0.1–0.2 eV larger than that of graphite.[62]

TABLE 9-1 Threshold electrical field values for different materials for a 10 mA/cm^2 current density[63]

Material	Threshold electrical field (V/μm)
Mo tips	50–100
Si tips	50–100
p-type semiconducting diamond	130
Undoped, defective CVD diamond	30–120
Amorphous diamond	20–40
Cs-coated diamond	20–30
Graphite powder(<1mm size)	17
Nanostructured diamond[a]	3–5 (unstable >30 mA/cm^2)
Carbon nanotubes[b]	1–3 (stable at 1 A/cm^2)

[a]Heat treated in H plasma

[b]random SWNT film

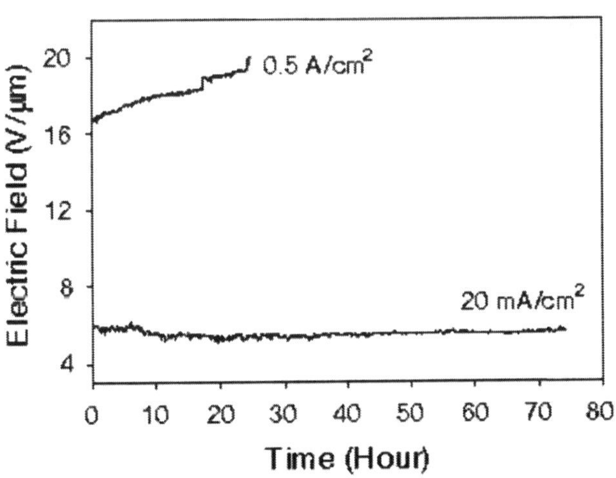

FIGURE 9-5 Stability test of a random laser-ablation-grown SWNT film showing stable emission at 20 mA/cm^2 (from Ref. [58]).

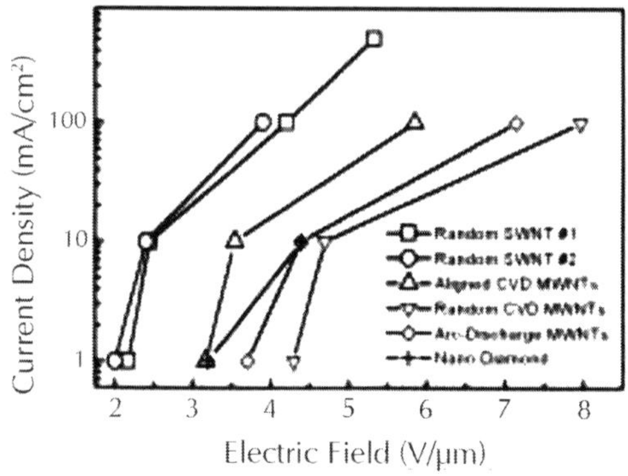

FIGURE 9-6 Current density versus electric field measured for various forms of carbon nanotubes.

SWNTs generally have a higher degree of structural perfection than either MWNTs or CVD-grown materials and have a capability for achieving higher current densities and have a longer lifetime. Stable emission above 20mA/cm^2 has been demonstrated in SWNT films deposited on Si substrates.[58] A current density above 4A/cm^2(measured by a 1mm local probe) was obtained from SWNTs produced by the laser ablation method.[58] Figure 9-7 is a CCD image of the setup for electron emission measurement, showing a Mo anode (1mm diameter) and the edge of the SWNT cathode in a vacuum chamber. The Mo anode is glowing due to bombarding from field emitted electrons, demonstrating the high current capability of the SWNTs. This particular image was taken at a current density of 0.9A/cm^2. The current densities observed from the carbon nanotubes are significantly higher than from conventional emitters, such asnano-diamonds which tend to fall below 30mA/cm^2 current density.[63] Carbon nanotube emitters are particularly attractive for a variety of applications including microwave amplifiers.

Although carbon nanotube emitters show clear advantageous properties over conventional emitters in terms of threshold electric field and current density, their emission site density (number of functioning emitters per unit area) is still too low for high resolution display applications. Films presently fabricated have typical emission site densities of 103-104/cm^2 at the turn-on field, and ~106/cm^2 is typically required for high resolution display devices.

9.5.3.4.2 Prototype electron emission devices based on carbon nanotubes

A. Cathode ray lighting elements: Cathode ray lighting elements have been fabricated with carbon nanotube materials as the field emitters by Ise Electronic Co. in Japan.[64] As illustrated in Figure 9-7, these nanotube-based lighting elements have a triode-type design. In the early models, cylindrical rods containing MWNTs, formed as a deposit by the arc-discharge method, were cut into thin disks and were glued to stainless steel plates by silver paste. In later models, nanotubes are now screen-printed onto the metal plates. A phosphor screen prints on the inner surfaces of a glass plate. Different colors are obtained by using different fluorescent materials. The luminance of the phosphor screens measured on the tube axis is 6.4×10^4 cd/cm^2 for green light on an anode current of 200μA, which is two times more intense than that of conventional thermionic cathode ray tube (CRT) lighting elements operated under similar conditions.[64]

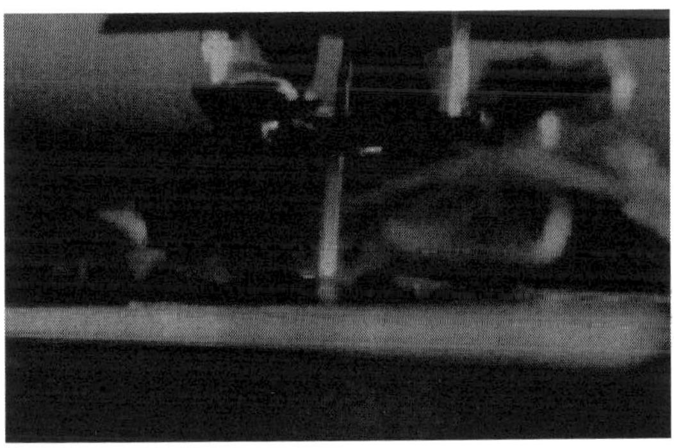

FIGURE 9-7 A CCD image showing a glowing Mo anode (1mm diameter) at an emission current density of 0.9 A/cm^2 from a SWNT cathode. Heating of the anode is due to field emitted electrons bombarding the Mo probe, thereby demonstrating a high current density (image provided by Dr. Wei Zhu of Bell Labs).

B. Flat panel display: Prototype matrix-addressable diode at panel displays have been fabricated using carbon nanotubes as the electron emission source. One demonstration (demo) structure constructed at Northwestern University consists of nanotube-epoxy stripes on the cathode glass plate and phosphor-coated ITO stripes on the anode plate. Pixels are formed at the intersection of cathode and anode stripes, as illustrated in Figure 9-8. At a cathode-anode gap distance of 30 μm, 230V is required to obtain the emission current density necessary to drive the diode display (~76 μmA/mm^2). The device is operated using the half-voltage off-pixel scheme. Pulses of ± 150V are switched among anode and cathode stripes, respectively, to produce an image.

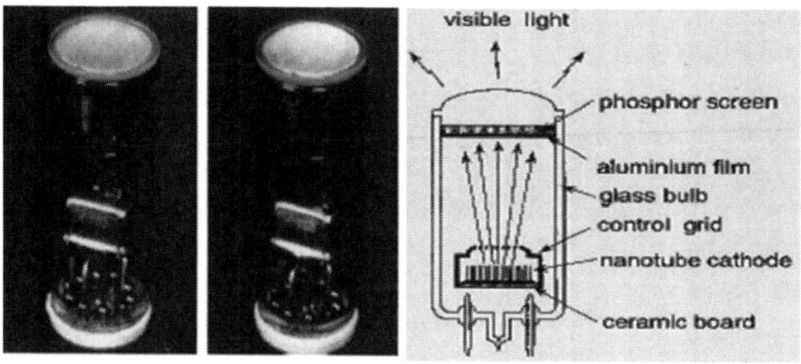

FIGURE 9-8 Demonstration field emission light source using carbon nanotubes as the cathodes (fabricated by Ise Electronic Co., Japan)[64].

Recently, a 4.5 inch diode-type field emission display has been fabricated by Samsung (Figure 9-9), with SWNT stripes on the cathode and phosphor-coated ITO stripes on the anode running orthogonally to the cathode stripes.[65] SWNTs synthesized by the arc-discharge method were dispersed in isopropyl alcohol and then mixed with an organic mixture of nitro cellulose. The paste was squeezed into sodalime glasses through a metal mesh, 20 μm in size, and then heat-treated to remove the organic binder. Y_2O_2S:Eu, ZnS:Cu,Al, and ZnS:Ag,Cl, phosphor-coated glass is used as the anode.

FIGURE 9-9 A prototype 4.5^{00} field emission display fabricated by Samsung using carbon nanotubes (image provided by Dr. W. Choi of Samsung Advanced Institute of Technologies).

C. Gas-discharge tubes in telecom networks: Gas discharge tube protectors, usually consisting of two electrodes parallel to each other in a sealed ceramic case filled with a mixture of noble gases, is one of the oldest methods used to protect against transient over-voltages in a circuit. They are widely used in telecom network interface device boxes and central office switching gear to provide protection from lightning and AC power cross faults on the telecom network. They are designed to be insulated under normal voltage and current row. Under large transient voltages, such as from lightning, a discharge is formed between the metal electrodes, creating a plasma breakdown of the noble gases inside the tube. In the plasma state, the gas tube becomes a conductor, essentially short-circuiting the system and thus protecting the electrical components from over-voltage damage. These devices are robust, moderately inexpensive, and have a relatively small shunt capacitance, so they do not limit the bandwidth of high-frequency circuits as much as other nonlinear shunt components. Compared to solid state protectors, GDTs can carry much higher currents. However, the current GDT protector units are unreliable from the standpoint of mean turn-on voltage and run-to-run variability.

Prototype GDT devices using carbon nanotube-coated electrodes have recently been fabricated and tested by a group from UNC and Raychem Co.[66]. Molybdenum electrodes with various interlayer materials were coated with SWNTs and analysed for both electron field emission and discharge properties. A mean DC breakdown voltage of 448.5V and a standard deviation of 4.8V over 100 surges were observed in nanotube-based GDTs with 1mm gap spacing between the electrodes. The breakdown reliability is a factor of 4–20 better and the breakdown voltage is ~30% lower than the two commercial products measured. The enhanced performance shows that nanotube based GDTs are attractive over-voltage protection units in advanced telecom networks such as an ADSL, where the tolerance is narrower than what can be provided by the current commercial GDTs.

9.5.3.4.3 Energy storage

Carbon nanotubes are being considered for energy production and storage. Graphite, carbonaceous materials and carbon fiber electrodes have been used for decades in fuel cells, battery, and several other electrochemical applications. Nanotubes are special because they have small dimensions, a smooth surface topology, and perfect surface specificity, since only the basal graphite planes are exposed in their structure. The rate of electron transfer at carbon electrodes ultimately determines the efficiency of fuel cells and this depends on various factors, such as the structure and morphology of the carbon material used in the electrodes. Several experiments have pointed out that compared to conventional carbon electrodes, the electron transfer kinetics take place fastest on nanotubes, following the ideal Nernstianbehavior.[67] Nanotube microelectrodes have been constructed using a binder and have been successfully used in bioelectrochemical reactions, for example, oxidation of dopamine. Their performance has been found to be superior to other carbon electrodes in terms of reaction rates and reversibility.[68] Pure MWNTs and MWNTs deposited with metal catalysts (Pd, Pt, Ag) have been used in electro-catalyze an oxygen reduction reaction, which is important for fuel cells.[69–71] It is seen from several studies that nanotubes could be excellent replacements for conventional carbon-based electrodes. Similarly, the improved selectivity of nanotube-based catalysts have been demonstrated in heterogeneous catalysis. Ru-supported nanotubes were found to be superior to the same metal on graphite and on other carbons in the liquid phase hydrogenation reaction of cinnamaldehyde.[71] The properties of catalytically grown carbon nanofibers, which are basically defective nanotubes, have been found to be desirable for high-power electrochemical capacitors.[72]

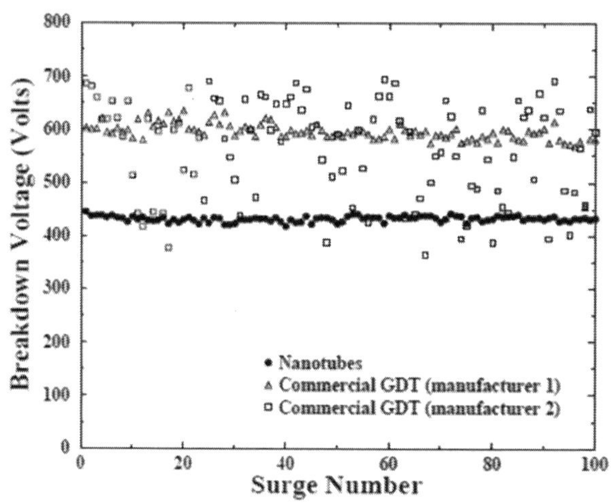

FIGURE 9-10 DC breakdown voltage of a gas discharge tube with SWNT-coated electrodes compared with commercial gas discharge tubes (GDTs) (from Rosen et al. in Ref. [66]).

9.5.3.4.4 Electrochemical system

A. Lithium-ion battery: The basic working mechanism of rechargeable lithium batteries is electrochemical intercalation and de-intercalation of lithium between two working electrodes. Current state-of-art lithium batteries use transition metal oxides (i.e., Li_xCoO_2 or $Li_xMn_2O_4$) as the cathodes and carbon materials (graphite or disordered carbon) as the anodes. It is desirable to have batteries with a high energy capacity, fast charging time, and long cycle time. The energy capacity is determined by the saturation lithium concentration of the electrode materials. For graphite, the thermodynamic equilibrium saturation concentration is LiC_6 which is equivalent to 372 mAh/g. Higher Li concentrations have been reported in disordered carbons (hard and soft carbon)[73,74] and metastable compounds formed under pressure.[75]

It has been speculated that a higher Li capacity may be obtained in carbon nanotubes if all the interstitial sites (intershell van der Waals spaces, intertube channels, and inner cores) are accessible for Li intercalation. The electrochemical intercalation of MWNTs [76, 77] and SWNTs [78, 79] has been investigated by several groups. Figure 9-11 (top) shows representative electrochemical intercalation data collected from an arc-discharge-grown MWNT sample using an electrochemical cell with a carbon nanotube film and a lithium foil as the two working electrodes.[79] A reversible capacity (C_{rev}) of 100–640 mAh/g has been reported, depending on the sample processing and annealing conditions.[76,77,79] In general, well-graphitized MWNTs such as those synthesized by the arc-discharge method have a lower C_{rev} than those prepared by the CVD method. Structural studies[80,81] have shown that alkali metals can be intercalated into the intershell spaces within the individual MWNTs through defect sites.

Single-walled nanotubes are shown to have both high reversible and irreversible capacities.[78,79] Two separate groups reported 400–650mAh/g reversible and ~1000mAh/g irreversible capacities in SWNTs produced by the laser ablation method. The exact locations of the Li ions in the intercalated SWNTs are still unknown. The intercalation and in-situ TEM and EELS measurements on individual SWNT bundles suggested that the intercalates reside in the interstitial sites between the SWNTs.[62] It is shown that the Li/C ratio can further be increased by ball-milling which fractures the SWNTs. A reversible capacity of 1000mAh/g[66] was reported in processed SWNTs. The large irreversible capacity is related to the large surface area of the SWNT films (~300m^2/g by BET characterization) and the formation of a solid electrolyte interface. The SWNTs are also found to perform well under high current rates. For example, 60% of the full capacity can be retained when the charge-discharge rate is increased from 50mA/h to 500mA/h.[78] The high capacity and high-rate performance warrant further studies on the potential of utilizing carbon nanotubes as battery electrodes. The large observed voltage hysteresis (see Figure 9-11) is undesirable for battery application. It is at least partially related to the kinetics of the intercalation reaction and can potentially be reduced/eliminated by processing, that is, cutting the nanotubes to short segments.

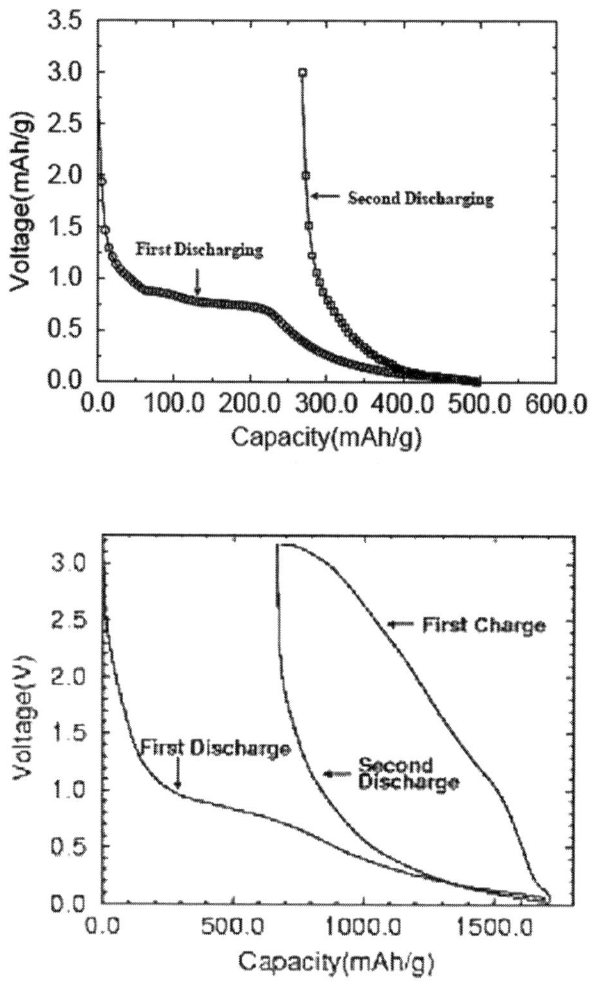

FIGURE 9-11 (Top): Electrochemical intercalation of MWNTs with lithium. (Bottom): Charge-discharge data of purified and processed SWNTs. (Figures are from B. Gao et al. in Ref. [66]).

The characteristics of a carbon nanotube when used as a filler in the electrodes of lithium-ion batteries can be summarized as follows:[82]

i. The small diameter of the nanotube makes it possible to distribute the nanotubes homogeneously in the thin electrode material and to introduce a larger surface area to react with the electrolyte.
ii. The improved electrical conductivity of the electrode is related to the high electrical conductivity of the tubes, and the function of the electrical bridge between graphite particles.
iii. The relatively high intercalation ability of nanotubes does not in itself lower the capacity of anode materials upon cycling.
iv. A high flexibility of the electrode is also achieved due to network formation of the nanotube in a tube-mat structure.
v. The electrode has high endurance due to the presence of nanotubes, which absorb the stress caused by intercalation of lithium ions.
vi. Improved penetration of the electrolyte due to the homogeneous distribution of the tubes surrounding the anode material.
vii. The cyclic efficiency of the lithium-ion battery was improved for a relatively long cycle when compared with that of carbon black.

B. Additives to the electrodes of lead-acid batteries: In order to increase the conductivity of electrodes in lead-acid batteries, different weight percents of carbon nanotubes are added to the active anode material with average diameters of ca. 2 5 mm of the positive electrode. The resistivity of the electrode is lowered for the case of 1.5% nanotube addition. When this sample (0.5–1 wt%) is incorporated in the negative electrode, the cycle characteristics are greatly improved compared with those of an electrode without additive (Figure 9-12) (Endo et al. 2001b). This is probably due to the ability of carbon nanotubes to act as a physical binder, resulting in electrodes that undergo less mechanical disintegration and shedding of their active material. Therefore, it is expected that the use of carbon nanotubes as an electrode's filler should produce an enhanced cyclic behaviour for electrodes in lead-acid batteries compared with electrodes using conventional graphite powder, because the unusual morphology of the carbon nanotube, such as the concentric orientation of their graphite crystallites along the fibre cross-section, induces a high resistance toward oxidation, and furthermore the nanotube network embedded in the polymer would enhance the reactivity of the electrode.[82]

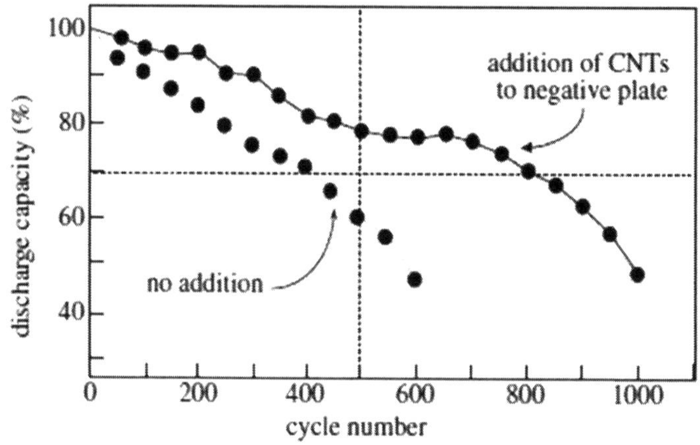

FIGURE 9-12 The effects of carbon nanotubes embedded in a conductive filler and compared with a carbon black/polymer composite.[82]

C. The electric double-layer capacitor: The merit of the EDLC is considered to be a high discharge rate (Conway 1999), which makes them applicable as a hybrid energy source for electric vehicles and portable electric devices (Miyadera 2002). EDLCs containing carbon nanotubes in the electrode exhibit relatively high capacitances resulting from the high surface area accessible to the electrolyte (Niu et al . 1997; An et al . 2001; Frackowiak&Beguin 2002). On the other hand, the most important factor in commercial EDLCs is considered to be the overall resistance. In this context, carbon nanotubes with strong electrical and mechanical properties can be used as an electrical conductive additive in the electrode of EDLC. It has been demonstrated that the addition of carbon nanotubes results in an enhanced capacity at higher current densities, when compared with electrodes containing carbon black (Takeda et al . 2001).[82]

D. Fuel cells: Fuel cells have been considered as the next generation of energy devices because this type of system transforms the chemical reaction energy from hydrogen and oxygen into electric energy (Williams 2001). Carbon nanotubes decorated with metal nanoparticles as the electrode have doubled the performance of the fuel cell, owing to the increased catalytic activity of nanotube-based electrodes (Britto et al. 1999; Che et al. 1999; Yoshitake et al. 2002). In this context, we have reported the efficient impregnation of Pt nanoparticles (outer diameter less than 3 nm) on cup-stacked type carbon nanotubes (Endo et al. 2003). The method involves dispersion of the fibres in H_2PtC_{16}, followed by low-temperature annealing. The Pt particle deposition is always homogeneous, and can be controlled selectively on the outer or inner core using the hydrophobic nature of the material. Since the Pt particle activity on the fibres is high, this material could find application as an efficient catalyst or biological device. It may be that carbon nanotube technology will contribute to the development of fuel cells as a catalyst support and also as a main component of bipolar systems. However, additional basic and applied research is necessary.[82]

E. Hydrogen storage: The area of hydrogen storage in carbon nanotubes remains active and controversial. Extraordinarily high and reversible hydrogen adsorption in SWNT-containing materials[83–85] and GNFs[86] has been reported and has attracted considerable interest in both academia and industry.

Materials with high hydrogen storage capacities are desirable for energy storage applications. Metal hydrides and cryo-adsorption are the two commonly used means to store hydrogen, typically at high pressure and/or low temperature. In metal hydrides, hydrogen is reversibly stored in the interstitial sites of the host lattice. The electrical energy is produced by direct electrochemical conversion. Hydrogen can also be stored in the gas phase in the metal hydrides. The relatively low gravimetric energy density has limited the application of metal hydride batteries. Because of their cylindrical and hollow geometry, and nanometer-scale diameters, it has been predicted that the carbon nanotubes can store liquid and gas in the inner cores through a capillary effect.[87] A temperature-programmed desorption (TPD) study on SWNT-containing material (0.1–0.2wt% SWNT) estimates a gravimetric storage density of 5–10wt% SWNT when H_2 exposures were carried out at 300 torr for 10 minutes at 277K followed by 3 minutes at 133K.[83] If all the hydrogen molecules are assumed to be inside the nanotubes, the reported density would imply a much higher packing density of H_2 inside the tubes than expected from the normal H_2–H_2 distance. The same group recently performed experiments on purified SWNTs and found essentially no H_2 absorption at 300K. Upon cutting (opening) the nanotubes by an oxidation process, the amount of absorbed H_2 molecules increased to 4–5 wt%. A separate study of higher purity materials reports ~8wt% of H_2 adsorption at 80 K, but using a much higher pressure of 100 atm,[88] suggesting that nanotubes have the highest hydrogen storage capacity of any carbon material. It is believed that hydrogen is first adsorbed on the outer surface of the crystalline ropes.

An even higher hydrogen uptake, up to 14–20wt%, at 20–400°C under ambient pressure was reported[84] in alkali-metal intercalated carbon nanotubes. It is believed that in the intercalated systems, the alkali metal ions act as a catalytic center for H_2 dissociative adsorption. FTIR measurements show strong alkali-H and C-H stretching modes. An electrochemical absorption and desorption of hydrogen experiment performed on SWNT-containing materials (MER Co, containing a few percent of SWNTs) reported a capacity of 110mAh/g at low discharge currents. The experiment was done in a half-cell configuration in 6M KOH electrolyte and using a nickel counter electrode. Experiments have also been performed on SWNTs synthesized by a hydrogen arc-discharge method.[85] Measurements performed on relatively large amount materials (~50% purity, 500 mg) showed a hydrogen storage capacity of 4.2wt% when the samples were exposed to 10MPa hydrogen at room temperature. About 80% of the absorbed H_2 could be released at room temperature.[85]

The potential of achieving/exceeding the benchmark of 6.5wt% H_2 to the system weight ratio set by the Department of Energy has generated considerable research activities in the universities, major automobile companies, and national laboratories. At this point, it is still not clear whether carbon nanotubes will have real technological applications in the hydrogen storage application area. The values reported in the literature will need to be verified on well-characterized materials under controlled conditions. What is also lacking is a detailed understanding of the storage mechanism and the effect of materials processing on hydrogen storage. Perhaps, the ongoing neutron scattering and proton nuclear magnetic resonance measurements will shed some light in this direction. In addition to hydrogen, carbon nanotubes readily absorb other gaseous species under ambient conditions which often leads to drastic changes in their electronic properties.[89–91]

This environmental sensitivity is a double-edged sword. From the technological point of view, it can potentially be utilized for gas detection.[48] On the other hand, it makes very difficult to deduce the intrinsic properties of the nanotubes, as demonstrated by the recent transport[90] and nuclear magnetic resonance[91] measurements. Care must be taken to remove the adsorbed species which typically require annealing the nanotubes at elevated temperatures under at least 10^{-6} torr dynamic vacuum.

9.5.3.4.5 Filled composites

The mechanical behavior of carbon nanotubes is exciting since nanotubes are seen as the "ultimate" carbon fiber ever made. The traditional carbon fibers[45] have about fifty times the specific strength (strength/density) of steel and are excellent load-bearing reinforcements in composites. Nanotubes should then be ideal candidates for structural applications. Carbon fibers have been used as reinforcements in high strength, light weight, high performance composites; one can typically find these in a range of products ranging from expensive tennis rackets to spacecraft and aircraft body parts. NASA has recently invested large amounts of money in developing carbon nanotube-based composites for applications such as the futuristic Mars mission.

Early theoretical work and recent experiments on individual nanotubes, mostly MWNTs have confirmed that nanotubes are one of the stiffest structures ever made.[92–95] Since C–C covalent bonds are one of the strongest in nature, a structure based on a perfect arrangement of these bonds oriented along the axis of nanotubes would produce an exceedingly strong material. Theoretical studies have suggested that SWNTs could have a Young's modulus as high as 1TPa,[93] which is basically the in-plane value of defect-free graphite. For MWNTs, the actual strength in practical situations would be further affected by the sliding of individual graphene cylinders with respect to each other. In fact, very recent experiments have evaluated the tensile strength of individual MWNTs using a nano-stressing stage located within a scanning electron microscope. The nanotubes broke by a sword-in-sheath failure mode.[45] This failure mode corresponds to the sliding of the layers within the concentric MWNT assembly and the breaking of individual cylinders independently. Such failure modes have been observed previously in vapor grown carbon fibers.[45] The observed tensile strength of individual MWNTs corresponded to <60GPa. Experiments on individual SWNT ropes are in progress and although a sword-in-sheath failure mode cannot occur in SWNT ropes, failure could occur in a very similar fashion. The individual tubes in a rope could pull out by shearing along the rope axis, resulting in the final breakup of the rope, in stresses much below the tensile strength of individual nanotubes. Although testing of individual nanotubes is challenging, and requires specially designed stages and nanosize loading devices, some clever experiments have provided valuable insights into the mechanical behavior of nanotubes and have provided values for their modulus and strength. For example, in one of the earlier experiments, nanotubes projecting out onto holes in a TEM specimen grid were assumed to be equivalent to clamp homogeneous cantilevers; the horizontal vibration amplitudes at the tube ends were measured from the blurring of the images of the nanotube tips and were then related to the Young's modulus.[94] Recent experiments have also used atomic force microscopy to bend nanotubes attached to substrates and thus obtain quantitative information about their mechanical properties.[95,96]

Most of the experiments done to date corroborate theoretical predictions suggesting the values of Young's modulus of nanotubes to be around 1 TPa (see Figure 11-12). Although the theoretical estimate for the tensile strength of individual SWNTs is about 300 GPa, the best experimental values on MWNTs are close to ~50 GPa, which is still an order of magnitude higher than that of carbon fibers.[45]

The fracture and deformation behavior of nanotubes is intriguing. Simulations on SWNTs have suggested very interesting deformation behavior; highly deformed nanotubes were seen to switch reversibly into different morphological patterns with abrupt releases of energy. Nanotubes gets attened, twisted and buckled as they deform (see Figure 9-12). They sustain large strains (40%) in tension without showing signs of fracture.[93] The reversibility of deformations, such as buckling, has been recorded directly for MWNT, under TEM observations.[37] Flexibility of MWNTs depends on the number of layers that make up the nanotube walls; tubes with thinner walls tend to twist and flatten more easily. This flexibility is related to the in-plane flexibility of a planar graphene sheet and the ability for the carbon atoms to rehybridize, with the degree of sp^2-sp^3 rehybridization depending on the strain. Such flexibility of nanotubes under mechanical loading is important for their potential application as nanoprobes, for example, for use as tips of SPMs.

Recently, an interesting mode of plastic behavior has been predicted in nanotubes.[97] It is suggested that pairs of 5–7 (pentagon-heptagon) pair defects, called a Stone–Wales defect,[98] in sp^2 carbon systems, are created at high strains in the nanotube lattice and that these defect pairs become mobile. This leads to a step-wise diameter reduction (localized necking) of the nanotube. These defect pairs become mobile. The separation of the defects creates local necking of the nanotube in the region where the defects have moved. In addition to localized necking, the region also changes lattice orientation (similar in effect to a dislocation passing through a crystal). This extraordinary behavior initiates necking, but also introduces changes in helicity in the region where the defects have moved (similar to a change in lattice orientation when a dislocation passes through a crystal). This extraordinary behavior could lead to a unique nanotube application: a new type of probe, which responds to mechanical stress by changing its electronic character. High temperature fracture of individual nanotubes under tensile loading has been studied by molecular dynamics simulations.[99] Elastic stretching elongates the hexagons until, at high strain, some bonds are broken. This local defect is then redistributed over the entire surface, by bond saturation and surface reconstruction. The final result of this is that instead of fracturing, the nanotube lattice unravels into a linear chain of carbon atoms. Such behavior is extremely unusual in crystals and could play a role in increasing the toughness by increasing the energy absorbed during deformation of nanotube-filled ceramic composites during high temperature loading.

The most important application of nanotubes based on their mechanical properties will be as reinforcements in composite materials. Although nanotube-filled polymer composites are an obvious materials application area, there have not been many successful experiments, which show the advantage by using nanotubes as fillers over traditional carbon fibers. The main problem is in creating a good interface between nanotubes and the polymer matrix and attaining good load transfer from the matrix to the nanotubes, during loading. The reason for this is essentially two-fold. First, nanotubes are atomically smooth and have nearly the same diameters and aspect ratios (length/diameter) as polymer chains. Second, nanotubes are almost always organized into aggregates which behave differently in response to a load, as compared to individual nanotubes. There have been conflicting reports on the interface strength in nanotube-polymer composites.[100–102] Depending on the polymer used and processing conditions, the measured strength seems to vary. In some cases, fragmentation of the tubes has been observed, which is an indication of a strong interface bonding. In some cases, the effect of sliding of layers of MWNTs and easy pull-out is seen, suggesting poor interface bonding. Micro-Raman spectroscopy has validated the latter, suggesting that sliding of individual layers in MWNTs and shearing of individual tubes in SWNT ropes could be limiting factors for good load transfer, which is essential for making high strength composites. To maximize the advantage of nanotubes as reinforcing structures in high strength composites, the aggregates needs to be broken up and dispersed or cross-linked to prevent slippage. In addition, the surfaces of nanotubes have to be chemically modified (functioned) to achieve strong interfaces between the surrounding polymer chains.

There are certain advantages that have been realized in using carbon nanotubes for structural polymer (e.g., epoxy) composites. Nanotube reinforcements will increase the toughness of the composites by absorbing energy during their highly flexible elastic behavior. This will be especially important for nanotube-based ceramic matrix composites. An increase in fracture toughness on the order of 25% has been seen in nano-crystalline alumina nanotube (5% weight fraction) composites, without compromising on hardness. Other interesting applications of nanotube-filled polymer films will be in adhesives where a decoration of nanotubes on the surface of the polymer films could alter the characteristics of the polymer chains due to interactions between the nanotubes and the polymer chains; the high surface area of the nanotube structures and their dimensions being nearly that of the linear dimensions of the polymer chains could give such nanocomposites new surface properties. The low density of the nanotubes will clearly be an advantage for nanotube-based polymer composites, in comparison to short carbon fiber reinforced random composites. Nanotubes would also offer multifunctionality, since carbon fibers are extremely brittle. Nanotubes will also offer better performance during compressive loading in comparison to traditional carbon fibers due to their flexibility and low propensity for carbon nanotubes to fracture under compressive loads.

Other than for structural composite applications, some of the unique properties of carbon nanotubes are being pursued by filling photo-active polymers with nanotubes. Recently, such a scheme has been demonstrated in a conjugated luminescent polymer, poly(m-phenylenevinylene-co-2,5-dioctoxyp-phenylenevinylene) (PPV), filled with MWNTs and SWNTs.[103] Nanotube/PPV composites have shown large increases in electrical conductivity (by nearly eight orders of magnitude) compared to the pristine polymer, with little loss in photoluminescence/electro-luminescence yield. In addition, the composite is far more robust than the pure polymer regarding mechanical strength and photo-bleaching properties (breakdown of the polymer structure due to thermal effects). Preliminary studies indicate that the host polymer interacts weakly with the embedded nanotubes, but that the nanotubes act as nano-metric heat sinks, which prevent the build up of large local heating thermal effects within the polymer matrix. While experimenting with the composites of conjugated polymers, such as PPV and nanotubes, a very interesting phenomenon has been recently observed;[104] it seems that the coiled morphology of the polymer chains helps to wrap around nanotubes suspended in dilute solutions of the polymer. This effect has been used to separate nanotubes from other carbonaceous material present in impure samples. Use of the nonlinear optical and optical limiting properties of nanotubes has been reported for designing nanotube-polymer systems for optical applications, including photo-voltaic applications. Functionalization of nanotubes and the doping of chemically modified nanotubes in low concentrations into photo-active polymers, such as PPV, have been shown to provide a means to alter the hole transport mechanism and hence the optical properties of the polymer. Small loadings of nanotubes are used in these polymer systems to tune the color of emission when used in organic light emitting devices. The interesting optical properties of nanotube-based composite systems arise from the low dimensionality and unique electronic band structure of nanotubes; such applications cannot be realized using larger micron-size carbon fibers.

There are other less-explored areas where nanotube-polymer composites could be useful. For example, nanotube filled polymers could be useful in electromagnetic induction (EMI) shielding applications where carbon fibers have been used extensively.[45] Membranes for molecular separations (especially biomolecules) could be built from nanotube-polycarbonate systems, making use of the remarkable small pores sizes that exist in nanotubes. Very recently, work done at RPI suggested that composites made from nanotubes (MWNTs) and a biodegradable polymer (polylactic acid; PLA) act more efficiently than carbon fibers for osteointegration (growth of bone cells), especially under electrical stimulation of the composite.

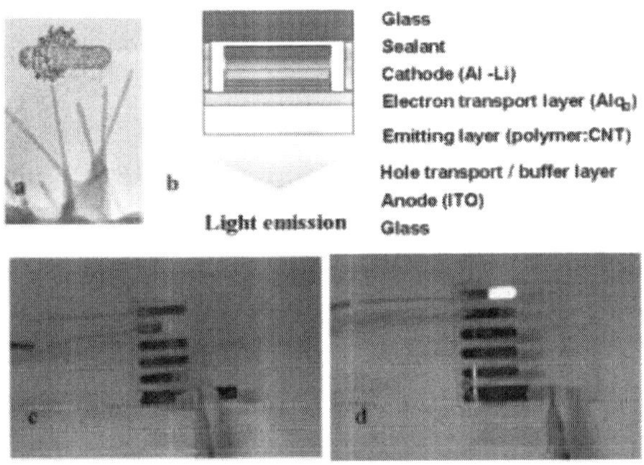

FIGURE 9-13 Results from the optical response of nanotube-doped polymers and their use in organic light emitting diodes (OLED). The construction of the OLED is shown in the schematic of (top). The bottom figure shows emission from OLED structures. Nanotube doping tunes the emission color. With SWNTs in the buffer layer, holes are blocked and recombination takes place in the transport layer and the emission color is red. Without nanotubes present in the buffer layer, the emission color is green (not shown in the figure). (Figures are Courtesy of Prof. David Carroll).

There are challenges to be overcome when processing nanotube composites. One of the biggest problems is dispersion. It is extremely difficult to separate individual nanotubes during mixing with polymers or ceramic materials and this creates poor dispersion and clumping together of nanotubes, resulting in a drastic decrease in the strength of composites. By using high power ultrasound mixers and using surfactants with nanotubes during processing, good nanotube dispersion may be achieved, although the strengths of nanotube composites reported to date have not seen any drastic improvements over high modulus carbon fiber composites. Another problem is the difficulty in fabricating high weight fraction nanotube composites, considering the high surface area for nanotubes, which results in a very high viscosity for nanotube-polymer mixtures. Notwithstanding all these drawbacks, it needs to be said that the presence of nanotubes stiffens the matrix (the role is especially crucial at higher temperatures) and could be very useful as a matrix modifier, particularly for fabricating improved matrices useful for carbon fiber composites. The real role of nanotubes as an efficient reinforcing fiber will have to wait until we know how to manipulate the nanotube surfaces chemically to make strong interfaces between individual nanotubes (which are really the strongest material ever made) and the matrix materials. In the meanwhile, novel and unconventional uses of nanotubes will have to take the center stage.

It has been shown that carbon nanotubes could behave as the ultimate one-dimensional material with remarkable mechanical properties (Lu 1997; Li et al. 2000). The density-based modulus and strength of highly crystalline SWNTs are 19 and 56 times that of steel. Based on a continuum shell model, the armchair tube exhibits larger stress–strain response than the zigzag tube under tensile loading (Figure 9-14).

Strong mechanical properties of carbon nanotubes due to a strong C–C covalent bond are highly dependent upon the atomic structure of nanotubes and the number of shells. Moreover, carbon nanotubes exhibit strong electrical and thermal conducting properties. Therefore, carbon nanotubes (single- and multi-walled) have been studied intensively as fillers in various matrices, especially polymers.[100,105–108]

The best use of the intrinsic properties of these fibrous nanocarbons in polymers can be achieved by optimizing the interface interaction of the nanotube surface and the polymer. Therefore, surface treatments via oxidation could be used in order to improve adhesion properties between the filler and the matrix. This results in a good stress transfer from the polymer to the nanotube. There are various surface oxidative processes, such as electrochemical, chemical, and plasma techniques. From the industrial point of view, ozone treatment is a very attractive technique. In addition, the dispersion of the nanotubes/nanofibres in the polymer should be uniform within the matrix.[82]

When cup-stacked-type carbon nanotubes are incorporated in polypropylene, the improvement of the tensile strength is very noticeable (up to 40%). This remarkable result can be explained by the particular morphology of cup-stacked-type carbon nanotubes. In other words, a large portion of edge sites on the outer surface of the nanotubes might act as nucleation sites, resulting in good adhesion between nanotubes and polymers (good stress transfer). Recently, various studies on the nucleation effect of nanotubes on the crystallization of semi-crystalline polymers have been reported (Valentini et al. 2003; Bhattacharyya et al. 2003; Cadek et al. 2003).[82]

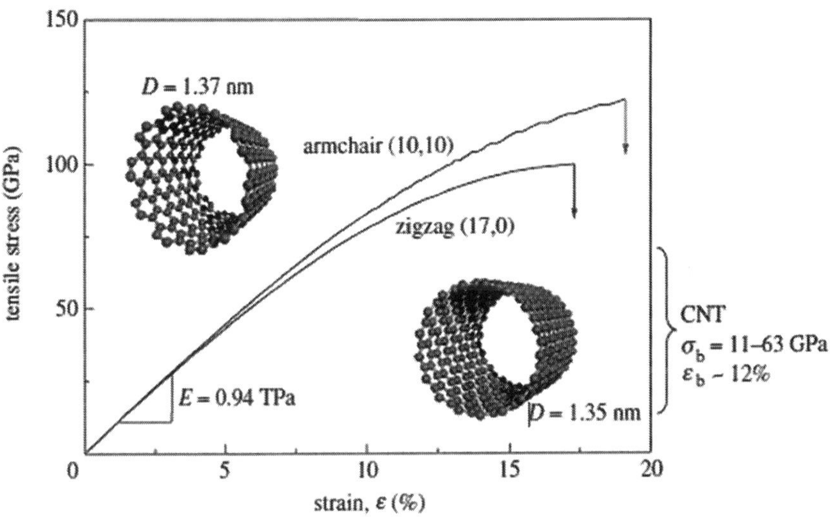

FIGURE 9-14 Stress–strain response of a carbon nanotube under tensile loading.

9.5.3.4.6 *Nanoprobes and sensors*

The small and uniform dimensions of the nanotubes produce some interesting applications. With extremely small sizes, high conductivity, high mechanical strength and flexibility (ability to easily bend elastically), nanotubes may ultimately become indispensable in their use as nanoprobes. One could think of such probes as being used in a variety of applications, such as high resolution imaging, nano-lithography, nanoelectrodes, drug delivery, sensors, and field emitters. The possibility of nanotube-based field emitting devices has been already discussed. Use of a single MWNT attached to the end of a scanning probe microscope tip for imaging has already been demonstrated (see Figure 9-15).[109] Since MWNT tips are conducting, they can be used in STM, AFM instruments as well as other scanning probe instruments, such as an electrostatic force microscope. The advantage of the nanotube tip is its slenderness and the possibility to image features (such as very small, deep surface cracks), which are almost impossible to probe using the larger, blunter etched Si or metal tips. Biological molecules, such as DNA, can also be imaged with higher resolution using nanotube tips, compared to conventional STM tips. MWNT and SWNT tips were used in a tapping mode to image biological molecules such as amyloid-b-protofibrils (related to Alzheimer's disease), with a resolution never achieved before.[110] In addition, due to the high elasticity of the nanotubes, the tips do not suffer from crashes on contact with the substrates. Any impact will cause buckling of the nanotube, which generally is reversible on the retraction of the tip from the substrate. Attaching individual nanotubes to the conventional tips of SPMs has been the real challenge. Bundles of nanotubes are typically pasted on to AFM tips and the ends are cleaved to expose individual nanotubes (see Figure 9-15). These tip attachments are not very controllable and will result in vibration problems and in instabilities during imaging, which decrease the image resolution. However, successful attempts have been made to grow individual nanotubes onto Si tips using CVD,[111] in which case the nanotubes are firmly anchored to the probe tips. Due to the longitudinal (high aspect) design of nanotubes, nanotube vibration still will remain an issue, unless short segments of nanotubes can be controllably grown (see Figure 9-15 reff-tip).

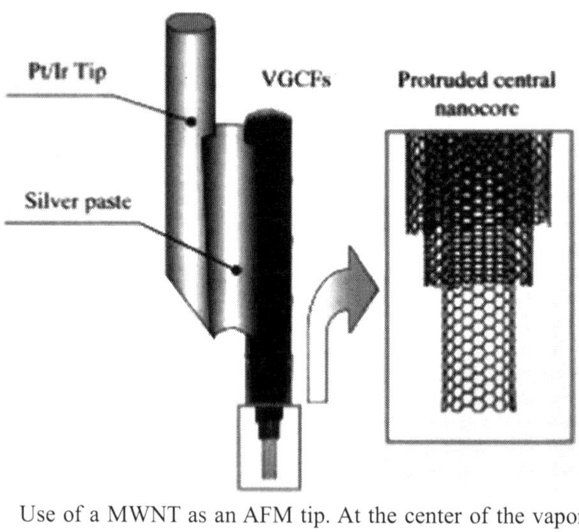

FIGURE 9-15 Use of a MWNT as an AFM tip. At the center of the vapor grown carbon fiber (VGCF) is a MWNT which forms the tip. The VGCF provides a convenient and robust technique for mounting the MWNT probe for use in a scanning probe instrument.

In addition to the use of nanotube tips for high resolution imaging, it is also possible to use nanotubes as active tools for surface manipulation. It has been shown that if a pair of nanotubes can be positioned appropriately on an AFM tip, they can be controlled like tweezers to pick up and release nanoscale structures on surfaces; the dual nanotube tip acts as a perfect nano-manipulator in this case.[112] It is also possible to use nanotube tips in AFM nano-lithography. Ten nanometer lines have been written on oxidized silicon substrates using nanotube tips at relatively high speeds,[113] a feat that can only be achieved with tips as small as nanotubes.

Since nanotube tips can be selectively modified chemically through the attachment of functional groups,[114] nanotubes can also be used as molecular probes, with potential applications in chemistry and biology. Open nanotubes with the attachment of acidic functionalities have been used for chemical and biological discrimination on surfaces.[115] Functioned nanotubes were used as AFM tips to perform local chemistry, to measure binding forces between protein-ligand pairs and for imaging chemically patterned substrates. These experiments open up a whole range of applications, for example, as probes for drug delivery, molecular recognition, chemically sensitive imaging, and local chemical patterning, based on nanotube tips that can be chemically modified in a variety of ways. The chemical functionalization of nanotubes is a major issue with far-reaching implications. The possibility to manipulate, chemically modify, and perhaps polymerize nanotubes in solution will set the stage for nanotube-based molecular engineering and many new nanotechnological applications.

Electromechanical actuators have been constructed using sheets of SWNTs. It was shown that small voltages (a few volts), applied to strips of laminated (with a polymer) nanotube sheets suspended in an electrolyte, bend the sheet to large strains, mimicking the actuator mechanism present in natural muscles. The nanotube actuators would be superior to conducting polymer-based devices, since in the former no ion intercalation (which limits actuator life) is required. This interesting behavior of the nanotube sheets in response to an applied voltage suggests several applications, including nanotube-based micro-cantilevers for medical catheter applications and as novel substitutes, especially at higher temperatures, for ferroelectrics.

Recent research has also shown that nanotubes can be used as advanced miniaturized chemical sensors.[89] The electrical resistivities of SWNTs were found to change sensitively on exposure to gaseous ambient containing molecules of NO_2, NH_3, and O_2. By monitoring the change in the conductance of nanotubes, the presence of gases could be precisely monitored. It was seen that the response times of nanotube sensors are at least an order of magnitude faster (a few seconds for a resistance change of one order of magnitude) than those based on presently available solid-state (metal-oxide and polymers) sensors. In addition, the small dimensions and high surface area offer special advantages for nanotube sensors, which could be operated at room temperature or at higher temperatures for sensing applications.

9.5.3.4.7 Templates

Since nanotubes have relatively straight and narrow channels in their cores, it was speculated from the beginning that it might be possible to fill these cavities with foreign materials to fabricate one-dimensional nanowires. Early calculations suggested that strong capillary forces exist in nanotubes, strong enough to hold gases and fluids inside them.[87] The first experimental proof was demonstrated in 1993, by the filling and solidification of molten lead inside the channels of MWNTs.[14] Wires can be fabricated by this method inside nanotubes. A large body of work now exists in the literature,42–44 to site a few examples, concerning the filling of nanotubes with metallic and ceramic materials. Thus, nanotubes have been used as templates to create nanowires of various compositions and structures (see Figure 9-16).

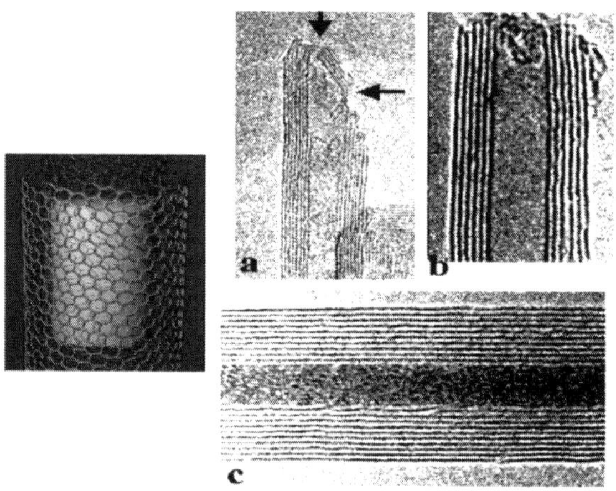

FIGURE 9-16 Results that show the use of nanotubes as templates. The left-hand figure is a schematic that shows the filling of the empty one-dimensional hollow core of nanotubes with foreign substances. (a) Shows a high resolution TEM image of a tube tip that has been attacked by oxidation; the preferential attack begins at locations where pentagonal defects were originally present (arrows) and serves to open the tube. (b) TEM image that shows a MWNT that has been completely opened by oxidation. (c) TEM image of a MWNT with its cavity filled uniformly with lead oxide. The filling was achieved by capillarity.[43]

The critical issue in the filling of nanotubes is the wetting characteristics of nanotubes, which seem to be quite different from that of planar graphite, because of the curvature of the tubes. Wetting of low melting alloys and solvents occurs quite readily in the internal high curvature pores of MWNTs and SWNTs. In the latter, since the pore sizes are very small, filling is more difficult and can be done only for a selected few compounds. It is intriguing that one could create one-dimensional nanostructures by utilizing the internal one-dimensional cavities of nanotubes. Liquids such as organic solvents wet nanotubes easily and it has been proposed that interesting chemical reactions could be performed inside nanotube cavities.[44]

A whole range of experiments remains to be performed inside these constrained one-dimensional spaces, which are accessible once the nanotubes can be opened. The topology of closed nanotubes provides a fascinating avenue to open them through the simple chemical method of oxidation.[42]

As in fullerenes, the pentagonal defects that are concentrated at the tips are more reactive than the hexagonal lattice of the cylindrical parts of the nanotubes. Hence, during oxidation, the caps are removed prior to any damage occurring to the tube body, thus easily creating open nanotubes. The opening of nanotubes by oxidation can be achieved by heating nanotubes in air (above 600°C) or in oxidizing solutions (e.g., acids). It is noted here that nanotubes are more stable to oxidation than graphite, as observed in TGA experiments, because the edge planes of graphite where reaction can initiate are conspicuous by their absence in nanotubes. After the first set of experiments, reporting the opening and filling of nanotubes whit air, simple chemical methods, based on the opening and filling nanotubes in solution, were discovered to develop generalized solution-based strategies to fill nanotubes with a range of materials.[43]

In these methods an acid is first used to open the nanotube tip and to act as a low surface tension carrier for solutes (metal-containing salts) to fill the nanotube hollows. Calcination of solvent-treated nanotubes leaves deposits of oxide material (e.g., NiO) inside nanotube cavities. The oxides can then be reduced to metals by annealing in reducing atmospheres. Observation of solidification inside the one-dimensional channels of nanotubes provides a fascinating study of phase stabilization under geometrical constraints. It is experimentally found that when the channel size gets smaller than a certain critical diameter, solidification results in new and oftentimes disordered phases (e.g., V_2O_5).[116] Crystalline bulk phases are formed in larger cavities. Numerous modeling studies are under way to understand the solidification behavior of materials inside nanotubes and the physical properties of these unique, filled nano-composite materials.

Filled nanotubes can also be synthesized in situ, during the growth of nanotubes in an electric arc or by laser ablation. During the electric arc formation of carbon species, encapsulated nanotubular structures are created in abundance. This technique generally produces encapsulated nanotubes with carbide nanowires, for example, transition metal carbides inside. Laser ablation also produces heterostructures containing carbon and metallic species. Multielement nanotube structures consisting of multiple phases, for example, coaxial nanotube structures containing SiC, SiO, BN, and C have been successfully synthesized by reactive laser ablation.[117] Similarly, postfabrication treatments can also be used to create heterojunctions between nanotubes and semiconducting carbides.[118,119] It is expected that these hybrid nanotubebased structures, which are combinations of metallic, semiconducting, and insulating nanostructures, will be useful in future nanoscale electronic device applications. Nanocomposite structures based on carbon nanotubes can also be built by coating nanotubes uniformly with organic or inorganic structures. These unique composites are expected to have interesting mechanical and electrical properties due to a combination of dimensional effects and interface properties. Finely-coated nanotubes with monolayers of layered oxides have been made and characterized, for example, vanadium pentoxide films.[116]

The interface formed between the nanotubes and the layered oxide is atomically at due to the absence of covalent bonds across the interface. It has been demonstrated that after the coating is made, the nanotubes can be removed by oxidation leaving behind freely-standing nanotubes made of oxides, with nanoscale wall thickness. These novel ceramic tubules, made using nanotubes as templates, could have interesting applications in catalysis. Recently, researchers have also found that nanotubes can be used as templates for the self-assembly of protein molecules. Dipping MWNTs in a solution containing proteins, results in monolayers of proteins covering nanotubes; what is interesting is that the organization of the protein molecules on nanotubes corresponds directly to the helicity of the nanotubes. It seems that nanotubes with controlled helicities could be used as unique probes for molecular recognition, based on the helicity and dimensions, which are recognized by organic molecules of comparable length scales.

There are other ways in which pristine nanotubes can be modified into composite structures. Chemical functionalization can be used to build macromolecular structures from fullerenes and nanotubes. The attachment of organic functional groups on the surface of nanotubes has been achieved, and with the recent success in breaking up SWNTs into shorter fragments, the possibility of functionalizing and building structures through chemistry has become a reality. Decoration of nanotubes with metal particles has been achieved for different purposes, most importantly for use in heterogeneous catalysis.[71] SWNT bundles have been doped with alkali metals, and with the halogens Br_2 and I_2, resulting in an order of magnitude increase in electrical conductivity.[120] In some cases, it is observed that the dopants form a linear chain and sit in the one-dimensional interstitial channels of the bundles. Similarly, Li intercalation inside nanotubes has been successfully carried out with possible impact on battery applications, which has already been discussed in a previous section. The intercalation and doping studies suggest that nanotube systems provide an effective host lattice for the creation of a range of carbon-based synthetic metallic structures.

The conversion of nanotubes through vapor chemistry can create unique nanocomposites with nanotubes as a backbone. When volatile gases such as halogenated compounds or SiO_x are then reacted with nanotubes, the tubes get converted into carbide nano-rods of similar dimensions. These reactions can be controlled, such that the outer nanotube layers can be converted to carbides, keeping the inner graphite layer structure intact. The carbide rods so produced (e.g., SiC, NbC) should have a wide range of interesting electrical and mechanical properties, which could be exploited for applications as reinforcements and nanoscaleelectrical devices.[13]

9.5.3.4.8 Challenges and potential for carbon nanotube applications

Carbon nanotubes have come a long way since their discovery in 1991. The structures that were first reported in 1991 were MWNTs with a range of diameters and lengths. These were essentially the distant relatives of the highly defective carbon nanofibers grown via catalytic chemical vapor deposition. The latter types of fibers (e.g., the lower quality carbon nanofibers made commercially by the Hyperion Corporation and more perfect nanotube structures revealed by Morinobu Endo in his 1975 Ph.D. thesis) had existed for more than a decade. The real molecular nanotubes arrived when they were found accidentally while a catalyst (Fe, Co) material was inserted in the anode during electric arc discharge synthesis. For the first time, there was hope that molecular fibers based purely on carbon could be synthesized and the excitement was tremendous, since many physical properties of such a fiber had already been predicted by theory. It was really the theoretical work proposed on SWNTs and the availability of nanoscale technology (in characterization and measurements) that made the field takes off in 1991.

The greatness of a SWNT is that it is a macro-molecule and a crystal at the same time. The dimensions correspond to extensions of fullerene molecules and the structure can be reduced to a unit cell picture, as in the case of perfect crystals. A new predictable (in terms of atomic structure-property relations) carbon fiber was born. The last decade of research has shown that indeed the physical properties of nanotubes are remarkable, as elaborated in the various chapters of this book. A carbon nanotube is an extremely versatile material: it is one of the strongest materials, yet highly elastic, highly conducting, small in size, but stable, and quite robust in most chemically harsh environments. It is hard to think of another material that can compete with nanotubes in versatility.

As a novel material, fullerenes failed to make much of an impact on applications. It seems from the progress made in recent research that the story of nanotubes is going to be very different. There are already real products based on nanotubes on the market, for example, the nanotube attached AFM tips used in metrology. The United States, Europe, and Japan have all invested heavily in developing nanotube applications. Nanotube-based electronics tops this list and it is comforting that the concepts of devices (such as room temperature field effect transistors based on individual nanotubes) have already been successfully demonstrated. As in the case of most products, especially in high technology areas, such as nano-electronics, the time lag between concept demonstration and real products could be several years to decades and one will have to wait and see how long it is going to take nanotube electronics to pervade high technology. Other more obvious and direct applications are some of the bulk uses, such as nanotube-based polymer composites and electrochemical devices. These, although very viable applications, face challenges, as detailed in this chapter. What is also interesting is that new and novel applications are emerging, as, for example, nanotubes affecting the transport of carriers and hence luminescence in polymer-based organic light emitting diodes, and nanotubes used as actuators in artificial muscles. It can very well be said that some of these newly found uses will have a positive impact on the early stages of nanotube product development.

There are also general challenges that face the development of nanotubes into functional devices and structures. First of all, the growth mechanism of nanotubes, similar to that of fullerenes, has remained a mystery. With this handicap, it is not really possible yet to controllably grow these structures in a controlled way. There have been some successes in growing nanotubes of certain diameter (and to a lesser extent, of predetermined helicity) by tuning the growth conditions by trial and error. Especially for electronic applications, which rely on the electronic structure of nanotubes, this inability to select the size and helicity of nanotubes during growth remains a drawback. More so, many predictions of device applicability are based on joining nanotubes via the incorporation of topological defects in their lattices. There is no controllable way, as of yet, of making connections between nanotubes. Some recent reports, however, suggest the possibility of constructing these interconnected structures by electron irradiation and by template mediated growth.

For bulk applications, such as fillers in composites, where the atomic structure (helicity) has a much smaller impact on the resulting properties, the quantities of nanotubes that can be manufactured still falls far short of what the industry would need. There are no available techniques that can produce nanotubes of reasonable purity and quality in kilogram quantities. The industry would need tonnage quantities of nanotubes for such applications. The market price of nanotubes is also too high presently for any realistic commercial application. But it should be noted that the starting prices for carbon fibers and fullerenes were also prohibitively high during their initial stages of development, but has come down significantly in time. Another challenge is in the manipulation of nanotubes. Nanotechnology is in its infancy and the revolution that is unfolding in this field relies strongly on the ability to manipulate structures at the atomic scale. Some recent reports, however, suggest the possibility of constructing these interconnected structures by electron irradiation, template mediated growth and nano-manipulation (e.g., use of an AFM tip to push nanotubes placed on a substrate). This will remain a major challenge in this field, among several others.

9.6 CONCLUSION

In recent decades, with the development of nanotechnology, carbon nanotubes due to the unique characteristics and remarkable electronic, mechanical, and thermal properties have attracted many attentions. This chapter has described several possible applications of carbon nanotubes, with emphasis on nanoscience and engineering-based applications.

According to the studies conducted in recent years, as reported, the carbon nanotubes have potential applications in vacuum microelectronics and prototype electron emission devices, which is included of cathode ray lighting elements, flat panel display, and gas-discharge tubes in telecom networks. In addition, carbon nanotubes have special applications in an electrochemical system like lithium-ion and additives to the electrodes of lead-acid batteries, the electric double-layer capacitors, and fuel cells. Also, these materials as fillers could give many fine properties to composites, and finally, as the last applications of the carbon nanotubes should refer to their use in nanoprobes, sensors, and templates.

It is obvious that the prospect of the large and growing applications of carbon nanotubes in the near future could open the way to many other nano-sciences and engineering sciences.

KEYWORDS

- **Carbon Nanotubes Applications**
- **Carbon Nanotubes**
- **Nanoscience**

REFERENCES

1. Haghi, A.; Zaikov, G. *POLYMERIC NANOSYSTEMS.* Handbook of Research on Functional Materials: Principles, Capabilities and Limitations, 2014: p. 105.
2. Allhoff, F.; Lin, P.; Moore, D. *What is nanotechnology and why does it matter: from science to ethics*. 2009: John Wiley & Sons.
3. Ramsden, J. *Applied nanotechnology: the conversion of research results to products*. 2013: William Andrew.
4. Feynman, R. P. *There's plenty of room at the bottom.* Engineering and science, 1960. **23**(5): p. 22–36.
5. Bonaccorsi, A.; Thoma, G. *Institutional complementarity and inventive performance in nano science and technology.* Research Policy, 2007. **36**(6): p. 813–831.
6. Binnig, G.; Rohrer, H. *Scanning tunneling microscope*, 1982, Google Patents.
7. Bennig, G. K. *Atomic force microscope and method for imaging surfaces with atomic resolution*, 1988, Google Patents.
8. Bonaccorsi, A. *Search regimes and the industrial dynamics of science.* Minerva, 2008. **46**(3): p. 285–315.
9. Taniguchi, N. *On the basic concept of nano-technology Proceedings of the International Conference on Production Engineering Tokyo Part II Japan Society of Precision Engineering.* 1974.
10. Young, R.; Ward, J.; Scire, F. *The topografiner: an instrument for measuring surface microtopography.* Review of Scientific Instruments, 1972. **43**(7): p. 999–1011.
11. Zhang, Y., et al. *Recent development of polymer nanofibers for biomedical and biotechnological applications.* Journal of Materials Science: Materials in Medicine, 2005. **16**(10): p. 933–946.
12. Ramakrishna, S., et al. *An introduction to electrospinning and nanofibers.* Vol. 90. 2005: World Scientific.

13. Dresselhaus, M. S., et al. *Carbon nanotubes*. 2000: Springer.
14. Bernholc, J.; Roland, C.; Yakobson, B. I. *Nanotubes*. Current opinion in solid state and materials science, 1997. **2**(6): p. 706–715.
15. Dresselhaus, M. S.; Dresselhaus, G.; Eklund, P. C. *Science of fullerenes and carbon nanotubes: their properties and applications*. 1996: Academic press.
16. Rao, C.; Sen, R.; Govindaraj, A. *Fullerenes and carbon nanotubes*. Current Opinion in Solid State and Materials Science, 1996. **1**(2): p. 279–284.
17. Iijima, S. *Helical microtubules of graphitic carbon*. Nature, 1991. **354**(6348): p. 56–58.
18. Iijima, S.; Ichihashi, T. *Single-shell carbon nanotubes of 1-nm diameter*. 1993.
19. Dekker, C. *Carbon nanotubes as molecular quantum wires*. Physics Today, 1999. **52**: p. 22–30.
20. Ebbesen, T. W.; Ajayan, P. M. *Large-scale synthesis of carbon nanotubes*. Nature, 1992. **358**(6383): p. 220–222.
21. Bethune, D., et al. *Cobalt-catalysed growth of carbon nanotubes with single-atomic-layer walls*. 1993.
22. Iijima, S.; Ajayan, P. M.; Ichihashi, T. *Growth model for carbon nanotubes*. Physical Review Letters, 1992. **69**(21): p. 3100.
23. Thess, A., et al. *Crystalline ropes of metallic carbon nanotubes*. Science-AAAS-Weekly Paper Edition, 1996. **273**(5274): p. 483–487.
24. Dai, H., et al. *Single-wall nanotubes produced by metal-catalyzed disproportionation of carbon monoxide*. Chemical Physics Letters, 1996. **260**(3): p. 471–475.
25. Li, W. Z., et al. *Large-scale synthesis of aligned carbon nanotubes*. Science, 1996. **274**(5293): p. 1701–1703.
26. Terrones, M., et al. *Controlled production of aligned-nanotube bundles*. Nature, 1997. **388**(6637): p. 52–55.
27. Beck, R.D., et al. *Resilience of all-carbon molecules C60, C70, and C84: a surface-scattering time-of-flight investigation*. The Journal of Physical Chemistry, 1991. **95**(21): p. 8402–8409.
28. Hirahara, K., et al. *One-dimensional metallofullerene crystal generated inside single-walled carbon nanotubes*. Physical Review Letters, 2000. **85**(25): p. 5384.
29. Dresselhaus, M. S.; Dresselhaus, G.; Saito, R. *Physics of carbon nanotubes*. Carbon, 1995. **33**(7): p. 883–891.
30. Ajayan, P. M., et al. *Growth morphologies during cobalt-catalyzed single-shell carbon nanotube synthesis*. Chemical Physics Letters, 1993. **215**(5): p. 509–517.
31. Kingston, C. T.; Simard, B. *Fabrication of carbon nanotubes*. Analytical Letters, 2003. **36**(15): p. 3119–3145.
32. Jorio, A., et al. *Linewidth of the Raman features of individual single-wall carbon nanotubes*. Physical Review B, 2002. **66**(11): p. 115411.
33. Hone, J. *Phonons and thermal properties of carbon nanotubes*, in *Carbon Nanotubes*. 2001, Springer. p. 273–286.
34. Kroto, H. W., et al. *C 60: buckminsterfullerene*. Nature, 1985. **318**(6042): p. 162-163.
35. Ebbesen, T. W. *Carbon nanotubes: preparation and properties*. 1996: CRC press.
36. Saito, R.; Dresselhaus, G.; Dresselhaus, M. S. *Physical properties of carbon nanotubes*. Vol. 4. 1998: World Scientific.
37. Ajayan, P. M. *Nanotubes from carbon*. Chemical Reviews, 1999. **99**(7): p. 1787–1800.
38. Ajayan, P. M.; Ichihashi, T.; Iijima, S. *Distribution of pentagons and shapes in carbon nano-tubes and nano-particles*. Chemical Physics Letters, 1993. **202**(5): p. 384–388.

39. Carroll, D. L., et al. *Electronic structure and localized states at carbon nanotube tips.* Physical Review Letters, 1997. **78**(14): p. 2811.
40. Kosaka, M., et al. *Electron spin resonance of carbon nanotubes.* Chemical Physics Letters, 1994. **225**(1): p. 161–164.
41. Liu, J., et al.; Rodriguez-Macias, F.; Shon, Y. S.; Lee, T. R.; Colbert, D. T.; Smalley, R. E. *Fullerene pipes.* Science, 1998. 280, p: 1253-1256.
42. Ajayan, P. M., et al. *Opening carbon nanotubes with oxygen and implications for filling.* Nature, 1993. **362**(6420): p. 522–525.
43. Tsang, S. C., et al. *A simple chemical method of opening and filling carbon nanotubes.* Nature, 1994. 372. p. 159–162.
44. Dujardin, E., et al. *Capillarity and wetting of carbon nanotubes.* Science, 1994. **265**(5180): p. 1850–1852.
45. Dresselhaus, M. S., et al. *Graphite fibers and filaments.* Vol. 5. 1988: Springer-Verlag Berlin.
46. Journet, C., et al. *Large-scale production of single-walled carbon nanotubes by the electric-arc technique.* Nature, 1997. **388**(6644): p. 756–758.
47. Ren, Z. F., et al. *Synthesis of large arrays of well-aligned carbon nanotubes on glass.* Science, 1998. **282**(5391): p. 1105–1107.
48. Kong, J., et al. *Synthesis of individual single-walled carbon nanotubes on patterned silicon wafers.* Nature, 1998. **395**(6705): p. 878–881.
49. Brodie, I.; Spindt, C. A. *Vacuum microelectronics.* Advances in electronics and electron physics, 1992. **83**: p. 1–106.
50. Castellano, J. A. *Handbook of display technology.* 1992: Elsevier.
51. Scott, A. *Understanding microwaves.* New York, John Wiley & Sons, Inc., 1993, 557.
52. Davis, J. J., et al. *Protein electrochemistry at carbon nanotube electrodes.* Journal of Electroanalytical Chemistry, 1997. **440**(1): p. 279–282.
53. Saito, Y., et al. *Field emission patterns from single-walled carbon nanotubes.* Japanese Journal of Applied Physics, 1997. **36**(10A): p. L1340.
54. Collins, P. G.; Zettl, A. *A simple and robust electron beam source from carbon nanotubes.* Applied Physics Letters, 1996. **69**(13): p. 1969–1971.
55. Wang, Q. H., et al. *Field emission from nanotube bundle emitters at low fields.* Applied Physics Letters, 1997. **70**(24): p. 3308–3310.
56. De Heer, W. A.; Chatelain, A.; Ugarte, D. *A carbon nanotube field-emission electron source.* Science, 1995. **270**(5239): p. 1179–1180.
57. Küttel, O. M., et al. *Electron field emission from phase pure nanotube films grown in a methane/hydrogen plasma.* Applied Physics Letters, 1998. **73**(15): p. 2113–2115.
58. Zhu, W., et al. *Large current density from carbon nanotube field emitters.* Applied Physics Letters, 1999. **75**(6): p. 873–875.
59. Dean, K. A.; Chalamala, B. R. *Field emission microscopy of carbon nanotube caps.* Journal of Applied Physics, 1999. **85**(7): p. 3832–3836.
60. Dean, K. A.; Chalamala, B. R. *Current saturation mechanisms in carbon nanotube field emitters.* Applied Physics Letters, 2000. **76**(3): p. 375–377.
61. Robertson, J. *Mechanisms of electron field emission from diamond, diamond-like carbon, and nanostructured carbon.* Journal of Vacuum Science & Technology B, 1999. **17**(2): p. 659–665.
62. Suzuki, S., et al. *Work functions and valence band states of pristine and Cs-intercalated single-walled carbon nanotube bundles.* Applied Physics Letters, 2000. **76**(26): p. 4007–4009.

63. Zhu, W.; Kochanski, G. P.; Jin, S. *Low-field electron emission from undoped nanostructured diamond.* Science, 1998. **282**(5393): p. 1471–1473.
64. Saito, Y.; Uemura, S.; Hamaguchi, K. *Cathode ray tube lighting elements with carbon nanotube field emitters.* Japanese journal of Applied Physics, 1998. **37**(3B): p. L346.
65. Choi, W. B., et al. *Fully sealed, high-brightness carbon-nanotube field-emission display.* Applied Physics Letters, 1999. **75**(20): p. 3129–3131.
66. Gao, B., et al. *Fabrication and electron field emission properties of carbon nanotube films by electrophoretic deposition.* Advanced Materials, 2001. **13**(23): p. 1770–1773.
67. Nugent, J. M., et al. *Fast electron transfer kinetics on multiwalled carbon nanotube microbundle electrodes.* Nano Letters, 2001. **1**(2): p. 87–91.
68. Britto, P. J.; Santhanam, K. S. V.; Ajayan, P. M. *Carbon nanotube electrode for oxidation of dopamine.* Bioelectrochemistry and Bioenergetics, 1996. **41**(1): p. 121–125.
69. Britto, P. J., et al. *Improved charge transfer at carbon nanotube electrodes.* Advanced Materials, 1999. **11**(2): p. 154–157.
70. Che, G., et al. *Carbon nanotubule membranes for electrochemical energy storage and production.* Nature, 1998. **393**(6683): p. 346–349.
71. Planeix, J. M., et al. *Application of carbon nanotubes as supports in heterogeneous catalysis.* Journal of the American Chemical Society, 1994. **116**(17): p. 7935–7936.
72. Niu, C., et al. *High power electrochemical capacitors based on carbon nanotube electrodes.* Applied Physics Letters, 1997. **70**(11): p. 1480–1482.
73. Ammundsen, B.; Paulsen, J. *Novel Lithium-Ion Cathode Materials Based on Layered Manganese Oxides.* Advanced Materials, 2001. **13**(12–13): p. 943–956.
74. Kaskhedikar, N. A.; Maier, J. *Lithium storage in carbon nanostructures.* Advanced Materials, 2009. **21**(25-26): p. 2664–2680.
75. Avdeev, V. V.; Nalimova, V. A.; Semenenko, K. N. *The alkali metals in graphite matrixes-new aspects of metallic state chemistry.* High Pressure Research, 1990. **6**(1): p. 11–25.
76. Frackowiak, E., et al. *Electrochemical storage of lithium in multiwalled carbon nanotubes.* Carbon, 1999. **37**(1): p. 61–69.
77. Wu, G. T., et al. *Structure and lithium insertion properties of carbon nanotubes.* Journal of the Electrochemical Society, 1999. **146**(5): p. 1696–1701.
78. Claye, A. S., et al. *Solid-State Electrochemistry of the Li Single Wall Carbon Nanotube System.* Journal of the Electrochemical Society, 2000. **147**(8): p. 2845–2852.
79. Shimoda, H., et al. *Lithium intercalation into etched single-wall carbon nanotubes.* Physica B: Condensed Matter, 2002. **323**(1): p. 133–134.
80. Shimoda, H., et al. *Lithium intercalation into opened single-wall carbon nanotubes: storage capacity and electronic properties.* Physical Review Letters, 2001. **88**(1): p. 015502.
81. Suzuki, S.; Tomita, M. *Observation of potassium-intercalated carbon nanotubes and their valence-band excitation spectra.* Journal of Applied Physics, 1996. **79**(7): p. 3739–3743.
82. Endo, M., et al. *Applications of carbon nanotubes in the twenty–first century.* Philosophical Transactions of the Royal Society of London. Series A: Mathematical, Physical and Engineering Sciences, 2004. **362**(1823): p. 2223–2238.
83. Dillon, A. C., et al. *Storage of hydrogen in single-walled carbon nanotubes.* Nature, 1997. **386**(6623): p. 377–379.
84. Chen, P., et al. *High H_2 uptake by alkali-doped carbon nanotubes under ambient pressure and moderate temperatures.* Science, 1999. **285**(5424): p. 91–93.

85. Liu, C., et al. *Hydrogen storage in single-walled carbon nanotubes at room temperature.* Science, 1999. **286**(5442): p. 1127–1129.
86. Chambers, A., et al. *Hydrogen storage in graphite nanofibers.* The Journal of Physical Chemistry B, 1998. **102**(22): p. 4253–4256.
87. Pederson, M. R.; Broughton, J. Q. *Nanocapillarity in fullerene tubules.* Physical Review Letters, 1992. **69**(18): p. 2689.
88. Ye, Y., et al. *Hydrogen adsorption and cohesive energy of single-walled carbon nanotubes.* Applied Physics Letters, 1999. **74**(16): p. 2307–2309.
89. Kong, J., et al. *Nanotube molecular wires as chemical sensors.* Science, 2000. **287**(5453): p. 622–625.
90. Collins, P. G., et al. *Extreme oxygen sensitivity of electronic properties of carbon nanotubes.* Science, 2000. **287**(5459): p. 1801–1804.
91. Tang, X. P., et al. *Electronic structures of single-walled carbon nanotubes determined by NMR.* Science, 2000. **288**(5465): p. 492–494.
92. Overney, G.; Zhong, W.; Tomanek, D. *Structural rigidity and low frequency vibrational modes of long carbon tubules.* Zeitschrift für Physik D Atoms, Molecules and Clusters, 1993. **27**(1): p. 93–96.
93. Yakobson, B. I.; Brabec, C. J.; Bernholc, J. *Nanomechanics of carbon tubes: instabilities beyond linear response.* Physical Review Letters, 1996. **76**(14): p. 2511.
94. Treacy, M. M. J.; Ebbesen, T. W.; Gibson, J. M. *Exceptionally high Young's modulus observed for individual carbon nanotubes.* 1996.
95. Wong, E. W.; Sheehan, P. E.; Lieber, C. M. *Nanobeam mechanics: elasticity, strength, and toughness of nanorods and nanotubes.* Science, 1997. **277**(5334): p. 1971–1975.
96. Falvo, M. R., et al. *Bending and buckling of carbon nanotubes under large strain.* Nature, 1997. **389**(6651): p. 582–584.
97. Yakobson, B. I. *Mechanical relaxation and "intramolecular plasticity" in carbon nanotubes.* Applied Physics Letters, 1998. **72**(8): p. 918–920.
98. Stone, A. J.; Wales, D. J.*Theoretical studies of icosahedral C< sub> 60</sub> and some related species.* Chemical Physics Letters, 1986. **128**(5): p. 501–503.
99. Yakobson, B. I.; Brabec, C. J.; Bernholc, J. *Structural mechanics of carbon nanotubes: From continuum elasticity to atomistic fracture.* Journal of Computer-Aided Materials Design, 1996. **3**(1–3): p. 173–182.
100. Ajayan, P. M., et al. *Aligned carbon nanotube arrays formed by cutting a polymer resin—nanotube composite.* Science, 1994. **265**(5176): p. 1212–1214.
101. Wagner, H. D., et al. *Stress-induced fragmentation of multiwall carbon nanotubes in a polymer matrix.* Applied Physics Letters, 1998. **72**(2): p. 188–190.
102. Jin, L.; Bower, C.; Zhou, O. *Alignment of carbon nanotubes in a polymer matrix by mechanical stretching.* Applied Physics Letters, 1998. **73**(9): p. 1197–1199.
103. Curran, S. A., et al. *A composite from poly(m-phenylenevinylene-co-2, 5-dioctoxy-p-phenylenevinylene) and carbon nanotubes: a novel material for molecular optoelectronics.* Advanced Materials, 1998. **10**(14): p. 1091–1093.
104. Coleman, J. N., et al. *Phase separation of carbon nanotubes and turbostratic graphite using a functional organic polymer.* Advanced Materials, 2000. **12**(3): p. 213–216.
105. Lau, A. K. T.; Hui, D. *The revolutionary creation of new advanced materials—carbon nanotube composites.* Composites Part B: Engineering, 2002. **33**(4): p. 263–277.
106. Calvert, P. *Nanotube composites: a recipe for strength.* Nature, 1999. **399**(6733): p. 210–211.

107. Mamedov, A. A., et al. *Molecular design of strong single-wall carbon nanotube/polyelectrolyte multilayer composites.* Nature Materials, 2002. **1**(3): p. 190–194.
108. Zhan, G. D., et al. *Single-wall carbon nanotubes as attractive toughening agents in alumina-based nanocomposites.* Nature Materials, 2003. **2**(1): p. 38–42.
109. Dai, H., et al. *Nanotubes as nanoprobes in scanning probe microscopy.* Nature, 1996. **384**(6605): p. 147–150.
110. Wong, S. S., et al. *Carbon nanotube tips: high-resolution probes for imaging biological systems.* Journal of the American Chemical Society, 1998. **120**(3): p. 603–604.
111. Hafner, J. H.; Cheung, C. L.; Lieber, C. M. *Growth of nanotubes for probe microscopy tips.* Nature, 1999. **398**(6730): p. 761–762.
112. Kim, P. and C.M. Lieber, *Nanotube nanotweezers.* Science, 1999. **286**(5447): p. 2148–2150.
113. Wong, S. S., et al. *Single-walled carbon nanotube probes for high-resolution nanostructure imaging.* Applied Physics Letters, 1998. **73**(23): p. 3465–3467.
114. Chen, J., et al. *Solution properties of single-walled carbon nanotubes.* Science, 1998. **282**(5386): p. 95–98.
115. Wong, S. S., et al. *Covalently functionalized nanotubes as nanometre-sized probes in chemistry and biology.* Nature, 1998. **394**(6688): p. 52–55.
116. Ajayan, P. M., et al. *Carbon nanotubes as removable templates for metal oxide nanocomposites and nanostructures.* 1995.
117. Zhang, Y., et al. *Coaxial nanocable: silicon carbide and silicon oxide sheathed with boron nitride and carbon.* Science, 1998. **281**(5379): p. 973–975.
118. Hu, J., et al. *Controlled growth and electrical properties of heterojunctions of carbon nanotubes and silicon nanowires.* Nature, 1999. **399**(6731): p. 48–51.
119. Zhang, Y., et al. *Heterostructures of single-walled carbon nanotubes and carbide nanorods.* Science, 1999. **285**(5434): p. 1719–1722.
120. Lee, R. S., et al. *Conductivity enhancement in single-walled carbon nanotube bundles doped with K and Br.* Nature, 1997. **388**(6639): p. 255–257.

CHAPTER 10

PATHWAYS IN PRODUCING ELECTROSPUN NANOFIBERS

S. PORESKANDAR, F. RAEISI, SH. MAGHSOODLOU, and A. K. HAGHI

Department of Textile Engineering, Faculty of Engineering, University of Guilan, P.O. Box: 3756, Rasht, Iran

CONTENTS

ABSTRACT

Aligned electrospun nanofibers have found importance in many engineering applications, such as tissue engineering, filtration. An external electrostatic field should be acted on polymer solution/melt for producing a charged polymer jet. Therefore, the most significant part of electrospinning is controlling the process. There are various ways to control aligning the depositing fiber. For this case, understanding the influence of varying parameters is so important. The aim of this chapter is to review the effects of these parameters. In addition, some commonly used techniques to align the fibers are discussed in this chapter.

Symptoms	Definitions
C_i	the morphological transition from bead-only structure
C_f	the morphological transition from bead-free structure
M_w	the polymer molecular weight
M_e	the entanglement molecular weight
ϕ_p	the polymer volume fraction
$[\eta]C$	Berry number
$[\eta]$	the intrinsic viscosity
C	the solution concentration
(C/C_e)	normalized concentration
C_e	the entanglement concentration
ρ	polymer density
rpm	round per minute
MES	magnetic electrospinning
PEO	polyethylene oxide
PVA	polyvinyl alcohol
PMMA	polymethyl methacrylate
PANI	polyaniline
CA	cellulose acetate
PVP	polyvinyl pyrrolidone
PAA	polyacrylic acid
PA6	polyamide-6
PEO	polyethylene oxide
PAA	polyacrylic acid
PU	polyurethane
PCL	polycapro lactone
PS	polystyrene
DMF	dimethylformamide
THF	tetrahydrofuran

10.1 ESSENTIAL PARAMETERS FOR CONTROLLING ELECTROSPINNING PROCESS

In recent decades, electrospinning of polymeric materials has gained much attention mainly because of its being the cheapest and the simplest of all methods.[1–5] In this procedure, as shown in Figure 10-1, the polymer solution receives electrical charges from a high voltage supply; when the repulsive force between the charged ions overcomes the fluid surface tension, an electrified liquid jet could be formed and elongated toward the collector. At end, formatting nanofibers jet are collected on the surface of screen when the solvent is evaporated.[6–7]

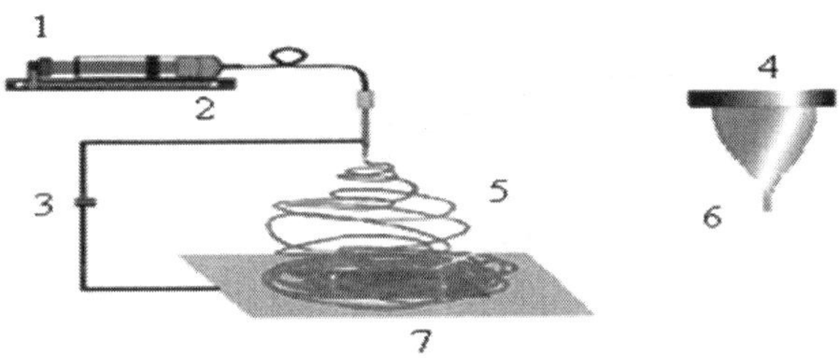

FIGURE 10-1 Sections of electrospinning process: 1) Polymer solution 2) Syringe 3) High voltage4) Taylor cone 5) Whipping instability 6) Nanofibers formation 7) Collector.

Many parameters affect the features of different parts of electrospun jet like the straight jet part, the instability region, and the jet path.[8]

The most significant challenge in this process is to attain uniform nanofibers consistently and reproducibly.[9–12] In addition, the mechanics of this process deserves a specific attention and necessary to predictive tools or a way for better understanding and optimization and controlling process.[13]

Also, Fiber diameter is an important characteristic for electrospinning, because of its direct influence on the properties of the produced webs.[12,14,15] Depending on several solution parameters, different results can be obtained using the same polymer and electrospinning setup.[16]

Many parameters effect fiber formation. These factors that are studied to have a primary effect on the formation of uniform fibers are the process parameters, environmental parameters, and solution parameters.[8,9,17–19]These dates are shown in the Figure 10-2.

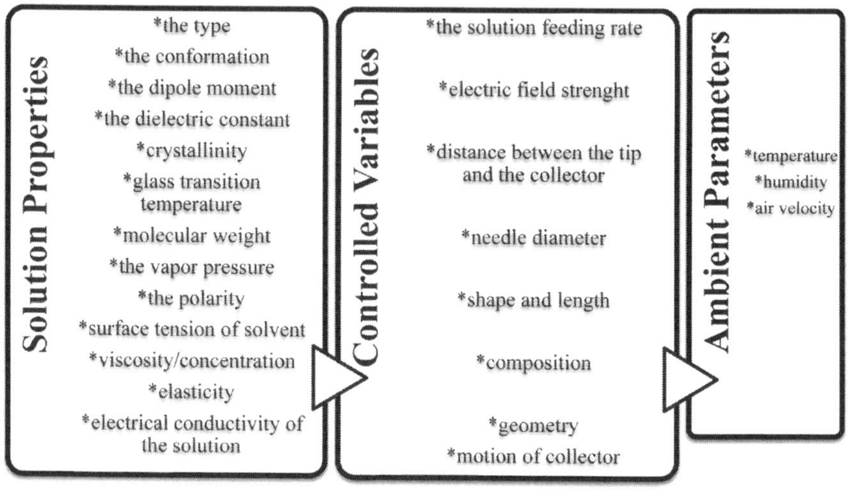

FIGURE 10-2 Parameters affect the morphology and size of electrospun nanofibers.

Several general relationships between these parameters and fiber morphology can be drawn; this relationship will differ for each polymer or solvent.[12,16,20,21]

In addition, many researchers' studies show the effect of these parameters on final fibers. A summary of the most important parameters with their effects are presented in Tables 10-1 and 10-2.

TABLE 10-1 Effects of some parameters on electrospinning nanofibers in researcher's studies

Parameters	Year	Name of Researcher	Effect	References
Needle to collector distance	1999	Fong et al	Inversely proportional to bead formation density Inverse to the electric field strength	[22]
	2003	Gupta & Wilkes	Inversely proportional to bead formation density Inversely proportional to fiber diameter	[23]
	2004	Theron et al	Exponentially inverse to the volume charge density Inverse to the electric field strength	[24]

TABLE 10-1 *(Continued)*

Parameters	Year	Name of Researcher	Effect	References
Flow rate	2004	Theron et al	Directly proportional to the electric current Inversely related to surface charge density Inversely related to volume charge density	[24]
	2005	Sawicka et al	Directly proportional to the fiber diameter	[25]
Voltage	2001	Deitzel et al	Direct effect on bead formation	[26]
	2003	Gupta & Wilkes	Inversely related to fiber diameter	[23]
	2004	Theron et al	Inversely proportional to surface charge density	[24]
	2004	Kessick et al	AC potential improved fiber uniformity	[27]
Concentration of polymer	2001	Deitzel et al	Power law relation to the fiber diameter	[26]
	2002	Demir et al	Cube of polymer concentration proportional to diameter	[28]
	2003	Gupta & Wilkes	Directly proportional to the fiber diameter	[23]
	2004	Hsu & Shivkumar	Parabolic-upper and lower limit relation to diameter	[29]
Ionic strength	2002	Zong et al	Directly proportional to charge density Inversely proportional to bead density	[30]
Temperature	2002	Demir et al	Inversely proportional to viscosity Uniform fibers with less beading	[28]
Solvent	2004	Theron et al	Effects volume charge density Directly related to the evaporation and solidification rate	[24]
Viscosity	2004	Hsu & Shivkumar	Parabolic relation to diameter and spinning ability	[29]

TABLE 10-2 Effects of some parameters on electrospinning nanofibers

	Parameters	Effect	References
Ambient Parameters	Humidity	When ambient humidity increases, the diameter decreases and finally results in beaded fibers in PEO The influence of humidity on the formation and the properties of nanofibers are studied using CA and PVP the humidity increases, the average fibre diameter of the CA nanofibers increases, while for PVP the average diameter decreases average diameter of nanofibers made by electrospinning change significantly through variation of humidity results in appearing circular pores on the fibers	[12, 20]
Ambient Parameters	Temperature	higher solution temperature produces thinner nanofibers because of the decrease in viscosity The influence of temperature on the formation and the properties of nanofibers are studied using CA and PVP solvent evaporation rate that increases with increasing temperature the viscosity of the polymer solution that decreases with increasing temperature average diameter of nanofibers made by electrospinning change significantly through variation of temperature	[12, 26]
Controlled Variables	Voltage	is strongly correlated with form bead defects in the fibers can conduct to form beads decreasing the electrical field decreases the bead density, regardless of concentrate the polymer in the solution	[20-21]
Controlled Variables	distance between the capillary and the fiber collector	affects the fiber drying and wants a minimum distance to give the fibers enough time to dry before achieving the fiber collector Increasing the distance, decreases the bead density, regardless of concentrate the polymer in the solution	[20]
Controlled Variables	geometry and composition of the collector	affect the surface of fiber, where the porous collector produces more porous fiber and metallic collector smooth fibers	[20]
Controlled Variables	fiber diameter	according to a power law relationship increasing with increasing solution concentration larger dielectric constant of solvent, result finer electrospun fibers A solvent with a larger dielectric constant has a higher net charge density in ejected jets. Increases with the increase of the concentration or viscosity of the polymer solution and higher conductivity yields smaller fibers in general, except for the technique using the polymers PAA and polyamide-6.	[20-21, 31]

TABLE 10-2 *(Continued)*

	Parameters	Effect	References
Solution Properties	addition of salts	adding small amounts of cationic or anionic polyelectrolyte into PEO solution could reduce the average diameter of electrospun PEO fibers and narrow their distributions by increasing the charge density in ejected jets	[31]
	surface tension	effect on the morphology of nanofiber and size of electrospun fiber have been found out, however there is no result on this issue yet	[20]
	solution viscosity	decreases with increasing temperature and reduces fiber diameter	[20]
	Solution concentration	has been found to most strongly affect fiber size	[21]
	pH	may affect the morphology and diameter of electrospun fibers the charge density in ejected jets affected in PVA solution	[31]

Processing conditions play a major character in the electrospinning technique because the electrospinning technique is governed by the external electric field produced by the applied voltage caused by charges on the jet surface. It can be divided into two sections; processing parameters and type of collector that are summarized in Figures 10-3 and 10-4. In addition, the processing conditions and external parameters also have a significant result on the diameter and morphology of the nanofibers. The electric field between the needle tip and the target can be controlled through these varying parameters:[8]

- The applied voltage
- The distance between needle tip and target
- The shape of the collector
- The diameter of the needle

Solution polymer parameters affect forming the nanofibers, causing defects in the fibers in the pattern of beads and junction by their low concentration or viscosity, low conductivity, and by their decrease of molecular weight. Ambient parameters play a role in the properties of the polymeric solution and so affect the morphology of the nanofibers.[20]

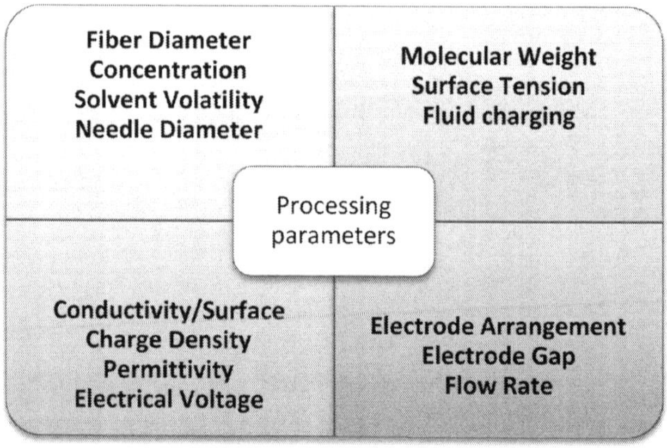

FIGURE 10-3 Important processing parameters.

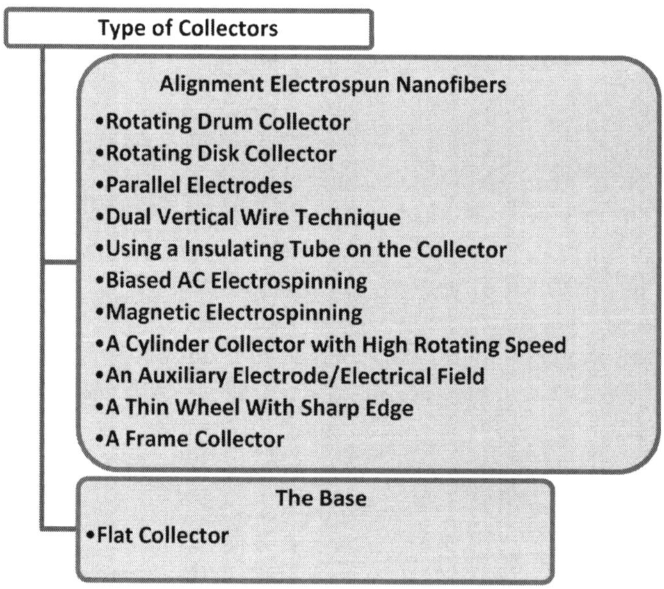

FIGURE 10-4 Type of collectors.

The solution must also have these feathers to prevent the gate from collapsing into droplets before the solvent has evaporated:[32]

- A surface tension, low enough
- A charge density, high enough
- A viscosity, high enough

In addition, major factors that control the diameter of the fibers are[33]

- Concentration of polymer in their solution
- Type of solvent used
- Conductivity of their solution
- Feeding rate of the solution

In the next part, we have discussed about these parameters.

10.1.1 CONCENTRATION

The concentrations of polymer solution play a significant role in the fiber formation during the electrospinning technique. Four critical concentrations from low to high should be remarked in these points:[10,21,30,34] As the concentration is low, polymeric micro or nano-particles will be produced. Now, electrospray occurs instead of electrospinning owing to the low viscosity and high surface tensions of the solution.

- As the concentration is little higher, a mixture of beads and fibers will be received.
- When the concentration is suitable, smooth nanofibers can be obtained.
- If the concentration is high, not nanoscale fibers, helix-shaped micro ribbons will be watched.

Normally, increasing solution concentration will increase the fiber diameter if the solution concentration is suitable for electrospinning. Also, solution viscosity can be also tuned by setting the solution concentration.[10]

In the electrospinning technique, a minimum solution concentration is needed for fiber formation. It has been found that at low concentration, beads and fibers are obtained. When the solution concentration increases, the shapes of beads change and finally uniform fibers are found. There is an optimum solution concentration for the electrospinning technique (At low concentrations, beads are formed, and at higher concentrations, continuous fibers are banned). Researchers have tried to determine a relationship between solution concentration and fiber diameter. They have found a power law relationship. Increasing concentrate solution causes increasing fiber diameter with gelatin in electrospinning. Solution surface tension and viscosity also play important roles in determining the range of concentrations from continuous fibers which can be obtained in electrospinning.[17] Polymer concentration determines the spinnability of a solution, that is to say, whether a fiber will form or not. The solution must have a high enough polymer concentration for chain entanglements to occur; however, the solution cannot be either too dilute or too centralized. On the other hand, polymer concentration influences both viscosity and surface tension of the solution technique. If the solution is to dilute, then the polymer fiber will give away up into droplets before reaching the collector (due to the effects of surface tension). If the solution is too concentrated, then fibers cannot be made (due to the high viscosity), because it is difficult to controlling solution flow rate through capillaries. Thus, an optimum range of polymer concentrations is called for.[16]

The molecular properties of polymer play a vital part in controlling fiber initiation and stabilization. Several investigators have attempted to establish optimum ranges for concentration and molecular weight in order to insure stable fiber formation. At any molecular weight, the effect of concentration on the breakdown of the solution jet can be distinguished by two critical concentrations, Ci and Cf. Below Ci, only beads may be developed (due to insufficient chain entanglements in the solution). Above Ci, a combination of beads and fibers is observed. When the concentration is increased above Cf, complete fibers are created. Ci is typically about the entanglement concentration Ce, at which chain entanglements in the solution become significant. Hence, Ci is a transition concentration at which fibers begin to come forth from the beads and Cf is the concentration at which a fibrous structure is stabilized. Shenoy et al. have developed the solution entanglement number, (ne) soln, for depicting the transition points for fiber initiation and complete fiber formation:[35]

$$(n_e)_{soln} = \frac{\phi_p M_w}{M_e} \tag{10-1}$$

Lyons et al. have reported that the fiber diameter varies exponentially with a molecular weight for melt electrospun polypropylene. The fiber diameter, normalized, with deference to the Berry number is plotted. A power law relationship is observed between D and [η]C as follows:[35]

$$D(nm) = 18.6\left([\eta]C\right)^{1.11} \tag{10-2}$$

Various investigators have also correlated the fiber diameter with a normalized concentration defined as[35]

$$C_e = \frac{\rho M_e}{M_w} \qquad (10\text{-}3)$$

10.1.2 SOLUTION VISCOSITY

One of the most significant parameters that influence the diameter and the morphology of the fiber is the viscosity of the resolution, which is indirectly affected by polymer characteristics such as molecular weight and concentration. When a polymer with a higher molecular weight is dissolved in a solvent, the viscosity of the polymer solution is higher than a solution of the same polymer with a lower molecular weight. Similarly, the viscosity of a polymer solution increases with an increased concentration of polymer in that solution. It is obvious that the viscosity of solution and polymer chain entanglements have a direct relationship. It has been proven that continuous and smooth fibers cannot be found in low viscosity, whereas high viscosity results in the hard ejection of jets from solution, namely there is a need of suitable viscosity for electrospinning. The viscosity range of different polymer solutions at electrospinning is different. It is important that viscosity, polymer concentration, and polymer molecular weight are linked to one another. For solution with low viscosity, surface tension is the dominant factor and just beads or beaded fiber formed. If the solution is of suitable viscosity, continuous fibers can be made. Also, the shape of the beads changes from spherical to elliptical when the viscosity of solution varies from low to high. Higher viscosity also results in larger diameter fibers and smaller deposition areas.[8,10,17,36]Taken together, these studies suggest there exist polymer-specific, optimal viscosity values for electrospinning and this property possesses a remarkable influence on the morphology of the fibers.[17]

10.1.3 MOLECULAR WEIGHT

Molecular weight of polymer also has an important impression on the morphology of electrospun fiber. In principle, molecular weight reflects entangling polymer chains in solutions, namely the solution viscosity. Keep the concentration fixed, lowering the molecular weight of polymer trends to form beads rather than smooth fiber. By increasing the molecular weight, smooth fiber will be obtained. Further by increasing molecular weight, micro ribbon will be received. Also, the authors establish that as the molecular weight is high, some patterned fibers can also be obtained at low concentration.[10]

It has been remarked that too low a molecular weight solution forms beads rather than fibers. A high molecular weight solution gives fibers with larger average diameters. Chain entanglement plays an important role in technique electrospinning. It has been discovered that high molecular weights are not always essential for the electrospinning technique if enough intermolecular interactions can provide a substitute for the interchange connectivity got through chain entanglements.[17]

In addition, this parameter plays a vital role in controlling fiber beginning and stabilization. Several investigators have tried to demonstrate the ideal ranges for concentration and molecular weight to insure stable fiber formation. The molecular weight of the polymer has a significant purpose of proving the structure in the electrospun polymer. At a constant concentration, the structure changes from beads, to beaded fibers, to complete fibers, and two flat ribbons as the molecular weight are increased.[35]

10.1.4 SURFACE TENSION

Surface tension is an important factor in electrospinning. Different solvents may contribute different surface tensions. With the concentration fixed, reducing the surface tension of the solution, beaded fibers can convert into smooth fibers. The surface tension and solution viscosity can be adjusted by varying the mass ratio of solvents mix and fiber morphologies. Surface tension decides the upper and lower boundaries of the electrospinning window if all other conditions specified.[10]A lower surface tension of the spinning solution helps electrospinning to occur at a lower electric field.[8,17,37] However, not necessarily a lower surface tension of a solvent will always be more suitable for electrospinning. Also, this parameter determines the upper and lower boundaries of the electrospinning window if all other variables are held constant.[17,37]

Adding a surfactant to the polymer solution also changes the surface tension. If all other variables are held constant, surface tension decides the upper and lower bounds of the electrospinning technique.[8]

It is caused by the attraction between the molecules in a liquid. In the most of liquid, each molecule is attracted equally in all directions by neighboring liquid molecules, resulting in a net force of zero. At the surface of the liquid, the molecules are subjected to a net inward force to balancing only by resisting liquid to compression. The net effecting causes the surface area to reduce it until controlling the possible lowest ratio of surface area to volume. In electrospinning, the charges on the polymer solution must be high enough to overwhelm the surface tension of the solution. As the electrician jet speeds up from the needle to the aim, the polymer jet is stretched. Then, surface tension of the solution may cause the jet to break up into droplets. In addition, if there is a lower concentration of polymer molecules, the surface tension causes beaded fibers formed.[8]

10.1.5 CONDUCTIVITY/SURFACE CHARGE DENSITY

Solution conductivity is mainly decided by the polymer type, solvent sort, and the salt. Usually, natural polymers are polyelectrolyte in nature, subjecting to higher tension under the electric field, resulting in the poor fiber formation. Also, the electrical conductivity of the solution can be tuned by adding the ionic salts. With the aid of ionic salts, nanofibers with small diameter can be produced. Sometimes, high solution conductivity can be also accomplished by using organic acid as the solvent. An increase in the solution conductivity favors forming thinner fibers.[10]

Also, solutions with high conductivity will cause a greater charge carrying capacity than solutions with low conductivity. Therefore, the fiber jet of conducive solutions will be subjected to a greater tensile force in the mean of an electric field than will a fiber jet from a solution with a low conductivity.[16]

The minimum voltage for electrospinning to occur can also be cut back if the conductivity of the polymer solution is increased. Also, higher solution conductivity results in greater bending instability and produces a larger deposition area of collecting fibers. It has been reported that the size of the ions in the solution has an important impact on the electrospun fiber diameter besides the charges carried by the jet. Ions with a smaller atomic radius have a higher charge density. Thus, a higher mobility under an external electric field is present.[8]

However, conductivity solution is unstable in the presence of strong electric fields, which results in a dramatic bending instability as well as a broad diameter distribution. Electrospun nanofibers with the smallest fiber diameter can be taken with the highest electrical conductivity and it has been found, there is a drop in the size of the fibers, which is because of the increased electrical conductivity. The ions increase the charge holding capacity of the jet with it subjecting it to higher tension. Thus, the fiber forming ability of the gelatin is less compared to the synthetic ones.[17] For example, Zong et al. have proved the effect of ions by adding ionic salt on the morphology and diameter of electrospun fibers.[17]

Stanger et al. found that an increase in charge density results in a reduction in the mass deposition rate and initial jet diameter during the electrospinning. In addition, a theory was proposed where they correlated reducing the curvature diameter of the Taylor cone with increasing charge density. Decreasing total electrostatic forces cause a smaller effective area. Similarly, other researchers have described the different behavior of the Taylor cone when compared to the Taylor's observation of ionic liquids.[8]

10.1.6 SOLVENT VOLATILITY

Selecting a suitable solvent or solvent as the carrier of a particular polymer is fundamental for optimizing electrospinning. On the other hand, it is critical in determining the critical minimum solution concentration to allow the transition from electrospraying to electrospinning, by significantly affecting solution spinnability and the morphology of the electrospun fibers.[34]

In addition, choice of solvent is too critical about whether the fibers are forming, as well as influencing fiber porosity. For enough solvent evaporation to happen between the capillary tip and the collector, a volatile solvent must be used. As the fiber jet travels through the air toward the collector, a phase separation occurs before the solid polymer fibers deposited, a technique that is influenced by the volatility of the solvent.[16]

In Megelski et al. studies the properties of polystyrene fibers electrospun from solutions containing various ratios of DMF and THF were examined. Electrospinning solutions from 100% THF (more volatile) explained a high density of pores, which increased the surface area of the fiber depending on the fiber diameter. Solutions electrospun from 100% DMF (less volatile) proved almost a loss of micro texture with forming smooth fibers. Between these two extremes, it was viewed that pore size increased with decreased pore depth (thus decreasing pore density) as the solvent volatility decreased. For volatile solvents, the region close to the fiber surface can be saturated with solvent in the vapor phase, which further limits penetrating nonsolvent. This can hinder skin formation leading to developing a porous surface morphology.[16]

10.1.7 FLUID CHARGING

In electrospinning, generation of charging within the fluid, usually occurs under contract with and flow across an electrode held at high (positive or negative) potential, referred to as induction charging. Depending on the nature of the fluid and polarity of the applied potential, free electrons, ions, or ion pairs may be produced as charge carriers in the fluid; the generation of charge carriers can be sensitive to solution impurities. Forming ions or ion pairs by induction results in forming an electrical double layer. Without flow, the double layer thickness is found by the ion mobility in the fluid; in the presence of flow, ions may be convected away from the electrode and the double layer continually renewed. Charging of the fluid in electrospinning is typically field-limited, with the breakdown field strength in dry air being between flat plates.[38]

10.1.8 PERMITTIVITY

As with conductivity, the permittivity of a solvent has an important influence on the electrospinning technique and fiber morphology. However, not much discussion has been published around these effects. Theron et al. describe a method for deciding the permittivity of an electrospinning solution by measuring the complex resistance of a small cylindrical volume of the fluid. Bead formation and the diameter of the resultant electrospun fibers can be diminished by using a solution with a higher permittivity. With bending instability and the traversed jet in the path, electrospinning jet increases with higher permittivity, which results in a reduction in the diameter of the fiber and a larger fiber deposition area. Solvents such as DMF can be utilized to increase the permittivity of polymer solutions.[8]

10.1.9 ELECTRICAL VOLTAGE

One of the major parameters which affects the fiber diameter to a remarkable extent is the applied electric potential. In general, a higher applied voltage ejects more fluid in a jet, resulting in a larger fiber diameter.[8,14]

If the applied electric potential is higher, a greater amount of charge will cause the jet to speed up faster, and more solution will be drawn out from the tip of the needle. At a critical voltage, the Taylor cone is no longer seen. The jet imminent directly from the nozzle with increasing applied voltage. The resultant electrical field between the needle and the target increases as well, which contributes to greater stretching of the solution because of the larger columbic force between the surface charges.[8]

An increase in the applied voltage therefore leads to a lessening in the diameter of the electrospun nanofibers. Similarly, Pawlowski et al. have proved the drier fibers can be generated if the voltage is increased because of the faster evaporation of the solvent that results. Zhao et al. have verified that a lower voltage leads to a weaker electrical field, which reduces speeding up the jet and increases the flight time of the electrospinning jet, thus producing thinner fibers. Therefore, they suggested that a voltage close to the minimum critical voltage needed for the onset of electrospinning might be efficient for getting thinner fibers. Higher voltages are related to a greater of bead formation, possibly because of the increased instability of the jet as the Taylor cone recedes into the syringe needle with the increased potential. The shape of the beads transforms from a spindle to a spherical shape with increased voltage, and sometimes the beads will get together to form thicker fibers because of the increased density of the beads on the other hand.[8]

Reneker and Chun have proved there is not much effect of electrical field on the diameter of electrospun PEO nanofibers. Several groups suggested that higher voltages simplified form large diameter fiber. For example, Zhang et al. explored the effect of voltage on morphologies and fiber diameter distribution with PVA or a water solution as a model. Several groups suggested that higher voltages can increase the electrostatic repulsive force on the charged jet, favoring the narrowing of fiber diameter. For example, Yuan et al. analyzed the effect voltage on morphologies and fiber alignment with PSF or DMAC or acetone as a model. Besides those phenomena, some groups also established that higher voltage offers the greatest chance of bead formation. Thus, we can find that voltage does the influence fiber diameter, but the meanings vary with the polymer solution concentration and on the distance between the tip and the collector.[10,30]

10.1.10 FLOW RATE

The flow rate decides the solution available for electrospinning. Keeping a stable Taylor cone needs a minimum solution flow rate for a given voltage and electrode gap. On the other hand, at low flow rate, the Taylor cone recedes into the needle, and the jet originates from the liquid surface within the needle. In contrast, if the solution flow rate is greater than the electrospinning rate, it causes solution droplets to come from the needle tip because of the lack of time for electrospinning the complete droplet to be an electrician. It has been viewed the diameter of the fiber and the size of the bead both increase with an increased flow rate.[16,17]

Also, at high flow rates significant numbers of bead defects were noticeable, because of the inability of fibers to dry before progressing to the collector. Incomplete fiber drying also leads to forming ribbon like (or flattened) fibers compared to fibers with a circular cross-section.[16]

A summary of researchers' work is shown in Figure 10-5.[16–17]

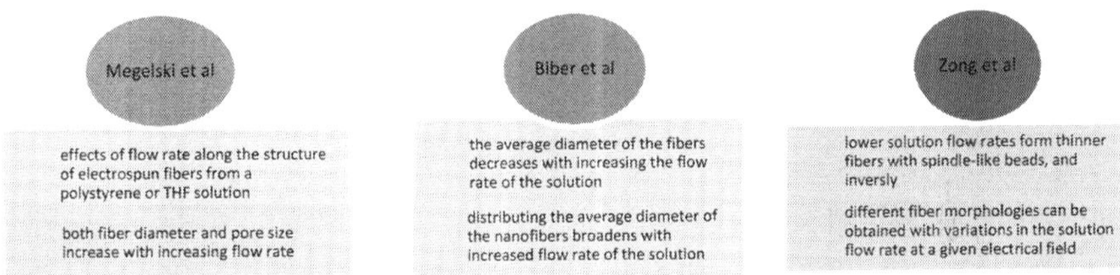

FIGURE 10-5 Summary of researchers' work on the flow rate parameter.

10.1.11 NEEDLE DIAMETER

The diameter of the needle has an effect on the electrospinning technique. A smaller needle diameter is found to reduce clogging at the tip of the needle as well as the number of beads in the collected nanofibers because of the lower exposure of the solution to the atmosphere during electrospinning. In addition, using smaller-diameter needles means the diameter of the electrospun nanofibers can also be smaller. The jet flying time of the solution between the needle and the collector plate can also be increased if the needle diameter is cut because the surface tension of the droplet is increased and the jet acceleration, decreased. Few studies have worked in this field that are summarized in Figure 10-6.[8]

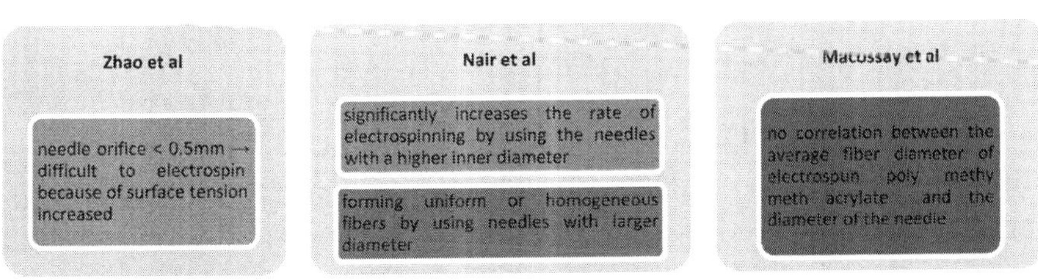

FIGURE 10-6 Summary of researchers' study.

10.1.12 DISTANCE BETWEEN THE COLLECTOR AND THE TIP OF THE SYRINGE

Varying the distance between the needle and the target causes a modification in the behavior of the electrospun jet and the morphology of the resultant nanofibers. Shortening the distance between the two electrodes causes an increase in the electrical field strength between the needle and the target and speeds up the electrospinning technique, therefore reducing the time available for evaporation. It has been also reported spreading the diameter of the nanofibers becomes narrower when the space between the two electrodes is increased. Conversely, in other cases, it was considered as the average diameter of the fiber increases with increased distances because of the decreased strength of the electric field.[8]

As a brief summarized, it has been proven the distance between the collector and the tip of the syringe can also affect the fiber diameter and morphology. In brief, if it is too short, the fiber will not have adequate time to solidify before reaching the collector, because dryness from the solvent is an important parameter on electrospun fiber. If the distance is too long, bead fiber can be obtained. It has been reported that flatter fibers can be brought out at closer distances. however, spinning distance is not important effect on fiber morphology.[16,17] Also, many researchers studied this parameter affect on fibers morphology and diameters as shown in figure 10-7.[16]

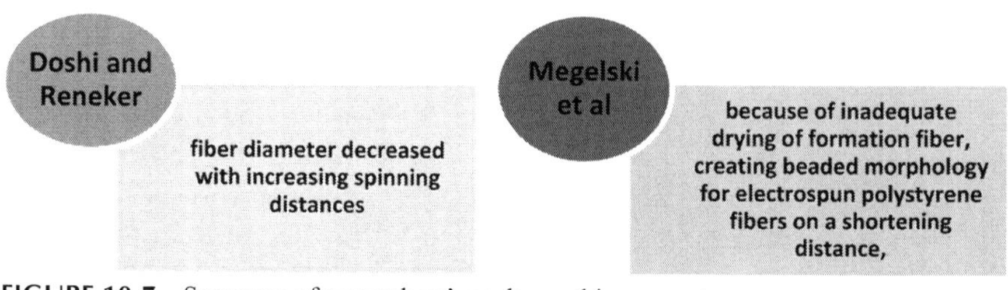

FIGURE 10-7 Summary of researchers' results on this parameter.

10.2 CONTROL ALIGNED FORMATION FIBERS

Aligned fibers have found importance in many engineering applications, such as tissue engineering, sensors, nano composites, filters, electronic devices, and so on. Some commonly used techniques to align the fibers are discussed in the subsections in the following.[33] There are various ways to control aligning the depositing fiber. One way is to use different collector like a rotating wheel instead of foil sheet collector.[39]

Recently, it was decided the nature and type of the collector influences significantly the morphological and the physical characteristics of spun fibers. The density of the fibers per unit area of the collector and fiber arrangement is affected by the degree of charge dissipation on fiber deposition. The most commonly used targets are the conductive metal plate that results in collection of randomly oriented fibers in the nonwoven form. The use of metal and conductive collectors helped dissipate the charges and reduced the repulsion among the fibers. Therefore, the fibers collected are smooth and densely compacted. However, the fibers collected on the nonconductive collectors do not fritter away the charges which repel one another. In addition, The fibers can also be collected on specially designed collector to get aligned fibers or arrays of fibers.[33]

In the next parts, we have discussed about the types of collectors.

10.2.1 TYPES OF COLLECTORS

In the former stages of electrospinning, researchers used a needle and a flat collector plate as electrodes. However, with developing electrospinning technology, many electrode arrangements have been tried as a means of changing the electric field and getting wanted nanofiber morphologies. Collector electrode arrangements have all been ground out as ways, summarized in Figure 10-8), to produce aligned fibers.[8]

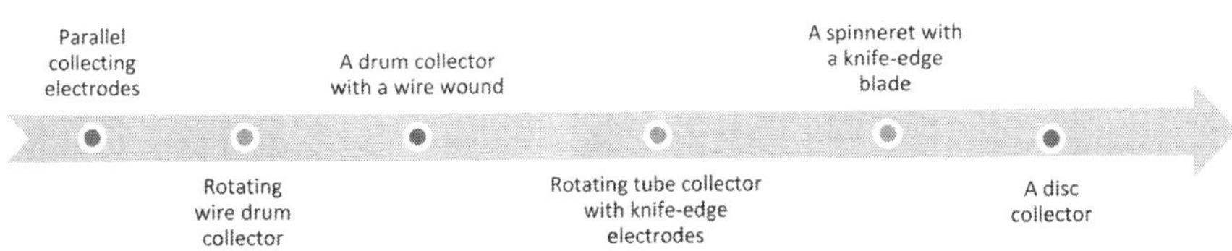

FIGURE 10-8 A summary of electrode arrangements.

One important aspect of the electrospinning technique is the type of collector used. In electrospinning technique, a collector serves as a conductive substrate where the nanofibers are collected. Aluminium foil is used as a collector, but because of difficulty in transferring of collecting fibers and with the need for aligning fibers for various applications, other collectors are also common types of collectors today as summarized in Figure 10-9.[17]

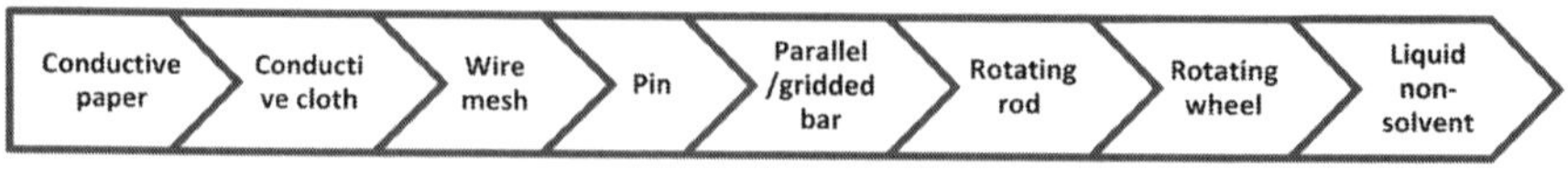

FIGURE 10-9 A summary on other common types of collectors.

In addition, the fiber alignment is found out by the type of the target or collector and its rotation speed. The created nanofibers are deposited in the collector as a random mass because of the bending instability of the charged jet. Nowadays, Several types of collectors, such as a rotating drum or a rotating wheel-like bobbin or metal frame, have been utilized for getting aligned nanofibers.[17] In the next section, we have discussed briefly about these types of collectors.

10.2.1.1 FLAT COLLECTOR

During the electrospinning technique, collectors usually acted as conductive substrates to collect the charged fibers. As usual, a foil is used as a collector.[10]

In addition, this is the most widely utilized methods of fiber collection. The collector can be either a solid metal, foil, or screen. Other materials can also be localized between the capillary and the collector.[39]

However, it is difficult to transfer the collected nanofibers to other substrates for several applications. With the need of fibers transferring, diverse collectors have been developed including[10]

- Wire mesh
- Pin or grids
- Parallel or gridded bar
- Rotating rods or wheel
- Liquid bath

10.2.1.2 ROTATING DRUM COLLECTOR

This method is normally used to collect an aligned array of fibers. Also, the diameter of the fiber can be controlled and tailored based on the rotational speed of the drum. The cylindrical drum is rotating at high speeds (a few 1000 rpm) and of orienting the fibers circumferentially (Figure 10-10). Ideally, the linear rate of the rotating drum should match the evaporation rate of the solvent; such fibers are deposited and held up on the surface of the drum. The alignment of the fibers is induced by the rotating drum and the degree of alignment improves with the rotational velocity. At rotational speeds slower than the fiber take-up speed, randomly oriented fibers are obtained along the drum. At higher speeds, a centrifugal force is prepared near the vicinity of the circumference of the rotating drum, which elongates the fibers before being collected on the drum. However, at much higher speed, the take-up velocity breaks the depositing fiber jet and continuous fibers are not taken.[33]

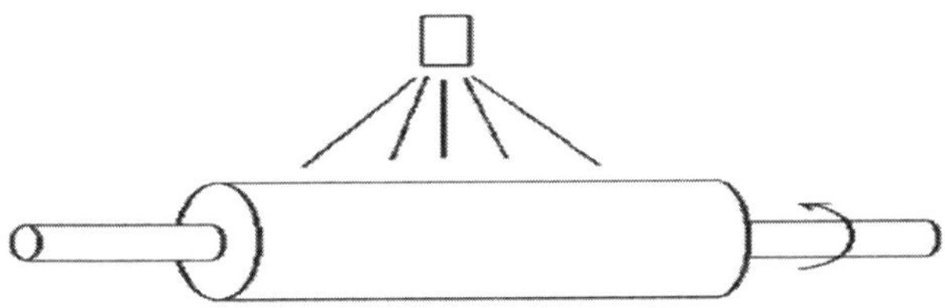

FIGURE 10-10 Rotating collector for electrospun nanofibers.

10.2.1.3 ROTATING DISK COLLECTOR

The rotating disk collector is a variation setup of the rotating drum collector and is practiced to obtain un-axially aligned fibers. The advantage of using a rotating disk collector over a drum collector is that most of the fibers deposited on the sharp-edged disk and are collected as aligned patterned nanofibers. The jet travels in a cone and inverse conical path with the utilization of the rotating disk collector as opposed to a conical path got when using a drum collector. During the first level, the jet follows the usual envelope cone path which is because of the instabilities influencing the jet. At a point above the disk, the diameter of the loop decreases as the conical form of the jet starts to shrink. This results in the inverted cone appearance, with the top of the cone resting on the disk. The electrical field applied concentrated on the tapered edge of the magnetic disc. Therefore, the charged polymer jet is pulled toward the edge of the wheel, which explains the inverted conical form of the jet at the disk edge. The fibers that are attracted to the edge of the disk are wound round the perimeter of the disk owing to the tangential force acting on the fibers produced from rotating the disk. This force further stretches the fibers and reduces their diameter. The quality of fiber alignment obtained using the disk is more beneficial than the rotating drum. However, only a small quantity of aligned fibers can be obtained since there is just a small area at the tip of the disk.[33]

10.2.1.4 PARALLEL ELECTRODES

There has already been several groups produced well-aligned nanofibers by using two grounded parallel electrodes, such as aluminium strips with a 1 cm gap used in Yi Xin's group and the metal frame in Dersch's group et al. This apparatus used in this method is uncomplicated. The same as the rotating drum method, it operates by varying the collectors. The drawbacks of both these two methods are taught:[39]

- They can only produce aligned fiber in a small area.
- Fibers fabricated by this method cannot be conveniently transferred to different types of substrates.

The advantage of utilizing this technique lies in the simplicity of the set-up and the ease of collecting single fibers for mechanical testing. Good alignment has been contracting with this technique. The air gap between electrodes creates residual electrostatic repulsion between spun fibers, which helps align fibers. Two nonconductive strips of materials are placed along a straight line and an aluminium foil is placed on each of the strips and connected to the ground. This technique enables fibers to be deposited at the end of the strips, so the fibers cling to the strips in an alternate fashion and collected as aligned arrays of fibers. A similar technique by Teo and Ramakrishna used double-edge steel blades on a line to collect aligned arrays of fibers. The fibers were deposited at the gap between the electrodes, however, few fibers were found to deposit along the blades. It was solved by applying a negative voltage between the blades, resulting in the deposition of fibers between the blades.[33]

10.2.1.5 DUAL VERTICAL WIRE TECHNIQUE

The dual vertical wire technique used in Surawut's study is a variance of the parallel electrode technique, comprised two stainless steel wires, used as the secondary target, and the grounded aluminium foil, used as the primary target. The two stainless steel wires are mounted vertically in parallel to each other along a center line between the top of the needle and the grounded aluminium foil. Both the needle and the foil were tilted about 45° from a vertical baseline. They also tried to utilize the secondary electrodes alone, but much smaller amounts of aligned fibers were good. This was because most of the fibers would instead deposit randomly around the first wire electrode. Based on this observation, both the primary and secondary electrodes were essential for making good-aligned fibers for the present set-up. The mechanism for the depositing fibers to extend across the wire electrodes is similar to the parallel electrodes described previously. Both aligned fibers between the parallel vertical wires and a randomly aligned fiber mat on the aluminium foil could be reached.[39]

10.2.1.6 USING AN INSULATING TUBE ON THE COLLECTOR

Ying Yang et al. produced a large area of oriented fibers by putting an insulating tube on the target for a long-time. The changed electrical field made the jet bends around the pipe, so the oriented fibers could be received with a suitable tube. Based on different height and diameter of the pipe, there were three kinds of collection of aligned fibers formed:[39]

- Only a round mat within the tube area
- A round belt outside the tube and a round mat within the tube area
- Only a ring belt outside the tube area

Still, since there is only one electrode as the target, the repelling force is not large enough on the tube to keep the jet falling around the tube all the time. In this font, the jet is not coaxial with the tube and results in disordered fibers.[39]

10.2.1.7 BIASED AC ELECTROSPINNING

Biased AC electrospinning is a new method used by Soumayajit Sarkar et al. It employs a combination of DC and AC potentials. This study aims at lessening the inherent instability of the fiber itself, compared to all techniques relying on lessening the fiber instability by using external forces on the fibers during electrospinning. By introducing a DC biased AC potential instead of either a pure AC or DC potential, alternating positively and negatively charged regions in the fiber result in a reduction of electrostatic repulsion and increase in fiber stability and stability can improve electrospun fiber quality.[39]

10.2.1.8 MAGNETIC ELECTROSPINNING

In this method, the polymer solution is magnetized by adding a few magnetic nanoparticles. The magnetic field stretches the fibers across the gap to make a parallel array as they land on the magnets. When the fibers fall down, the parts of the fibers close to the magnets are attracted to the surface of the magnets, finally the fibers land on the two magnets and suspend over the gap. This method is fundamentally different from all previously discussed methods in preparing aligned fibers, because the driving force is the magnetic field in it while electrostatic interaction plays the function as driving force in the other methods. In addition, this method has several advantages:[39]

- The magnetic field can be manipulated accurately.
- The resultant nanofibers can be transferred on to any substrate with full retention.
- The area of the aligned fibers is large compared to other techniques.

10.2.1.9 A CYLINDER COLLECTOR WITH HIGH ROTATING SPEED

It has been suggested that by rotating a cylinder collector at a high-speed up to thousands of rpm, electrospun nanofibers could be oriented circumferentially. Researchers from Virginia Commonwealth University have utilized this technique summarized in Figure 10-11.[14]

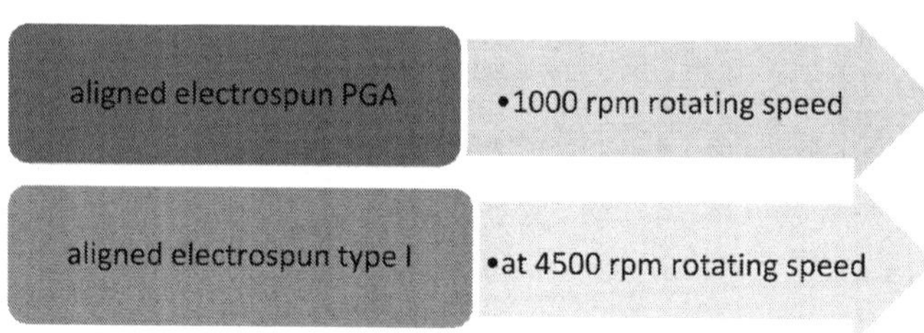

FIGURE 10-11 The speed of this parameter for aligned fibers.

When the linear velocity of rotating cylinder surface matches that of evaporated jet depositions, fibers are taken up on the surface of the cylinder tightly circumferentially, resulting in a fair alignment. Such a speed can be predicted as an alignment speed. If the surface velocity of the cylinder is slower than the alignment speed, randomly deposited fibers will be collected, as it is the fast chaos motions of jets control the final deposition manner. On the other hand, there must be a limit rotating speed above which continuous fibers cannot be collected from the over fast take-up speed will stop the fiber jet. The reasons a perfect alignment is difficult to achieve can be applied to the fact the chaotic motions of polymer jets are not probable to be consistent and are less controllable.[14]

10.2.1.10 A THIN WHEEL WITH SHARP EDGE

A significant advancement in collecting aligned electrospun nanofibers has been recently constructed. The tip-like edge substantially concentrates the electrical field, so the as-spun nanofibers are almost all attracted to and can be continuously wound on the bobbin edge of the revolving wheel. It was explained that before getting to the electrically grounded target, the nanofibers keep enough residual charges to repel each other. As a result, once a nanofiber is attached to the wheel tip, it will exert a repulsive force on the next fiber attracted to the tip. This repulsion is from one another results in a detachment between the deposited nanofibers. The variation in the separation distances is because of varying repulsive forces related to nanofiber diameters and residual charges.[14]

10.2.1.11 A FRAME COLLECTOR

In order to make an individual nanofiber for experimental characterizations, we have recently developed another approach to fiber alignment by simply placing a rectangular frame under the spinning jet. In addition, different frame materials result in different fiber alignments (i.e., aluminium frame favors better fiber alignments than a wooden form). More investigation has been undergoing to understand the alignment characteristics in varying the configuration and size of frame rods, the distance between the frame rods, and the inclination angle of a single frame. These will be useful in deciding how many positions would be best suitable making a polygonal multiframe structure.[14]

10.3 CONCLUDING REMARKS

Producing nanofibers by electrospinning is a simple and widely utilized method for varied applications. As mention earlier, aligned fibers have found importance in many applications. The most significant part of electrospinning is how to control the process. Therefore, we must know the behavior of every part of process and control instabilities were made in it. Some commonly used techniques to align the fibers have been discussed in this chapter. In the next chapter, we have discussed about the behavior formation of nanofiber jet. In addition, the most important tools for better controlling process are modeling and simulating. In near future, we review them.

KEYWORDS

- **Aligned Nanofibers**
- **Controlling Process**
- **Electrospun Nanofibers**
- **Process Parameters**

REFERENCES

1. Šimko, M.; Erhart, J.; Lukáš, D. *A Mathematical Model of External Electrostatic Field of a Special Collector for Electrospinning of Nanofibers.* Journal of Electrostatics, 2014. **72**(2): p. 161–165.
2. Brooks, H.; Tucker, N. *Electrospinning Predictions Using Artificial Neural Networks.* Polymer, 2015. **58**: p. 22–29.
3. Fridrikh, S. V., et al. *Controlling The Fiber Diameter during Electrospinning*, in *Physical Review Letters*. 2003, American Physical Society. p. 144502–144502.
4. Zeng, Y., et al. *Numerical Simulation of Whipping Process in Electrospinning.* in *WSEAS International Conference. Proceedings. Mathematics and Computers in Science and Engineering*. 2009: World Scientific and Engineering Academy and Society.
5. Ciechańska, D. *Multifunctional Bacterial Cellulose/Chitosan Composite Materials for Medical Applications.* Fibres & Textiles in Eastern Europe, 2004. **12**(4): p. 69–72.
6. Ghochaghi, N. *Experimental Development of Advanced Air Filtration Media Based on Electrospun Polymer Fibers*, in *Mechnical and Nuclear Engineering*. 2014, Virginia Commonwealth. p. 1–165.
7. Ziabari, M.; Mottaghitalab, V.; Haghi, A. K. *Evaluation of Electrospun Nanofiber Pore Structure Parameters.* Korean Journal of Chemical Engineering, 2008. **25**(4): p. 923–932.
8. Angammana, C. J. *A Study of the Effects of Solution and Process Parameters on the Electrospinning Process and Nanofibre Morphology*. 2011, University of Waterloo.
9. Lu, P.; Ding, B. *Applications of Electrospun Fibers.* Recent Patents on Nanotechnology, 2008. **2**(3): p. 169–182.

10. Li, Z.; Wang, C. *Effects of Working Parameters on Electrospinning*, in *One-Dimensional Nanostructures*. 2013, Springer. p. 15–28.
11. Bognitzki, M., et al. *Nanostructured Fibers via Electrospinning*. Advanced Materials, 2001. **13**(1): p. 70–72.
12. De, V. S., et al., *The Effect of Temperature and Humidity on Electrospinning*. Journal of Materials Science, 2009. **44**(5): p. 1357–1362.
13. Yarin, A. L.; Koombhongse, S.; Reneker, D. H. *Bending Instability in Electrospinning of Nanofibers*. Journal of Applied Physics, 2001. **89**(5): p. 3018–3026.
14. Huang, Z. M., et al. *A review on Polymer Nanofibers by Electrospinning and Their Applications in Nanocomposites*. Composites Science and Technology, 2003. **63**: p. 2223–2253.
15. Haghi, A. K., *Electrospun Nanofiber Process Control*. Cellulose Chemistry & Technology, 2010. **44**(9): p. 343–352.
16. Sill, T. J.; von. R, H. A. *Electrospinning: Applications in Drug Delivery and Tissue Engineering*. Biomaterials, 2008. **29**(13): p. 1989–2006.
17. Bhardwaj, N.; Kundu, S. C. *Electrospinning: A Fascinating Fiber Fabrication Technique*. Biotechnology Advances, 2010. **28**(3): p. 325–347.
18. Rafiei, S., et al. *New Horizons in Modeling and Simulation of Electrospun Nanofibers: A Detailed Review*. Cellulose Chemistry and Technology, 2014. **48**(5–6): p. 401–424.
19. Tan, S. H., et al. *Systematic Parameter Study for Ultra-Fine Fiber Fabrication via Electrospinning Process*. Polymer, 2005. **46**(16): p. 6128–6134.
20. Zanin, M. H. A.; Cerize, N. N. P.; de. O, A. M. *Production of Nanofibers by Electrospinning Technology: Overview and Application in Cosmetics*, in *Nanocosmetics and Nanomedicines*. 2011, Springer. p. 311–332.
21. Deitzel, J. M., et al. *The Effect of Processing Variables on the Morphology of Electrospun Nanofibers and Textiles*. Polymer, 2001. **42**(1): p. 261–272.
22. Fong, H.; Chun, I.; Reneker, D. H. *Beaded Nanofibers Formed during Electrospinning*. Polymer, 1999. **40**(16): p. 4585–4592.
23. Gupta, P.; Wilkes, G. L. *Some Investigations on the Fiber Formation by Utilizing a Side-by-Side Bicomponent Electrospinning Approach*. Polymer, 2003. **44**(20): p. 6353–6359.
24. Theron, S. A.; Zussman, E.; Yarin, A. L. *Experimental Investigation of the Governing Parameters in the Electrospinning of Polymer Solutions*. Polymer, 2004. **45**(6): p. 2017–2030.
25. Sawicka, K.; Gouma, P.; Simon, S. *Electrospun Biocomposite Nanofibers for Urea Biosensing*. Sensors and Actuators B: Chemical, 2005. **108**(1): p. 585–588.
26. Deitzel, J. M., et al. *The Effect of Processing Variables on the Morphology of Electrospun Nanofibers and Textiles*. Polymer, 2001. **42**(1): p. 261–272.
27. Kessick, R.; Fenn, J.; Tepper, G. *The Use of AC Potentials in Electrospraying and Electrospinning Processes*. Polymer, 2004. **45**(9): p. 2981–2984.
28. Demir, M. M., et al. *Electrospinning of Polyurethane Fibers*. Polymer, 2002. **43**(11): p. 3303–3309.
29. Hsu, C. M.; Shivkumar, S. *Nano-Sized Beads and Porous Fiber Constructs of Poly(ε-Caprolactone) Produced by Electrospinning*. Journal of Materials Science, 2004. **39**(9): p. 3003–3013.
30. Zong, X., et al. *Structure and Process Relationship of Electrospun Bioabsorbable Nanofiber Membranes*. Polymer, 2002. **43**(16): p. 4403–4412.
31. Keun, S. W., et al. *Effect of Ph on Electrospinning of Poly (Vinyl Alcohol)*. Materials Letters, 2005. **59**(12): p. 1571–1575.

32. Frenot, A.; Chronakis, I. S. *Polymer Nanofibers Assembled by Electrospinning.* Current Opinion in Colloid & Interface Science, 2003. **8**(1): p. 64–75.
33. Baji, A., et al. *Electrospinning of Polymer Nanofibers: Effects on Oriented Morphology, Structures and Tensile Properties.* Composites Science and Technology, 2010. **70**(5): p. 703–718.
34. Luo, C. J.; Nangrejo, M.; Edirisinghe, M. *A Novel Method of Selecting Solvents for Polymer Electrospinning.* Polymer, 2010. **51**(7): p. 1654–1662.
35. Tao, J.; Shivkumar, S. *Molecular Weight Dependent Structural Regimes during The Electrospinning of PVA.* Materials Letters, 2007. **61**(11): p. 2325–2328.
36. Huang, Z. M., et al. *A Review on Polymer Nanofibers by Electrospinning and their Applications in Nanocomposites.* Composites Science and Technology, 2003. **63**(15): p. 2223–2253.
37. Reneker, D. H.; Yarin, A. L. *Electrospinning Jets and Polymer Nanofibers.* Polymer, 2008. **49**(10): p. 2387–2425.
38. Rutledge, G. C.; Fridrikh, S. V. *Formation of Fibers by Electrospinning.* Advanced Drug Delivery Reviews, 2007. **59**(14): p. 1384–1391.
39. Zhang, S. *Mechanical and Physical Properties of Electrospun Nanofibers.* 2009: p. 1–83.

CHAPTER 11

A DETAILED REVIEW AND UPDATE ON NANOFIBERS PRODUCTION AND APPLICATIONS

S. PORESKANDAR, F. RAEISI, SH. MAGHSOODLOU, and A. K. HAGHI

Department of Textile Engineering, Faculty of Engineering, University of Guilan, P.O. Box: 3756, Rasht, Iran

CONTENTS

ABSTRACT

When the diameters of fiber materials reduce from micrometers to nanometers, compared to other known form of the material in many research fields, several characteristics are varied. For these cases, researchers become more interested in analyzing the unique properties of these materials. In this chapter, the general and new methods of nanofibers production are studied, at first, and then the electrospinning process is investigated. Finally, the most important applications of these nanofibers are reviewed.

Symptoms	Definitions
V_c	Critical voltage
H	Distance between the capillary exit and the ground
L	Length of the capillary
R	Radius
γ	Surface tension of the liquid/ solution

Abbreviations	Symbols
Polyethylene oxide	PEO
Polyvinyl alcohol	PVA
Polymethyl methacrylate	PMMA
Polyaniline	PANI
Cellulose acetate	CA
Polyvinyl pyrrolidone	PVP
Polyacrylic acid	PAA
Polyamide-6	PA6
Polyethylene oxide	PEO
Polyacrylic acid	PAA
Polyurethane	PU
Polycapro lactone	PCL
Polystyrene	PS
Dimethylformamide	DMF
Tetrahydrofuran	THF
Multifunctional polymethylsilsesquioxane	PMSQ

11.1 AN INTRODUCTION TO THE IMPORTANCE OF NANOTECHNOLOGY

Nanotechnology has called attention of many scientists all over the world for various novel applications in recent years with its unique physical, chemical, and biological properties[1–3] (Tables 11-1 and 11-2). Therefore, researchers have started to analyze these properties.[3,4] When the diameters of polymer fiber materials are reduced from micrometers to nanometers, several characteristics change compared to other known form of materials in many research fields. These characteristics are as follows:[5]

- A large surface area to volume ratio
- Flexibility in surface functionalities
- High porosity
- Superior mechanical performance (e.g., Stiffness and tensile force)
- Nanofibers can produce from a spacious range of polymers.[2,6,7]

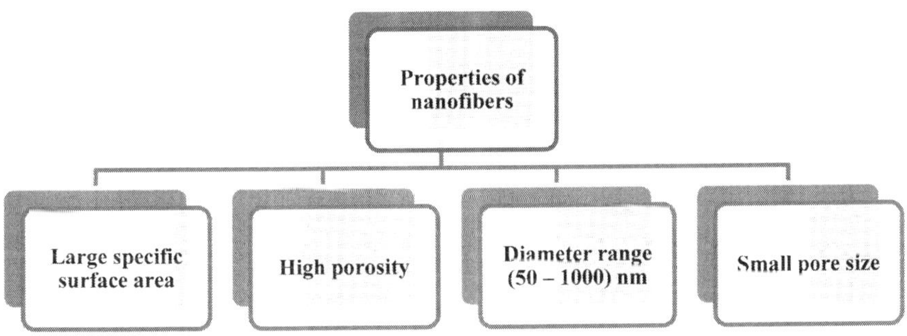

FIGURE 11-1 Properties of nanofibers.

These desirable properties make the polymer nanofibers best candidates for many important applications:[1–3,6–23]

- Environmental engineering and Biotechnology
  - Medical science
    - Tissue engineering
      - Wound healing
      - Tissue template

- Drug delivery
- Release control

- Composites
- Defense

  - Protective clothing

TABLE 11-1 Utilized polymer fibers in electrospinning process in different applications for tissue engineering[24]

Fiber diameter	Solvent	Polymer	Application
200–350 nm	2,2,2-trifluoroethanol Water	Poly(ε-caprolactone) (shell)+ Poly(ethylene glycol) (core)	
1–5 μm	Chloroform DMF Water	Poly(ε-caprolactone) and poly(ethylene glycol) (shell) Dextran (core)	
500–700 nm	Chloroform DMF Water	Poly(ε-caprolactone) (shell) Poly(ethylene glycol) (core	**Drug Delivery System**
~4 μm	DCM PBS	Poly(ε-caprolactone-co-ethyl ethylene phosphate)	
260–350 nm	DMF	Poly(D,L-lactic-co-glycolic acid), PEG-b-PLA، PLA	
1–10 μm	DCM	Poly(D,L-lactic-co-glycolic acid)	
690–1350 nm	Chloroform	Poly(L-lactide-co-glycolide) and PEG-PLLA	
2–10 nm	Chloroform methanol	Poly(ε-caprolactone)	
500–900 nm	Chloroform DMF	Poly(ε-caprolactone) (core)+ Zein (shell)	
500 nm	2,2,2-trifluoroethanol	Poly(ε-caprolactone) (core) + Collagen (shell)	
500–800 nm	DMF THF	Poly(D,L-lactic-co-glycolic acid) and PLGA-b-PEG-NH2	
1–4 mm	DMF acetone	Poly(ethylene glycol-co-lactide)	**General Tissue Engineering**
0.2–8.0 mm	2-propanol and water Water	Poly(ethylene-co-vinyl alcohol)	
180–250 nm	HFP	Collagen	
0.29–9.10 mm	2,2,2-trifluoroethanol	Gelatin	
120–610 μm	HFP	Fibrinogen	
130–380 nm	HFP	Poly(glycolic acid) and chitin	

TABLE 11-1 *(Continued)*

Fiber diameter	Solvent	Polymer	Application
0.2–1 nm	Chloroform DMF	Poly(ε-caprolactone)	**Vascular Tissue Engineering**
200–800 nm	Acetone	Poly(L-lactide-co-ε-caprolactone)	
5 μm	Chloroform	Poly(propylene carbonate)	
300 nm	1,4-dioxane DCM	Poly(L-lactic acid) and hydroxylapatite	
0.163–8.77 nm	HFP	Chitin	

11.2 USUAL METHODS OF PRODUCING NANOFIBERS

Various common techniques can be used for preparing polymer nanofibers such as refs. [7, 10, 15]:

- Drawing
- Template synthesis
- Phase separation
- Self-assembly
- Electrospinning

At first these techniques are introduced, and then special ways for producing nanofibers are proceeding.

11.2.1 DRAWING

The drawing technique is associated with evaporating the solvent from viscous polymer liquids directly, leading to solidification of the fiber. In this method, the nanofiber has an order of microns.[7,10,23] Figure 11-2 shows a drawing technique, each fiber is made from a micro droplet of polymer solution using a micropipette.[7,25]

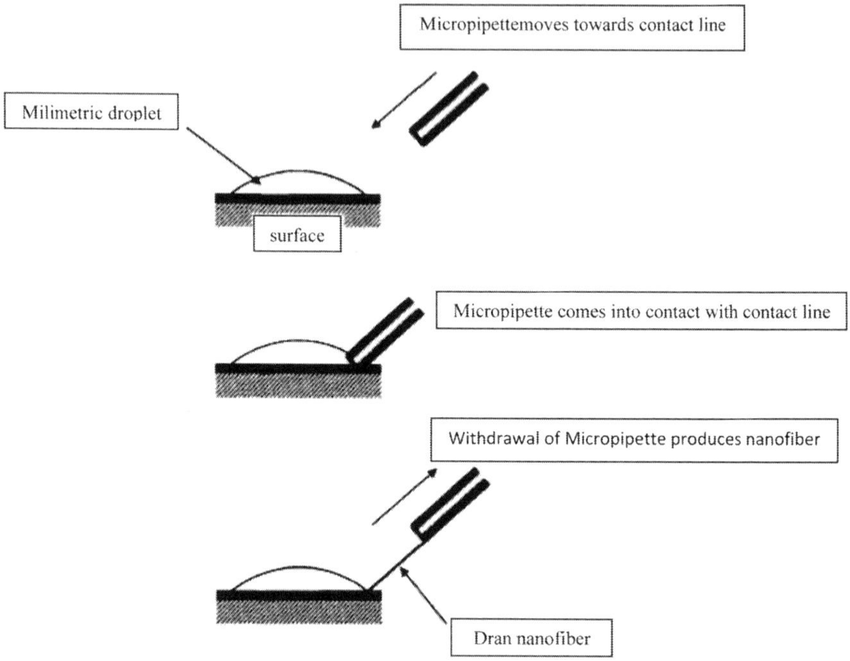

FIGURE 11-2 Schematic of drawing technique (each fiber is drawn from a microdroplet of polymer solution using a micropipette).

11.2.2 TEMPLATE SYNTHESIS

The template synthesis uses templates with pores such as membranes to make solid or hollow forms of nanofiber. This technique is similar to the extrusion in manufacturing.[7,25] This method cannot produce continuous nanofibers one-by-one. The most significant advantage of this method is that various materials are used for making nanofibers. These materials are[8,15]

- Conducting polymers
- Metals
- Semiconductors
- Carbons

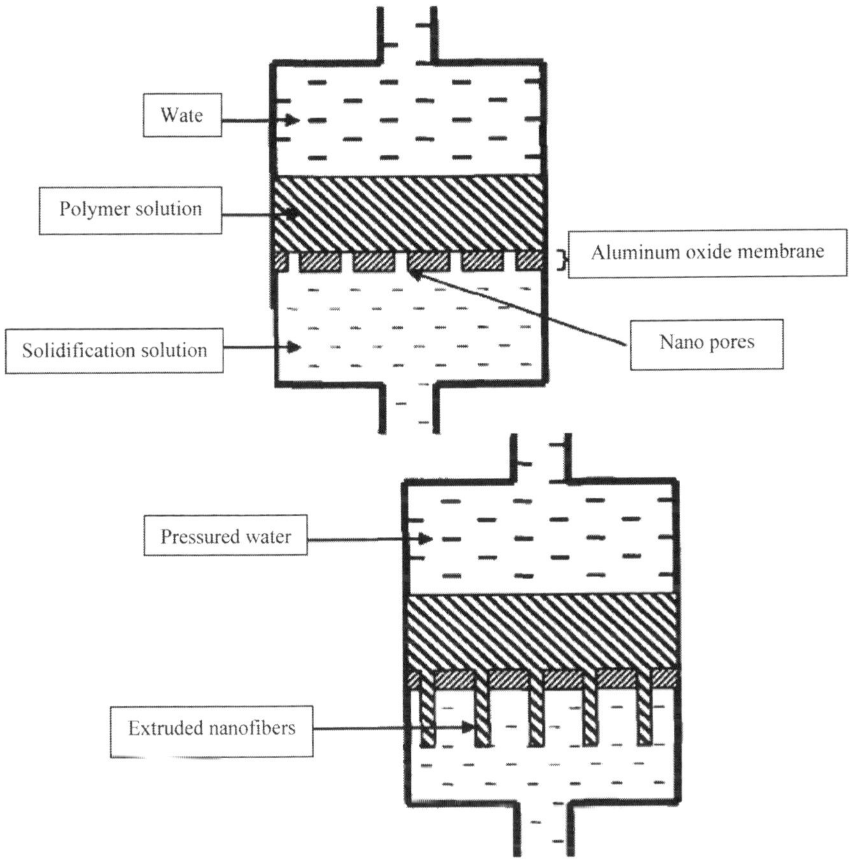

FIGURE 11-3 Schematic of template technique. Polymer extrudes through a nanoporous template by applying pressure.[7,25]

11.2.3 PHASE SEPARATION

The phase separation involves four levels:

1. Prepare a solution of polymer in solvent
2. Do polymer gelatination with low temperature
3. Get rid of solvent by immersion in water
4. Do freezing and freeze-drying

This technique calls for so long time. Figure 11-4 shows forming nanofiber by phase separation.[8,10,15,23,25]

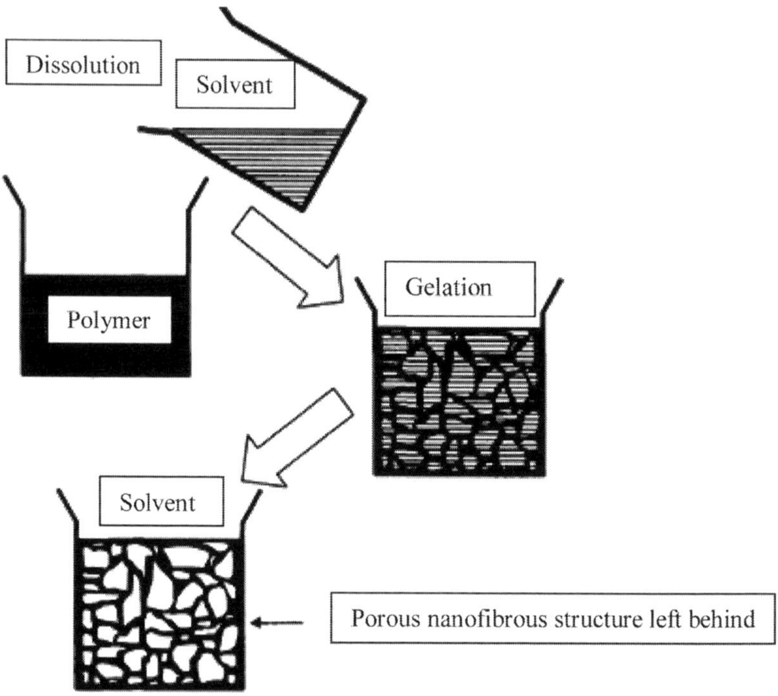

FIGURE 11-4 Formation of nanofiber by phase separation.[10,25]

11.2.4 SELF-ASSEMBLY

The self-assembly is another technique of producing nanofibers. In this technique, pre-existing items make up into favorable patterns. Although this technique is similar to the phase separation technique, the best feature of this technique is time-consuming for producing continuous polymer nanofibers. In the self-assembly technique, nanofibers are hung up molecule by molecule to bring out specific structures and functions.[7,8,10,15,23,25]

FIGURE 11-5 A) Molecular structure B) Nanostructure of self-assembling peptide-amphiphile nanofiber network.[10]

11.2.5 ELECTROSPINNING

Electrospinning is the most favorite technique for creating more efficient nanofibers.[12,26,27] This technique is simply a cheap and straightforward method to produce nanofibers.[5–6,28–30] It:

a) creates continuous fibers with diameters in nano range,
b) is applicable for a broad range of materials (e.g., synthetic, natural polymers, metals as well as ceramics and composite),
c) and prepares nanofibers with low cost

An ordinary electrospinning setup contains mainly three parts:[1,10]

- A high-power supply voltage
- A syringe with a needle and a pump
- A collector

Likewise, this technique is distinguished mainly by four sections:[21]

- Taylor Cone
- Steady Jet
- Instability part
- Base part

Nanofibers are formed from polymer solution or melt with a high-potential power source. Then this liquid is passed from capillary and collected along the collector.[10,31]

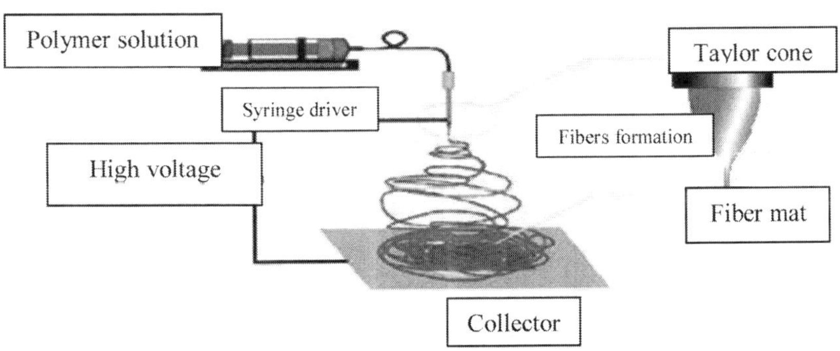

FIGURE 11-6 Standard electrospinning setup.

TABLE 11-2 Compression of common technology for producing nanofibers

Repeatability	Controllability	Simplicity	Scalability	Technology	Technique
No	Yes	Yes	No	Laboratory scale	Drawing
Yes	Yes	Yes	Yes	Laboratory scale	Template synthesis
No	Yes	Yes	No	Laboratory scale	Phase separation
No	No	Yes	No	Laboratory scale	Self-assembly
Yes	Yes	Yes	Yes	Industrial process	Electrospinning

11.3 NEW METHODS OF PRODUCING NANOFIBERS

11.3.1 GELATION TECHNIQUE

Initially, a gel is made using predetermined amounts of polymer and solvent followed by phase separation and gel formation. Finally, nanofiber forms when the gel is frozen and freeze-dried.[7]

11.3.2 BACTERIAL CELLULOSE TECHNIQUE

Cellulose nanofibers produced by bacteria have been used for long in diverse applications, including biomedical.[10,27,32]. Cellulose synthesis by Acetobacter involves polymerization of glucose residues into chains, followed by extra-cellular secretion, assembly, and crystallization of the chains into hierarchically comprised ribbons (Figure 11-7).[10,27,32]

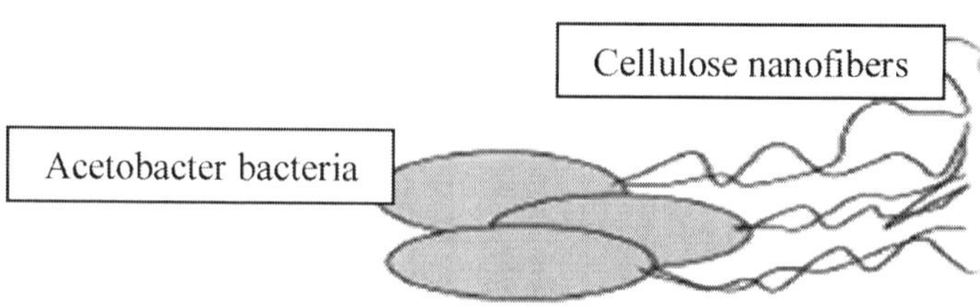

FIGURE 11-7 Acetobacter bacteria cells depositing cellulose nanofibers.[10]

Networks of cellulose nanofibers with diameters less than 100 nm are readily made. Fibers with different characteristics may be developed by different strains of bacteria. Copolymers have been created by adding polymers to the growth media of the cellulose producing bacteria.[10,27,32] Bacterial cellulose is mixed by the acetic bacterium *Acetobacter xylinum*. The fibrous structure of bacterial cellulose consists of a three-dimensional network of micro fibrils containing glucanchains bound by hydrogen bonds.[27,32]

11.3.3 EXTRACTION TECHNIQUE

Nanofibers can be extracted from natural materials using chemical and mechanical treatments. Cellulose fibrils can be sorted out from plant cell walls. In one example, cellulose nanofibers were extracted from wheat straw and soy hull with diameters ranging from 10 to 120 nm and lengths up to a few thousand nanometers. Invertebrates have also been utilized as a source for extracting nanofibers.[10]

11.3.4 VAPOR-PHASE POLYMERIZATION TECHNIQUE

Polymer nanofibers have also been made from vapor-phase polymerization. Plasma-induced polymerization of vapor phase vinyltrichlorosilane produced organosiloxane fibers with diameters around 25 nm and typical lengths of 400–600 nm and cyanoacrylate fibers with diameters from 100 to 400 nm and lengths of hundreds of microns.[10]

11.3.5 KINETICALLY CONTROLLED SOLUTION SYNTHESIS TECHNIQUE

Nanofibers and nanowires have been created in solution using linear aligned substrates as template agents such as iron-cation absorbed reverse cylindrical micelles and silver micelles. PVA - polymethyl methacrylate nanofibers were produced using silver nanoparticle that was linearly aligned in solution with vigorous magnetic stirring. These nanoparticle chain assemblies acted as a template for further polymerization of nanofibers with diameters from 10 to 30nm and lengths up to 60µm.[10]

11.3.6 CONVENTIONAL CHEMICAL OXIDATIVE POLYMERIZATION OF ANILINE TECHNIQUE

Chemical oxidation polymerization of aniline is a traditional method for synthesizing poly(aniline), and during the former stages of this synthesis technique, poly(aniline) nanofibers are formed. Optimization of polymerization conditions such as temperature, mixing speed, and mechanical agitation allows the end-stage formation of poly(aniline) nanofibers with diameters in the range of 30–120 nm.[10]

11.4 HISTORY OF ELECTROSPINNING AND NANOFIBERS

The word "fiber" has its root in "fibra" and the "nano" term comes from the definition that has been discussed generously. When the diameter of polymer fiber is reduced to nanoscale, the nanofibers become important in applications.[33] The electrospinning attracts more attention as ultrafine fibers of varied polymers with lower diameters to nanometers in nanotechnology in the recent years.[4,13,18] Employing electrostatic forces to deform materials in the liquid state goes back many centuries,[26,34] but the origin of electrospinning as fiber spinning technique comes back to 100 years ago.[9,16] Many researchers work on electrospinning set up and effective factor in this technique. The patents characterize an experimental set-up for producing polymers between 1934 and 1944.[34,35] Subjection Formhals work, the focus shifted to developing a better understanding process technique of the electrospinning. Here, we get a summary of electrospinning histories in Table 11-3. Several research groups, such as Dr Darrell Reneker and his research group, have expressed further interest in electrospinning with a series of papers published starting in early to mid 1990's and continuing today up to engage. This renewed interest spread quickly and many secondary academic groups became interested in the field of the electrospinning.[1,14,23]

The number of publications and patents in nanofibers fields and electrospinning have grown significantly in recent years.[1,14,18] Secondary academic groups have gained interest in the subject area of the electrospinning.[1,14,23]

TABLE 11-3 History of Electrospinning

Name of Researcher	Year	Subject	References
Lord Rayleigh	19th century	Understood the technique of electrospinning	[9]
William Gilbert	1600	Discovered first record of the electrostatic attraction of a liquid	[5]
Zeleny	1914	Introduced one of the earliest studies of electrified jetting phenomenon	[4]
Formhals	1934	Invented the experimental setup for the practical production of polymer filaments with an electrostatic force	[3]
Vonnegut and Neubauer	1952	Produce streams of uniform droplets and invented a simple tool for the electrical atomization	[2]
Drozin	1955	Examine the dispersion of a series of liquids into aerosols under high electric potentials	[1]
Simons	1966	Patented a tool for producing nonwoven fabrics of ultrathin and weightless	[6]
Taylor	1969	Published his work on the shape of the polymer droplet at the tip of the needle applying an electric field	[7]
Baumgarten	1971	Made a tool for electrospinning acrylic fibers with a stainless steel capillary tube and a high-voltage DC current. Estimated the jet speed by using energy balance when a critical voltage was applied	[15]
Larrondo and Mandley	1981	Produced polyethylene and polypropylene fibers by melting electrospinning successfully	[8]
Hayati *et al*	1987	To study effective factors on jet stability and technique of electrospinning	[28]
Reneker and Chun	1996	Has shown the possibility of electrospinning polymer solutions	[26]

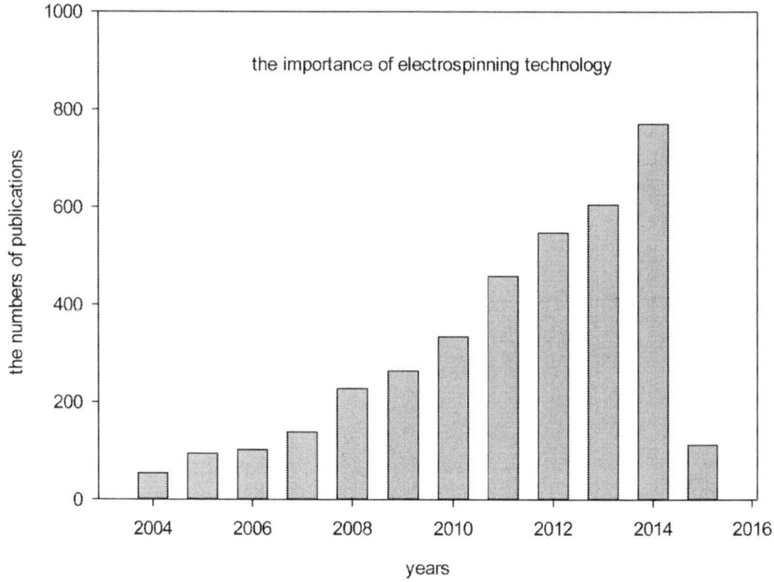

FIGURE 11-8 Numbers of publications about electrospinning.

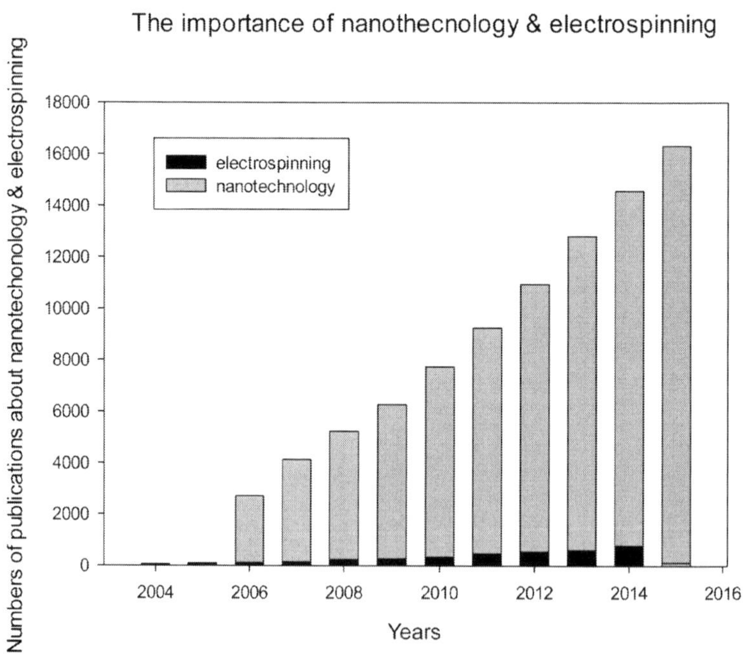

FIGURE 11-9 The importance of nanotechnology and electrospinning.

11.5 RULES FOR ELECTROSPINNING OF NANOFIBERS

As it was mentioned before, electrospinning is an efficient and the simplest technique for producing nanofibers with different structures and functionality.[22,33,36–41] The advantages of electrospinning technology are as follows:[42]

- Producing a hi'gh rate of nanofiber
- Simplest setup and low costs of production

A common electrospinning set-up includes the following:[11,37,41]

- A high-voltage power supply
- A syringe
- A needle
- A grounded collector screen

The technique of electrospinning includes several parts:[17,22,38]

- Charging of the fluid
- Formation of the cone-jet (Taylor cone)
- Thinning of the jet with an electric field
- Instability of the jet
- Collection of the jet on target

Here, a simple schematic of the electrospinning technique is shown in particular.

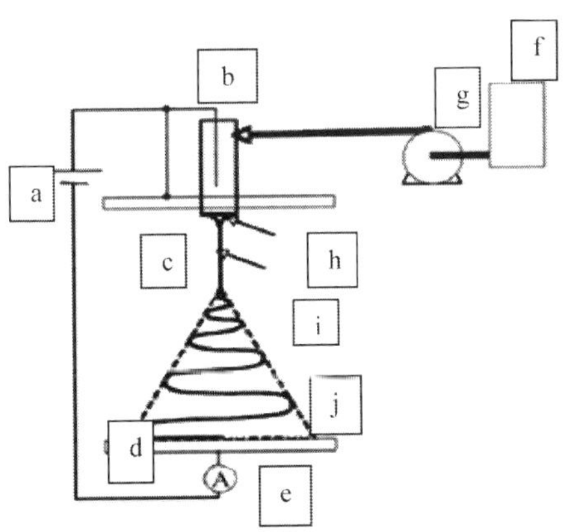

FIGURE 11-10 A simple schematic of electrospinning technique a) high-voltage power supply; b) charging devices; c) high-potential electrode (e.g., flat plate); d) collector electrode (e.g., flat plate); e) current measurement device; f) fluid reservoir; g) flow rate control; h) cone; i) thinning jet; j) instability region.[43]

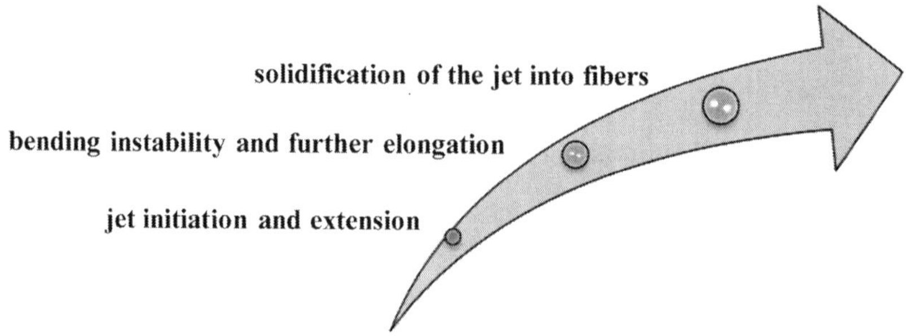

FIGURE 11-11 Basic principle of electrospinning.

Instead of formal methods of fiber formation (e.g. Dry or wet spinning), electrospinning makes nanofibers by electrostatic forces.[17] Ion migrates in the solution or melts with an electric field. When the potential comes into a critical value, the stream of jet starts formation to throw. This jet moves straight toward the collector then, bending instability develops into a series of loops expanding with time. The solvent evaporates during the jet moves. Finally, nanofibers are collected on plate.[31,44,45] It is important to observe that it is possible to electrospin all polymers into nanofibers, provided the molecular weight of the polymers is large enough and the solvent can be evaporated quickly enough during the technique.[8,42] The mechanics of this technique deserve a specific attention and necessary to predictive tools or direction for better understanding and optimization and controlling technique. It has been identified that during travelling a solution jet from the tip to collector, the primary jet may show instability during the pathSeveral videos, graphic and laser light scattering methods for watching over the three-dimensional path of jets in flight, and for seeing the diameter and rate of parts were developed.[37] On the other hand, as in any liquid, the surface tension reduces the entire surface of the jet, thus reduces the free energy of the liquid. If the viscosity is not enough to hold the jet as a continuous shape, what usually occurs is an instability that causes the jet to break up into droplets. This effect is known as Rayleigh instability. Which of these two opposing effects prevails depends on the nature of the fluid, especially its viscosity and surface tension. If the viscosity is enough high with good cohesiveness, the charged jet undergoes a straight jet stage and whipping instability takes place, the amplitude depends on the material and solvent, then dry thin fibers are gathered. Although the setup is straightforward, but controlling of electrospinning is complicated. Some studies are conducted by Taylor on the initial jet formation of electrospinning technique. He gained condition for critical electric potential where surface tension is in equipoise with the electrical force:[22,46]

$$V_C^2 = 4\frac{H^2}{L^2}\left(\ln\frac{2L}{R} - \frac{3}{2}\right)(0.117\pi\gamma R)$$

Although Taylor cone has been viewed in many subjects, the exact shape and the angle of the cone are not fixed and only applicable to slight conducive, monomeric fluids. Researchers studied the initial jet formation through computer simulation and compared with the experimental outcomes. They found that thinning the jet in the initial stage is determined by many features. Viscoelasticity is found to be the key element in the initial jet thinning behavior. Fluid with higher viscoelasticity is thicker. Studies of the whipping motion revealed that in the envelope of the cone it only controls a single jet. The jet undergoes a fast whipping motion and the whipping is so tight that the conventional camera cannot distinguish the splaying with whipping. The bending instability of electrified jet caused by repulsive forces between the charges carried by the jet.[13,31] The jet remains axisymmetric for some length. Then bending or whipping instability starts. At the onset of this instability, the jet follows a spiral path. As the jet spirals toward the collector, higher order instabilities reveal themselves. This instability makes the jet to loop in spirals with increasing radius.[7,29] The envelope of this closed circuit is a cone. Further, the electric field speeds up the jet. So, the jet rate increases. This leads to a decrease in the jet diameter. In addition, the electrostatic repulsion between excess charges in the solution stretches the jet. This stretching also decreases the jet diameter.[7,17]

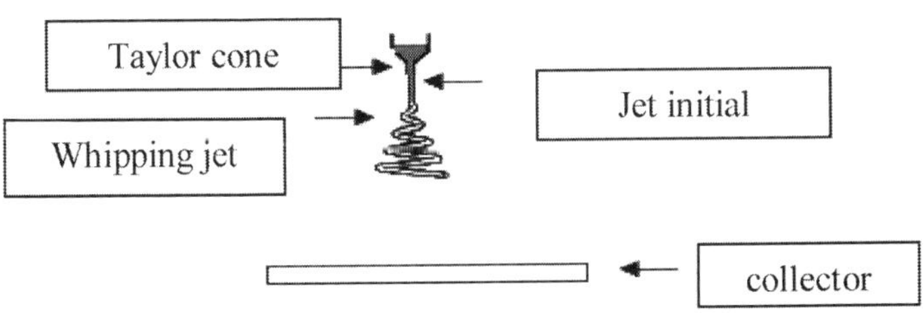

FIGURE 11-12 Whipping instability of jet in electrospinning technique.

11.6 THE USE OF NANOFIBERS IN VARIOUS SCIENCE SEARCHES

The fine electrospun nanofibers make them useful in a wide range of innovative applications.[22,47] Many materials are used for electrospinning.[8,42]

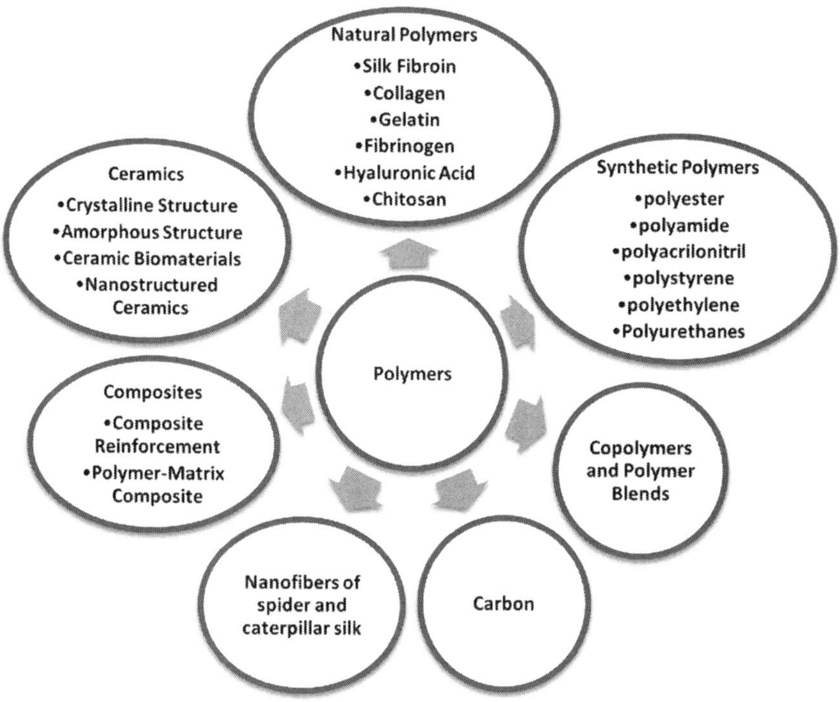

FIGURE 11-13 Varieties of polymers in electrospinning.

TABLE 11-4 Classes **of polymers with solvents**

Polymer class	Polymer	Solvent
	Polymides	Phenol
High performance polymers	Polyamic acid	m-cresol
	polyetherimide	methylene chioride
	Polyaramid	Sulphuric acid
Liquid crystalline polymers	Polygamma-benyzyl-glumate	dimethylformamide
	Polyp-phenylene terephthalamide	Sulphuric acid

TABLE 11-4 *(Continued)*

Polymer class	Polymer	Solvent
Copolymers	Nylon 6-polyimide	Formic acid
		Dimethylformamide
	Polyacrylonitrile	Trifuoroacetic acid and dichloromethane melt in vacuum
	Polyethylene terephthalate	
Textile fibre polymers	Naylon	Formic acid
	Polyvinyl alcohol	Water
Electrically conducting polymer	Polyaniline	Sulphuric acid
	DNA	Water
Biopolymers	Polyhydroxy butyrate-valerate	Chloroform
	Polycapro lactone	m-Cresol, Chlorophenol, Formic acid

Also new applications have been explored for these fibers continuously. Main application fields are shown in Figure 11-14:[39,48,49]

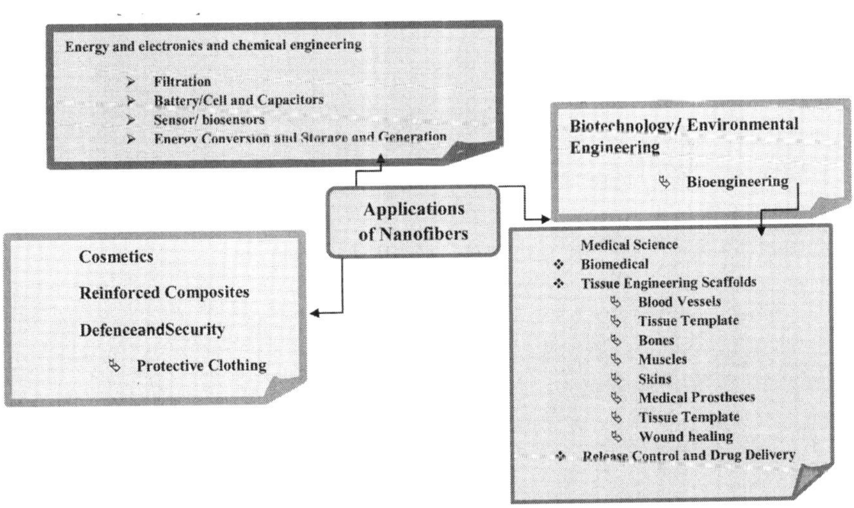

FIGURE11-14 **Potential** applications of electrospun fibers.

For selected applications, it is desirable to control not only the fiber diameter, but as well the internal morphology. Porous fibers are of interest for applications such as filtration or preparation of nanotubes by fiber templates.[35,38,45,50] Besides, small pore size and high surface area inherent in nanofiber has implications in biomedical applications such as scaffoldings for tissue growth.[35,51] Also, researchers have spun a fiber from a compound naturally present in the blood. This nanofiber can be used as forms of medical applications such as bandages or sutures that ultimately dissolve into the body. This nanofiber minimizes infection rate, blood loss and is also taken up by the body.[42] One of the most significant applications of nanofibers is to be used as reinforcements in composite developments. With these reinforcements, the composite materials can offer superior properties such as high modulus and strength to weight ratios, which cannot be achieved by other engineered monolithic materials alone. Information on the fabrication and structure-property relationship characterization of such nano composites is believed to be utilitarian. Such continuous carbon nanofiber composite also has possible applications as filters for[15]

- Separation of small particles from gas or liquid
- Supports for high temperature catalysts
- Heat management materials in aircraft and semiconductor devices
- Rechargeable batteries
- Super capacitors

11.6.1 BIOTECHNOLOGY/ENVIRONMENTAL ENGINEERING

Nonwoven electrospun nanofiber meshes are an excellent material for membrane preparation, particularly in biotechnology and environmental engineering applications for the following reasons:[17]

- High porosity
- Interconnectivity
- Micro-scale interstitial space
- A large surface to volume ratio

Biomacromolecules or cells can be tied to the nanofiber membrane for these applications:[17]

- In protein purification and waste water treatment (affinity membranes)
- Enzymatic catalysis or synthesis (membrane bioreactors)
- Chemical analysis and diagnostics (biosensors)

Electrospun nanofibers can form an effective size exclusion membrane for particulate removal from wastewater.[17] Affinity membranes are a broad class of membranes that selectively captures specific target molecules by immobilizing a specific capturing agent onto the membrane surface. In biotechnology, affinity membranes have applications in protein purification and toxin removal from bioproducts. In the environmental industry, affinity membranes have applications in organic waste removal and heavy metal removal in water treatment. To be used as affinity membranes, electrospun nanofibers must be surface functionalized with ligands. Mostly, the ligand molecules should be covalently attached to the membrane to prevent leaching of the ligands. Also, water pollution has been continuously emerging as a critical global issue. One important class of inorganic pollutant of great physiological significance is heavy metals, for example, Hg, Pb, Cu, and Cd. Distributing these metals in the environment is mainly applied to the release of metal containing waste-water from industries. For example, copper smelters may release high quantities of Cd, one of the most mobile and toxic among the trace elements, into nearby waterways. It is impossible to eliminate some classes of environmental contaminants, such as metals, by conventional water purification methods. Affinity membranes will play a critical role in wastewater treatment to remove (or recycle) heavy metal ions in the future. Polymer nanofibers functioned with a ceramic nanomaterial, mentioned in the following, could be suitable materials for fabrication of affinity membranes for water industry applications:[17]

- Hydrated alumina hydroxide
- Alumina hydroxide
- Iron oxides

The polymer nanofiber membrane acts as a bearer of the reactive nanomaterial that can attract toxic heavy metal ions, such As, Cr, and Pb, by adsorption or chemisorption and electrostatic attraction mechanisms. Again, affinity membranes provide an alternative access for removing organic molecules from wastewater.[17]

11.6.1.1 BIOENGINEERING

In biological viewpoint, almost entirely human tissues and organs are deposited in nanofibrous forms or structures. Some examples include the following:[15]

- Bone
- Dentin
- Collagen
- Cartilage
- Skin

All of them are characterized by well organized fibrous structures realigning in nanometer scale. Current research in electrospun polymer nanofibers has focused one of their major applications on bioengineering. We can easily find their promising potential in various biomedical fields.[15]

11.6.1.1.1 Medical Science

Nanofibers are used in medical applications, which include drug and gene delivery, artificial blood vessels, artificial organs, and medical face masks. For example, carbon fiber hollow Nanotubes, smaller than blood cells, have the potential to transport drugs into blood cells.[42]

Biomedical Application

Biomedical field is one of the important application areas among others, using the technique of electrospinning like[11,52]

- Filtration material
- Protective material
- Electrical applications
- Optical applications
- Sensors
- Nanofiber reinforced composites

Current medical practice is based almost on treatment regimes. However, it is envisaged that medicine in the future will be based heavily on early detection and prevention before disease expression. With nanotechnology, new treatment will emerge that will significantly reduce medical costs. With recent developments in electrospinning, both synthetic and natural polymers can be produced as nanofibers with diameters ranging from decades to hundreds of nanometers with controlled morphology. The potential of these electrospun nanofibers in human healthcare applications is promising, for example:[17]

- In tissue or organ repair and regeneration
- As vectors to deliver drugs and therapeutics
- As biocompatible and biodegradable medical implant devices
- In medical diagnostics and instrumentation
- As protective fabrics against environmental and infectious agents in hospitals and general surroundings
- In cosmetic and dental applications.

Tissue or organ repair and positive feedback are new avenues for potential treatment, avoiding the need for donor tissues and organs in transplantation and reconstructive surgery. In this advance, a scaffold is usually needed that can be fabricated from either natural or synthetic polymers by many techniques including electrospinning and phase separation. An animal model is utilized to study the biocompatibility of the scaffold in a biological system before the scaffold is introduced into patients for tissue-regeneration applications. Nanofibers scaffolds are suited to tissue engineering. These can be made up and shaped to fill anatomical defects. Its architecture can be designed to supply the mechanical properties necessary to support cell growth, differentiation, and motility. Also, it can be organized to provide growth factors, drugs, therapeutics, and genes to stimulate tissue regeneration. An inherent property of nanofibers is that they mimic ECM of tissues and organs. The ECM is a complex composite of fibrous proteins, such as collagen and fibronectin, glycoproteins, proteoglycans, soluble proteins such as growth factors, and other bioactive molecules that support cell adhesion and growth. One of the aim is to create electrospun polymer nanofiber scaffolds for engineering blood vessels, nerves, skin, and bone. In the pharmaceutical and cosmetic industry, nanofibers are promising tools for controlling these aims:[17,42,52]

- Delivery of drugs
- Therapeutics
- Molecular medicines
- Body-care supplements

Tissue Engineering Scaffolds

Successful tissue engineering needs synthetic scaffolds to bear similar chemical compositions, morphological, and surface functional groups to their natural counterparts. Natural scaffolds for tissue growth are three-dimensional networks of nanometer-sized fibers made of several proteins. Nonwoven membranes of electrospun nanofibers are well-known for their interconnected, 3D porous structures and large surface areas, which provide a class of ideal materials to mimic the natural ECM needed for tissue engineering. The electrospun nanofibrous support was treated with the cell solution and the nanofiber-cell was cultured in a rotating bioreactor to create the cartilage which controlled compressive strength similar to natural cartilage. The tissue engineered cartilages could be applied in treating cartilage degenerative diseases. The scaffold was applied as biomimic ECM, enzyme, gene, and medicine to revive skin, cartilage, blood vessel, and nerve. The scaffold was helpful in biocompatibility, mechanical property, porosity, and degradability in the human physical structure. The electrospun nanofibers showed moderate porosity, excellent mechanical property, and biocompatibility, which could be utilized to repair blood vessels, skin, and nervous tissue.[1] Tissue engineering is an emerging interdisciplinary and multidisciplinary research study. It involves the utilization of living cells, manipulated through their extracellular environment or genetically to develop biological substitutes for implantation into the body or to foster remodeling of tissues in some active manners. The purpose of tissue engineering is to renovate, replace, say, or improve the function of a particular tissue or organ. For a functional scaffold, a few basic needs have to be satisfied:

- A scaffold should control a high degree of porosity, with a suitable pore size distribution.
- A large surface area is needed.
- Biodegradability is often needed, with the degradation rate matching the rate of neotissue formation.
- The scaffold must control the needed structural integrity to prevent the pores of the scaffold from collapsing during neotissue formation, with the suitable mechanical properties.
- The scaffold should be nontoxic to cells and biocompatible, positively interacting with the cells to promote cell adhesion, growth, migration, and distinguished cell function.

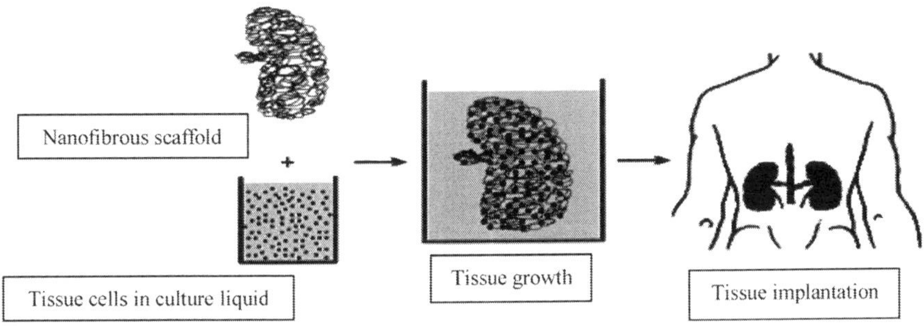

FIGURE 11-15 Principle of tissue engineering

Among all biomedical materials under evaluation, electrospun nanofibrous scaffolds have presented great performances in cell attachment, increase, and penetration.[6] One of the most promising potential applications is tissue scaffolding. The nonwoven electrospun mat has a high surface area and a high porosity. It contains an empty space between the fibers that is approximately the size of cells. The mechanical property, the topographical layout, and the surface chemistry in the nonwoven mat may have a direct effect on cell growth and migration.[31] Ultra-fine fibers of biodegradable polymers produced by electrospinning have found potential applications in tissue engineering because of their high surface area to volume ratios and high porosity of the fibers. However, the flexibility of seeding stem cells and human cells on the fibers makes electrospun materials most suited for tissue engineering applications. The fibers produced can be used systematically to design the structures that they perform not only mimic the properties of ECM, but also control high strength and high toughness. For instance, nonwoven fabrics show isotropic properties and support neotissue formation. These mats resemble the ECM matrix and can be applied as a skin-scaffold and wound dressing materials where the materials are needed to be more elastic than stiff. Many natural polymers like collagen, starch, chitin, and chitosan and synthetic biodegradable polymers like PCL, PLA, and PLGA have been widely investigated for potential applications in developing tissue scaffolds. These results confirm that electrospinning of natural or synthetic polymers for tissue engineering applications are promising.[9] Tissue engineering is one of the most exciting interdisciplinary and multidisciplinary research fields today, and there has been exponential growth in the number of research publications in this area in recent years. It involves the utilization of living cells, manipulated through their extracellular environment or genetically to develop biological substitutes for implantation into the body or to foster remodeling of tissues in some active manners. The purpose is to repair, replace, maintain, or increase the use of a particular tissue or organ. The core technologies intrinsic to this effort can be organized into three fields:[4]

- Cell technology
- Scaffold frame technology
- Technologies for *in vivo* integration

The scaffold frame technology focuses on these objectives:[4]

- Designing
- Manufacturing
- Characterizing three-dimensional scaffolds for cell seeding
- In vitro or in vivoculturing

Blood Vessels

Blood vessels vary in sizes, mechanical and biochemical properties, cellar content, and ultra structural organization, depending on their location and specific role. It is needed that the vascular grafts engineered should have wanted characteristics. Blood vessel replacement, a fine blood vessel (diameter<6 mm), has stayed a great challenge. Because the electrospun nanofiber mats can give good support during the initial development of vascular smooth muscle cells, smooth film combining with electrospun nanofiber mat could form a good 3D scaffold for blood vessel tissue engineering.[4,6]

Muscles

Collagen nanofibers were first applied to assess the feasibility of culturing smooth muscle cell. The cell growth on the collagen nanofibers was promoted and the cells were easily integrated into the nanofiber network after 7 days of seeding. Smooth muscle cells also adhered and proliferated well on another polymer nanofiber mats blended with collagen, incorporating collagen into nanofibers was observed to improve fiber elasticity and tensile strength, and increase the cell adhesion. The fiber surface wet ability influences cell attachment. The alignment of nanofibers can induce cell orientation and promote skeletal muscle cell morphologenesis and aligned formation.[4]

Medical Prostheses

Polymer nanofibers fabricated by electrospinning have been offered for several soft tissue prosthesis applications such as blood vessel, vascular, breast, and so on. In addition, electrospun biocompatible polymer nanofibers can also be deposited as a slender, porous film onto a hard tissue prosthetic device designed to be implanted into the human body. This coating film with a fibrous structure works as an interface between the prosthetic device and the host tissues. It is anticipated to reduce efficiently the stiffness mismatch at the tissue interphase and hence prevents the device failure after the implantation.[15]

Tissue Template

For treating tissues or organs in malfunction in a human body, one of the challenges in the area of tissue engineering or biomaterials is the design of ideal scaffolds or synthetic matrices. They can mimic the structure and biological functions of the natural ECM. Human cells can attach and organize well around fibers with diameters smaller than those of the cellular phones. Nanoscale fibrous scaffolds can provide an ideal template for cells to seed, migrate, and produce. A successful regeneration of biological tissues and organs calls for developing fibrous structures with fiber architectures useful for cell deposition and cell growth. Of particular interest in tissue engineering is creating reproducible and biocompatible three-dimensional scaffolds for cell growth resulting in biometrics composites for various tissue repair and replacement processes. Recently, people have begun to pay attention to making such scaffolds with synthetic polymers or biodegradable polymer nanofibers. It is believed that converting biopolymers into fibers and networks that mimic native structures will eventually improve the usefulness of these materials as large diameter fibers do not mimic the morphological characteristics of the native fibrils.[15,42]

Wound Healing

Wound healing is a native technique of regenerating dermal and epidermal tissues. When an individual is wounded, a set of complex biochemical actions take place in a closely orchestrated cascade to repair the harm. These events can be sorted into four groups:

1. Inflammatory
2. Proliferative
3. Remodeling phases
4. Epithelialization

Ordinarily, the body cannot heal a deep dermal injury. In full thickness burn or deep ulcers, there is no origin of cells remaining for regeneration, except from the wound edges. Dressings for the wound healing role to protect the wound, exude extra body fluids from the wound area, decontaminate the exogenous micro-organism improve the appearance and sometimes speed up the healing technique. For these functions, a wound dressing material should provide a physical barrier to a wound, but be permeable to moisture and oxygen. For a full thickness dermal injury, when an "artificial dermal layer" adhesion and integration consisting of a 3D tissue scaffold with well cultured dermal fibroblasts will aid there-epithelialization. Nanofiber membrane is a good wound dressing candidate because of its unique properties like:[4]

- the porous membrane structure
- well interconnected pores

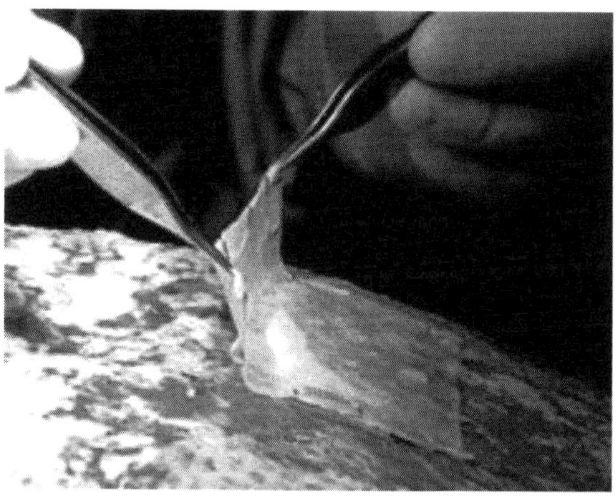

FIGURE 11-16 Nanofiber mats used for medical dressing.

They are important for exuding fluid from the wound. The small pores and high specific surface area not only inhibit the exogenous micro-organism invasions but also assist the control of fluid drainage. In addition, the electrospinning provides a simple path to add drugs into the nanofibers for any possible medical treatment and antibacterial purposes.[4]

For wound healing, an ideal dressing should have certain features:

1. Haemostatic ability
2. Efficiency as bacterial barrier
3. Absorption ability of excess exudates (wound fluid or pus)
4. Suitable water vapor transmission rate
5. Enough gaseous exchange ability
6. Ability to conform to the contour of the wound area
7. Functional adhesion
8. Painless to patient
9. Ease of removal
10. Low cost

Current efforts using nanofibrous membranes as a medical dressing are still in its early childhood, but electrospun materials meet most of the needs outlined for wound-healing polymer. Because their micro fibrous and nanofibrous provide the nonwoven textile with desirable properties.[40] Polymer nanofibers can also be utilized for the treatment of wounds or burns of a human skin, as well as designed for hemostatic devices with some unique characteristics. Fine fibers of biodegradable polymers can spray/spun on to the injured location of the skin to make a fibrous mat dressing. They let wounds heal by encouraging forming a normal skin development and remove form scar tissue, which would occur in a traditional treatment. Nonwoven nanofibrous membrane mats for wound dressing usually have pore sizes ranging from 500 nm to 1 mm, small enough to protect the wound from bacterial penetration by aerosol particle capturing mechanisms. High surface area of 5–100 m^2/g is efficient for fluid absorption and dermal delivery.[15,42] The electrospun nanofibers have been utilized in treating wounds or burns of human skin because of their high porosity which allows gas exchange and a fibrous structure that protects wounds from infection and dehydration. Nonwoven electrospun nanofibrous membranes for wound dressing usually have pore sizes in the range of 500 to 1,000 mm which is low enough to protect the wound from bacterial penetration. High surface area of electrospun nanofibers is efficient for fluid absorption and dermal delivery. Chong invented a composite containing a semi-permeable barrier and a scaffold filter layer of skin cells in wound healing by electrospinning.[1] Electrospinning could create scaffold with more homogeneity besides meeting other needs like oxygen permeation and protection of wound from infection and dehydration for use as a wound-dressing materials. Many other synthetic and natural polymers, like carboxyethyl, chitosan or PVA, collagen or chitosan, silk fibroin, have been electrician to advise them for wound-dressing applications.[11]

Release Control

Controlled release is an effective technique of delivering drugs in medical therapy. It can balance these features:

1. The delivery kinetics
2. Minimize the toxicity
3. Side effects
4. Improve patient convenience

In a controlled release system, the active substance is loaded into a carrier or device first, and then releases at a predictable rate in vivo when governed by an injected or noninjected route. As a potential drug delivery carrier, electrospun nanofibers have showed many advantages. The drug loading is easy to implement by electrospinning technique, and the high applied voltage used in the electrospinning technique had little influence on the drug activity. The high specific surface area and short diffusion passage length give the nanofiber drug system higher overall release rate than the bulk material (e.g., film). The release profile can be finely controlled by modulating of nanofiber morphology, porosity and composition. Nanofibers for drug release systems mainly come from biodegradable polymers, such as PLA, PCL, PDLA, PLLA, PLGA, and hydrophilic polymers such as PVA, PEG, and PEO. Nonbiodegradable polymers, such as PEU, were likewise found out.[4] Nanofiber systems for the release of drugs are needed to fill diverse roles. The mattress should be capable to protect the compound from decomposition and should allow for controlled release in the targeted tissue, over a needed period of time at a constant release rate.[13] Drug release and tissue engineering are closely related regions. Sometimes release of therapeutic causes can increase the efficiency of tissue engineering. Various nanostructured materials is applicable in tissue engineering. Electrospun fiber mats provide the advantage of increased drug release compared to roll-films because of the increased surface area.[11,35]

Drug Delivery and Pharmaceutical Composition

Delivery of drug or pharmaceuticals to patients in the most physiologically acceptable manner has always been an important concern in medicine. In general, the smaller the dimensions of the drug and the coating material wanted to encapsulate the drug, the better the drug to be assimilated by human being. Drug delivery with polymer nanofibers is based on the rule that the dissolution rate of a particulate drug increases with increasing surface area of both the drug and the similar carrier if needed. As the drug and carrier materials can be mixed for electrospinning of nanofibers, the likely modes of the drug in the resulting nanostructure products are as follows:

- Drug as particles attached to the surface of the carrier which is in the form of nanofibers.
- Both drug and carrier are nanofiber-form, therefore the product will be the two kinds of nanofibers interlaced together.
- The blend of drug and carrier materials integrated into one fiber containing both sections.
- The carrier material is electrospun into a tubular frame in which the drug particles are encapsulated.

However, as the drug delivery in the form of nanofibers is still in the early stage exploration, a real delivery mode after production and efficiency has yet to be determined in the future.[51] Drug delivery with electrospun nanofibers is based along the principle that drug releasing rate increases with increasing surface area of both the drug and the similar carrier used. The increased surface area of drug improved the bioavailability of the poor water-soluble drug. Various drugs such as avandia, eprosartan, carvedilol, hydrochloridethiazide, aspirin, naproxen, nifedipine, indomethacin, and ketoprofen were entrapped into PVP to form pharmaceutical compositions which provided controllable releasing. Not only synthetic polymers, but also natural polymers can be applied for modeling drug delivery system.[5] Controlled drug release over a definite period of time is possible with biocompatible delivery matrices of polymers and biodegradable polymers. They are mostly used as drug delivery systems to deliver therapeutic agents because they can be well designed for programmed distribution in a controlled fashion. Nanofiber mats are applied as drug carriers in drug delivery system because of their high functional characteristics. The drug delivery system relies on the rule that the dissolution rate of a particulate drug increases with increasing surface area of both the drug and the similar carrier. Importantly, the large surface area associated with nano spun fabrics allows for quick and efficient solvent evaporation, which provides the incorporated drug limited time to recrystallize which favors forming amorphous dispersions or solid solutions. Depending on the polymer carrier used, the release of pharmaceutical dosage can be designed as rapid, immediate, delayed, or varied dissolution. Many researchers successfully encapsulate drugs within electrospun fibers by mixing the drugs in the polymer solution to be electrospun. Various solutions containing low molecular weight drugs have been electrospun, including lipophilic drugs such as ibuprofen, cefazolin, rifampin, paclitaxel, and Itraconazole and hydrophilic drugs such as mefoxin and tetracycline hydrochloride. However, they have encapsulated proteins in electrospun polymer fibers. Besides the normal electrospinning process, another path to develop drug-loaded polymer nanofibers for controlling drug release is to use coaxial electrospinning and research has successfully encapsulated two kinds of medicinal pure drugs through this process.[40] Electrospinning affords great flexibility in selecting materials for drug delivery applications. Either biodegradable or nondegradable materials can be utilized to control whether drug release occurs by diffusion alone or diffusion and scaffold degradation. Also, because of the flexibility in material selection many drugs can be delivered including

- Antibiotics
- Anticancer drugs
- Proteins
- DNA

Using the various electrospinning techniques, many different drug loading methods can also be applied:

1. Coatings
2. Embedded drug
3. Encapsulated drug (coaxial and emulsion electrospinning)

However, as the drug delivery in the form of nanofibers is still in the early stage exploration, a real delivery mode after production and efficiency has yet to be found in the future.[41]

11.6.2 COSMETICS

The current skin masks applied as topical creams, lotions or ointments. They may include dusts or liquid sprays and more likely than fibrous materials to migrate into sensitive areas of the body, such as the nose and eyes where the skin mask is being utilized to the face. Electrospun polymer nanofibers have been tried as a cosmetic skin care mask for treating skin healing, skin cleaning, or other therapeutic or medical properties with or without various additives. This nanofibrous skin mask with small interstices and high surface area can make easy far greater utilization and speed up the rate of transfer of the additives to the skin for the fullest potential of the additive. The cosmetic skin mask from the electrospun nanofibers can be applied gently and painlessly as well as directly to the three-dimensional topography of the skin to provide healing or cure treatment to the skin.[51] Electrospun nanofibers have been aimed for use in cosmetic cares such as treating skin healing and skin cleaning with or without various additives in recent years. Despite the growth in the number of electrospun polymer nanofiber publications in the recent years, there is a rare work, including scientific papers and patents, in the cosmetic field about the use of electrospun nanofibers. Developing nanofibers in this field have been focused on skin treatment applications, such as care mask, skin healing, and skin cleaning, with active agents (cosmetics) with controlled release from time to time. The cosmetic application be included in the biomedicine application, which admits the drug delivery system employing the active agents used in cosmetics, body care supplements. Therefore, the cosmetic and drug delivery are closely interrelated areas. The electrospun nanofibers provide the advantage of the increasing drug release when compared to cast-films because of the increased surface area. Besides, it is agreeable in processing different polymers such as natural, synthetic and blends, according to their solubility or melting point. It is significant that although most of the researches in polymeric nanofiber by electrospinning consider the technique simple, cost-effective, and easily scalable from laboratory to commercial production, only a limited number of companies have commercially performed electrospun fibers.[5]

11.6.3 ENERGY AND ELECTRONICS AND CHEMICAL ENGINEERING

The demand for energy use of goods and services has been increasing every year throughout the world. However, it was reported the estimated reserve amounts of petroleum and natural gas in the world are only 41 years and 67 years, respectively. To solve this problem, new, clean, renewable and sustainable energies have to be ground and used to replace the current nonsustainable energies. Wind generator, solar power generator, hydrogen battery, and polymer battery are among the most popular alternatives to produce new energies. In recent years, electrospun nanofibers have presented their potential in these applications:

- Super capacitors
- Lithium cells
- Fuel cells
- Solar cells
- Transistors

Further, electrospun nanofibers with electrical and electrooptical have also got much interest recently because of their potential applications in creating nanoscale electronic and optoelectronic devices.[5]

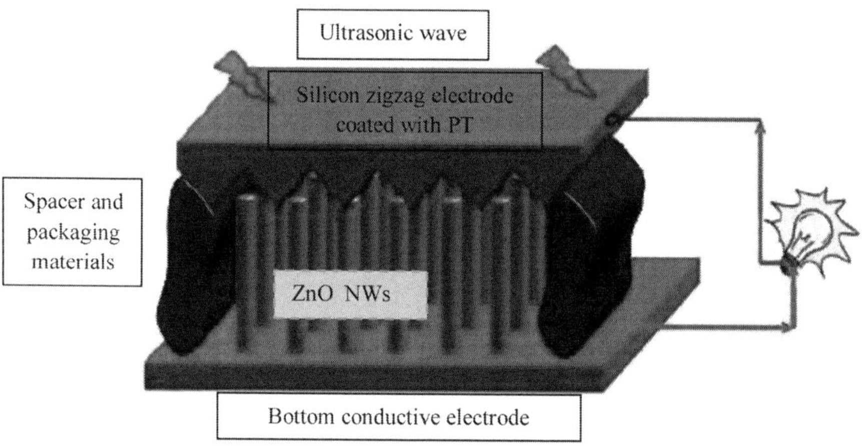

FIGURE 11-17 ZnO nanofibers in energy application.

11.6.3.1. FILTRATION APPLICATION

Filtration is necessary in many engineering fields. It was estimated that future filtration market would be up to US 700 billion US dollars by the year 2020.[1,15,42]

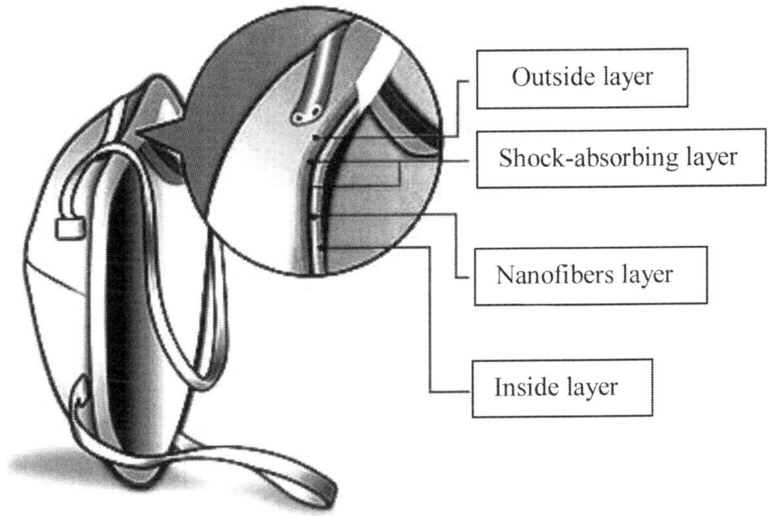

FIGURE 11-18 Applications of nanofibers in filtration.

Fibrous materials used for filter media provide advantages of high filtration efficiency and low air resistance. Filtration efficiency, which is closely related with the fiber fineness, is one of the most important concerns for the filter performance. One direct way of developing high efficient and effective filter media are by using nanometer-sized fibers in the filter structure.[15] With outstanding of polymeric nanofibers properties, such as high specific surface area, high porosity, and excellent surface adhesion, they are suited to be made into filtering media for filtering out particles in the sub-micron range. Also, this filter system could be used for processing waste water containing active sludge.[1,9] Electrospun fibers are being widely studied for aerosol filtration, air cleaning applications in industry and for particle collection in clean rooms. The advantage of using electrospun fibers in the filtration media is the fiber diameters that can be easily controlled and can produce an impact in high efficiency particulate air filtrations (HEPA).[9] The filtration efficiency is commonly influenced by these parameters:[4]

- The filter physical structure like
  - Fiber fineness
  - Matrix structure
  - Thickness
  - Pore size

    - Fiber surface electronic properties
    - Its surface chemical characteristic
- Surface free energy

The particle collecting capability is also associated with the size range of particles being collected. Besides the filtration efficiency, other properties such as pressure drop and flux resistance are also important factors to be assessed for a filter media.[4] Filter efficiency increases linearly with the decrease of thickness of filter membrane and applied pressure increase.[40]

11.6.4 REINFORCED COMPOSITES/ REINFORCEMENT

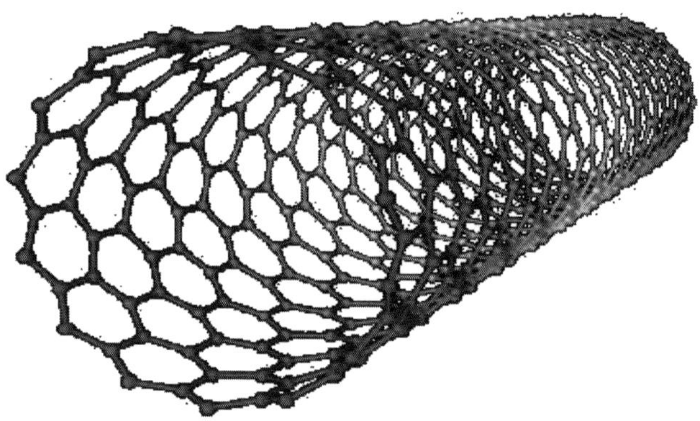

FIGURE 11-19 Carbone nanotubes composite.

Although electrospun fiber reinforced composites have significant potential for the development of high intensity or high toughness materials and materials with good thermal and electrical conductivity, few studies have found out the use of electrospun fibers in composites. Traditional reinforcements in polymer matrices can create stress concentration sites because of their irregular shapes and cracks spread by burning through the fillers or travelling up, down, and around the particles. However, electrospun fibers have various advantages over traditional fillers. The reinforcing effects of fibers are influenced essentially by fiber size. Smaller size fibers give more efficient support. Fibers with finer diameters have a preferential orientation of polymer chains along the fiber axis. The orientation of macromolecules in the fibers improves with decreasing in diameter, making finer diameter fibers strong. Therefore, the use of nanometer-sized fibers can significantly raise the mechanical integrity of polymer matrix compared to micron-sized fibers. However, the high percentage of porosity and irregular pores between fibers can contribute to an interpenetrated structure when spread in the matrix, which also improves the mechanical strength because of the interlocking mechanism. These characteristic features of nanofibers enable the transfer of applied stress to the fiber-matrix in a more serious fashion than most of the commonly used filler materials. Current issues related to the use of electrospun nanofibers as reinforcement materials are the control of dispersion and orientation of the fibers in the polymer matrix. To achieve better reinforcement, electrospun nanofibers may require to be collected as an aligned yarn instead of a randomly distributed felt, so the post-electrospinning stretching process could be applied to further improve the mechanical properties. Further, if the crack growth is transverse to the fiber orientation, the crack toughness of the composite can be optimized. Therefore, the interfacial adhesion between fibers and matrix material needs to be controlled in such a way that the fibers are deflecting the cracks by fiber-matrix interface debonding and fiber pullout. The interfacial adhesion should not be excessively strong or too weak. Ideal control can only be obtained by careful selective fiber surface treatment. Spreading electrospun mats in the matrix can be improved by cutting down the fibers to shorter fragments. This can be accomplished, if the electrospun fibers are collected as aligned bundles (instead of nonwoven network), which can then be optically or mechanically trimmed to get fiber fragments of several 100 nm in length.[9] Early studies on electrospun nanofibers also included reinforcement of polymers. As electrospun nanofiber mats have a large specific surface area and an irregular pore structure, mechanical interlocking among the nanofibers should occur.[4] One of the most significant applications of traditional fibers, especially engineering fibers such as carbon, glass, and Kevlar fibers, is to be used as reinforcements in composite developments. With these reinforcements, the composite materials can provide superior structural properties such as high modulus and strength to weight ratios, which cannot be attained by other engineered monolithic materials alone. Nanofibers will also eventually find important applications in making nano composites.This is because nanofibers can have even better mechanical properties than microfibers of the same materials, and therefore the superior structural properties of nanocomposites can be anticipated. However, nanofiber reinforced composites may control some extra merits which cannot be shared by traditional (microfiber) composites. For instance, if there is a difference in refractive indices between fiber and ground substance, the resulting composite becomes opaque or nontransparent because of light scattering. This limit, however, can be avoided when the fiber diameters become significantly smaller than the wavelength of visible illumination.[42]

11.6.5 DEFENCE AND SECURITY

Military, firefighter, law enforcement, and medical personal need high-level protection in many environments ranging from combat to urban, agricultural, and industrial, when dealing with chemical and biological threats like[17]

- Nerve agents
- Mustard gas
- Blood agents such as cyanides
- Biological toxins such as bacterial spores, viruses, and rickettsiae

Nanostructures with their minuscule size, large surface area, and light weight will improve, by orders of magnitude, our capability to[17]

- Detect chemical and biological warfare agents with sensitivity and selectivity
- Protect through filtration and destructive decomposition of harmful toxins
- Provide site-specific naturally prophylaxis

Polymer nanofibers are considered as excellent membrane materials owing to their lightweight, high surface area, and breathable (porous) nature. The high sensitivity of nanofibers toward warfare agents makes them excellent candidates as sensing of chemical and biological toxins in concentration levels of parts per billion. Governments across the globe are investing in strengthening the protection levels offered to soldiers in the battlefield. Various methods of varying nanofiber surfaces to improve their capture and decontamination capacity of warfare agents are under investigation. Nanofiber membranes may be employed to replace the activated charcoal in adsorbing toxins from the atmosphere. Active reagents can be planted in the nanofiber membrane by chemical functionalization, post-spinning variation, or through using nano particle polymer composites. There are many avenues for future research in nanofibers from the defense perspective. As well as serving protection and decontamination roles, nanofiber membranes will also suffer to provide the durability, wash ability, resistance to intrusion of all liquids, and tear strength needed of battledress fabrics.[17]

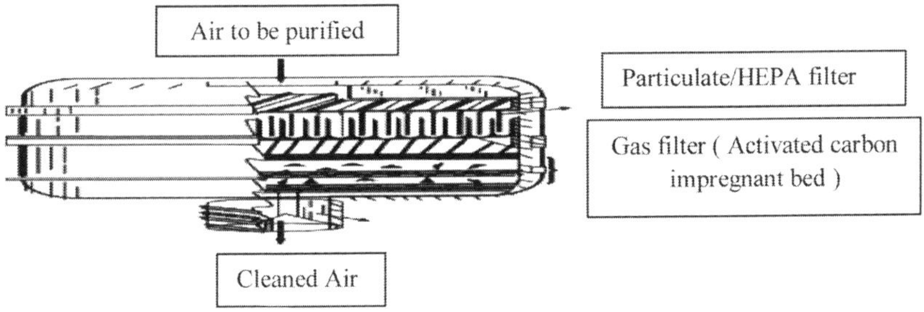

FIGURE 11-20 Cross-sectioning of a facemask canister used for protection from chemical and biological warfare agents.

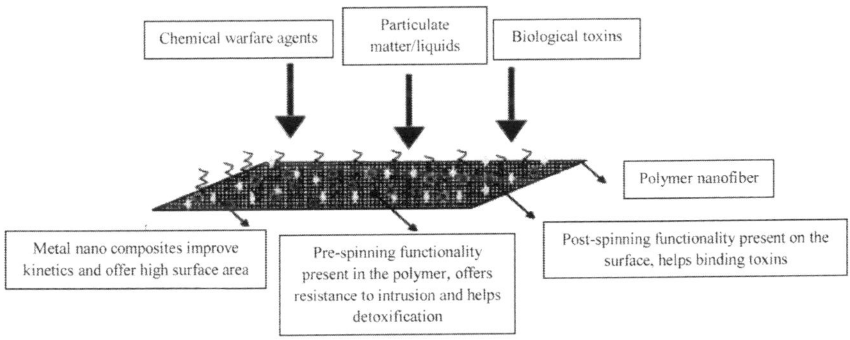

FIGURE 11-21 Incorporating functional groups into a polymer nanofiber mesh.

11.6.5.1 PROTECTIVE CLOTHING

The protective clothing in the military is largely expected to help increase the suitability, sustainability, and combat effectiveness of the individual soldier system against extreme climate, ballistics, and NBC warfare. In peace ages, breathing apparatus and protective clothing with the particular role of against chemical warfare agents such as sarin, soman, tabun become a special concern for combatants in conflicts and civilian populations in terrorist attacks. Current protective clothing containing charcoal absorbents has its terminal points for water permeability, extra weight-imposed to the article of clothing. A lightweight and breathable fabric, which is permeable to both air and water vapor, insoluble in all solvents and reactive with nerve gases and other deadly chemical agents, is worthy. Because of their large surface area, nanofiber fabrics are neutralizing chemical agents and without impedance of the air and water vapor permeability to the clothing. Electrospinning results in nanofibers lay down in a layer that has high porosity but small pore size, offering good resistance to penetrating chemical harms agents in aerosol form. Preliminary investigations indicate that compared to conventional textiles the electrospun nanofibers present both small impedance to moisture vapor diffusion and efficiency in trapping aerosol particles, as well as show strong promises as ideal protective clothing. Conductive nanofibers are expected to be utilized in fabricating tiny electronic or machines such as Schottky junctions, sensors, and actuators. Conduct (of electrical, ionic, and photoelectric) membranes also have potential for applications including electrostatic dissipation, corrosion protection, electromagnetic interference shielding, photovoltaic device, and so on.[51].

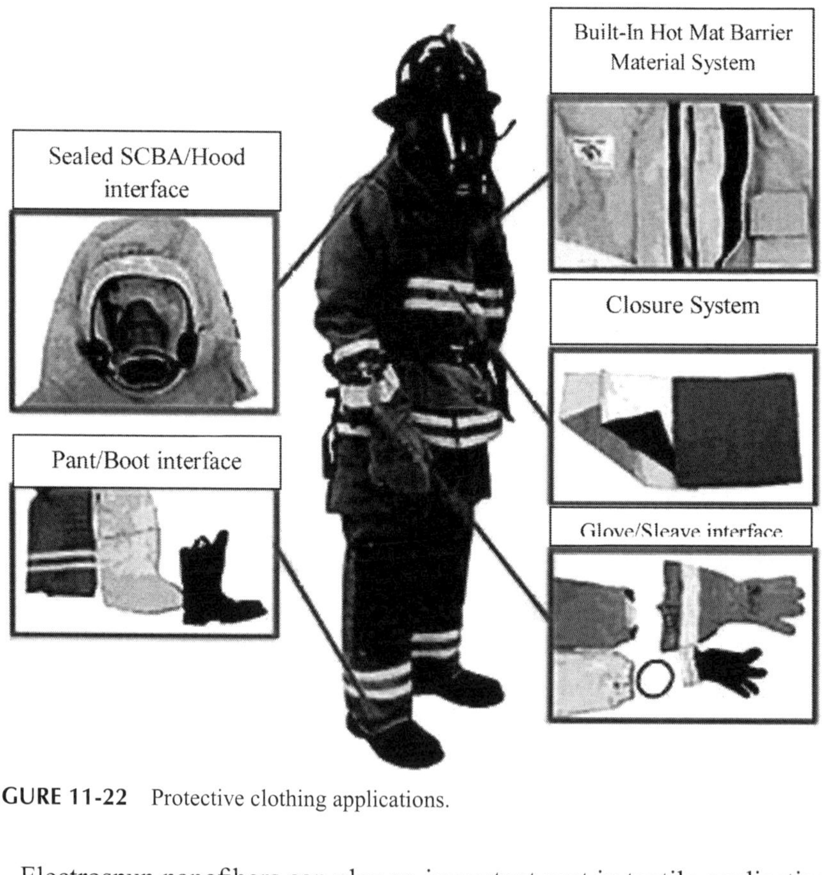

FIGURE 11-22 Protective clothing applications.

Electrospun nanofibers can play an important part in textile applications as protective clothing and other functional fabric materials. The electrospun nanofibrous membranes are capable of neutralizing chemical agents without impedance of the air and water vapor permeability to the clothing because of their high specific surface area and high porosity but small pore size. Preliminary investigations suggest the electrospun nanofibers control both minimal impedance to moisture vapor diffusion and efficiency in trapping aerosol particles compared with conventional textiles. Smith prepared a fabric comprising electrospun PEI nanofibers as lightweight protective clothing which was captured and neutralizing chemical warfare agents. This formed fabric also could be used in protective breathing apparatuses because PEI provides multiple amine sites for the nucleophilic decomposition of mustard gases and fluorophosphates nerve gases. A protective mask was constructed by attaching PC/PS electrospun nanofibrous layer to one side of a moist fabric composed of cellulose and wool. The diameter of the nanofibers in the protective layer was in the range of 100–10,000 nm. A nonwoven fabric composed of a submicrosized fiber, which receives a PC shell and a polyurethane core, was made by co-axial electrospinning. The resultant fabric combines the filtration efficiency of the PC and the mechanical effectiveness of polyurethane, which is useful in exposure suits and aviation clothing. A water-resistant and air-permeable laminated fabric was manufactured by utilizing hot-melt polyester as dots onto the surface of the electrospun nylon nonwoven fabric.[5] Ideally, protective clothing should have close to essential properties, such as lightweight, breathable fabric, air and water vapor permeability, insoluble in all solvents and improved toxic chemical resistance. Electrospun nanofiber membranes recognized as potential candidates for protective clothing applications for these causes:

- They're lightweight
- Large surface area
- High porosity (breathable nature)
- Great filtration efficiency
- Resistant to penetration of harmful chemical agents in aerosol form
- Their ability to neutralize the chemical agents without impedance of the air
- Water vapor permeability to the clothing

Various methods for the variation of nanofiber surfaces have been examined to improve protection against toxins. One protection method that has been used includes chemical surface variation and attachment of reactive groups such as axioms, cyclodextrins, and chloramines that bind and detoxify warfare agents.[40]

11.7 CONCLUDING REMARKS

Nanotechnology with its unique properties has become successful to interest many scientists all over the world for novel applications in recent years. Various common techniques can be used for preparing polymer nanofibers. Also, special ways such as gelation and bacterial cellulose can be utilized for producing nanofibers. Among these methods, electrospinning has been widely used as a novel technique for generating nanoscale fibers. Therefore, electrospun nanofibers are utilized in a wide range of applications. Also, this work has analyzed the recent advances of this technology in tissue engineering, drug delivery, and so on.

KEYWORDS

- **Electrospinning**
- **Nanofibers**
- **Nanofibers Applications**
- **Nanofibers Production**
- **Nanotechnology**

REFERENCES

1. Lu, P.; Ding, B. *Applications of Electrospun Fibers.* Recent patents on nanotechnology, 2008. **2**(3): p. 169–182.
2. Reneker, D. H., et al. *Electrospinning of Nanofibers from Polymer Solutions and Melts.* Advances in applied mechanics, 2007. **41**: p. 43–346.
3. Vonch, J.; Yarin, A.; Megaridis, C. M. *Electrospinning: A Study in the Formation of Nanofibers.* Journal of Undergraduate Research, 2007. **1**: p. 1–6.
4. Fang, J., et al. *Applications of Electrospun Nanofibers.* Chinese Science Bulletin, 2008. **53**(15): p. 2265–2286.
5. Rafiei, S., et al. *Mathematical Modeling in Electrospinning Process of Nanofibers: A Detailed Review.* Cellulose Chemistry and Technology, 2013. **47**(5–6): p. 323–338.
6. Fang, J.; Wang, X.; Lin, T. *Functional Applications of Electrospun Nanofibers.* Nanofibers—Production, Properties and Functional Applications, 2011: p. 287–326.
7. Karra, S. *Modeling Electrospinning Process and a Numerical Scheme using Lattice Boltzmann Method to Simulate Viscoelastic Fluid Flows.* 2007, Texas A&M University.
8. Angammana, C. J. *A Study of the Effects of Solution and Process Parameters on the Electrospinning Process and Nanofibre Morphology.* 2011, University of Waterloo.

9. Baji, A., et al. *Electrospinning of Polymer Nanofibers: Effects on Oriented Morphology, Structures and Tensile Properties.* Composites Science and Technology, 2010. **70**(5): p. 703–718.
10. Beachley, V.; Wen, X. *Polymer Nanofibrous Structures: Fabrication, Biofunctionalization, and Cell Interactions.* Progress in Polymer Science, 2010. **35**(7): p. 868–892.
11. Agarwal, S.; Wendorff, J. H.; Greiner, A. *Use of Electrospinning Technique for Biomedical Applications.* Polymer, 2008. **49**(26): p. 5603–5621.
12. Fridrikh, S. V., et al., *Controlling the Fiber Diameter during Electrospinning.* Physical Review Letters, 2003. **90**(14): p. 144502–144502.
13. Garg, K.; Bowlin, G. L. *Electrospinning Jets and Nanofibrous Structures.* Biomicrofluidics, 2011. **5**(1): p. 013403-1 - 013403-19.
14. Haghi, A. K. *Electrospun Nanofiber Process Control.* Cellulose Chemistry & Technology, 2010. **44**(9): p. 343–352.
15. Huang, Z. M., et al. *A Review on Polymer Nanofibers by Electrospinning and their Applications in Nanocomposites.* Composites Science and Technology, 2003. **63**(15): p. 2223–2253.
16. Kowalewski, T. A.; Blonski, S.; Barral, S. *Experiments and Modelling of Electrospinning Process.* Technical Sciences, 2005. **53**(4): p. 385–394.
17. Ramakrishna, S., et al. *Electrospun Nanofibers: Solving Global Issues.* Materials Today, 2006. **9**(3): p. 40–50.
18. Reneker, D. H.; Chun, I. *Nanometre Diameter Fibres of Polymer, Produced by Electrospinning.* Nanotechnology, 1996. **7**(3): p. 216–223.
19. Wang, H. S.; Fu, G. D. Li, X. S. *Functional Polymeric Nanofibers from Electrospinning.* Recent Patents on Nanotechnology, 2009. **3**(1): p. 21–31.
20. Zhang, C.; Ding, X.; Wu, S. *The Microstructure Characterization and the Mechanical Properties of Electrospun Polyacrylonitrile-Based Nanofibers.* Nanofibers—Production, Properties and Functional Applications, 2011: p. 177–196.
21. Zhang, S. *Mechanical and Physical Properties of Electrospun Nanofibers.* 2009: p. 1–83.
22. Zhou, H. *Electrospun Fibers from Both Solution and Melt: Processing, Structure and Property.* 2007, Cornell University.
23. Zanin, M. H. A.; Cerize, N. N. P.; Oliveira, A. M. de. *Production of Nanofibers by Electrospinning Technology: Overview and Application in Cosmetics. Nanocosmetics and Nanomedicines.* 2011, Springer. p. 311–332.
24. Khan, N. *Applications of Electrospun Nanofibers in the Biomedical Field.* Studies by Undergraduate Researchers at Guelph, 2012. **5**(2): p. 63–73.
25. Ramakrishna, S.; Fujihara, K. *An Introduction to Electrospinning and Nanofibers.* 2005. 1–383.
26. Zeng, Y., et al. *Numerical Simulation of Whipping Process in Electrospinning.* in *WSEAS International Conference. Proceedings. Mathematics and Computers in Science and Engineering.* 2009: World Scientific and Engineering Academy and Society.
27. Ciechańska, D. *Multifunctional Bacterial Cellulose/Chitosan Composite Materials for Medical Applications.* Fibres & Textiles in Eastern Europe, 2004. **12**(4): p. 69–72.
28. Stanger, J. J., et al. *Effect of Charge Density on the Taylor Cone in Electrospinning.* International Journal of Modern Physics B, 2009. **23**(06): p. 254–268.
29. Chronakis, I. S. *Novel Nanocomposites and Nanoceramics Based on Polymer Nanofibers Using Electrospinning Process—A Review.* Journal of Materials Processing Technology, 2005. **167**(2): p. 283–293.

30. Wu, Y., et al. *Controlling Stability of the Electrospun Fiber by Magnetic Field.* Chaos, Solitons & Fractals, 2007. **32**(1): p. 5–7.
31. Li, W. J., et al. *Electrospun Nanofibrous Structure: A Novel Scaffold for Tissue Engineering.* Journal of biomedical materials research, 2002. **60**(4): p. 613–621.
32. Brown, E. E.; Laborie, M. P. G. *Bioengineering Bacterial Cellulose/Poly(Ethylene Oxide) Nanocomposites.* Biomacromolecules, 2007. **8**(10): p. 3074–3081.
33. De, V. S., et al., *The Effect of Temperature and Humidity on Electrospinning.* Journal of Materials Science, 2009. **44**(5): p. 1357–1362.
34. Tao, J.; Shivkumar, S. *Molecular Weight Dependent Structural Regimes during the Electrospinning of PVA.* Materials Letters, 2007. **61**(11): p. 2325–2328.
35. Zong, X., et al. *Structure and Process Relationship of Electrospun Bioabsorbable Nanofiber Membranes.* Polymer, 2002. **43**(16): p. 4403–4412.
36. Lyons, J.; Li, C.; Ko, F. *Melt-Electrospinning Part I: Processing Parameters and Geometric Properties.* Polymer, 2004. **45**(22): p. 7597–7603.
37. Reneker, D. H.; Yarin, A. L. *Electrospinning Jets and Polymer Nanofibers.* Polymer, 2008. **49**(10): p. 2387–2425.
38. Bognitzki, M., et al. *Nanostructured Fibers via Electrospinning.* Advanced Materials, 2001. **13**(1): p. 70–72.
39. Deitzel, J. M., et al. *The Effect of Processing Variables on the Morphology of Electrospun Nanofibers and Textiles.* Polymer, 2001. **42**(1): p. 261–272.
40. Bhardwaj, N.; Kundu, S. C. *Electrospinning: A Fascinating Fiber Fabrication Technique.* Biotechnology Advances, 2010. **28**(3): p. 325–347.
41. Sill, T. J.; von Recum, H. A. *Electrospinning: Applications in Drug Delivery and Tissue Engineering.* Biomaterials, 2008. **29**(13): p. 1989–2006.
42. Patan, A.K., et al. *Nanofibers—A New Trend in Nano Drug Delivery Systems.* International Journal of Pharmaceutical Research & Analysis, 2013. **3**: p. 47–55.
43. Rutledge, G. C.; Fridrikh, S. V. *Formation of Fibers by Electrospinning.* Advanced Drug Delivery Reviews, 2007. **59**(14): p. 1384–1391.
44. Sawicka, K. M.; Gouma, P. *Electrospun Composite Nanofibers for Functional Applications.* Journal of Nanoparticle Research, 2006. **8**(6): p. 769–781.
45. Yousefzadeh, M., et al. *A Note on the 3D Structural Design of Electrospun Nanofibers.* Journal of Engineered Fabrics & Fibers (JEFF), 2012. **7**(2): p. 17–23.
46. Yarin, A.L.; Koombhongse, S; Reneker, D. H. *Bending Instability in Electrospinning of Nanofibers.* Journal of Applied Physics, 2001. **89**(5): p. 3018–3026.
47. Keun, S. W., et al. *Effect of PH on Electrospinning of Poly(Vinyl Alcohol).* Materials Letters, 2005. **59**(12): p. 1571–1575.
48. Feng, J. J. *The Stretching of an Electrified Non-Newtonian Jet: A Model for Electrospinning.* Physics of Fluids (1994–present), 2002. **14**(11): p. 3912–3926.
49. Maleki, M.; Latifi, M.; Amani-Tehran, M. *Optimizing Electrospinning Parameters for Finest Diameter of Nano Fibers.* World Academy of Science, Engineering and Technology, 2010. **40**: p. 389–392.
50. Thompson, C. J. *An Analysis of Variable Effects on a Theoretical Model of the Electrospin Process for Making Nanofibers.* 2006, University of Akron.
51. Luo, C. J.; Nangrejo, M.; Edirisinghe, M. *A Novel Method of Selecting Solvents for Polymer Electrospinning.* Polymer, 2010. **51**(7): p. 1654–1662.
52. Frenot, A.; Chronakis, I. S. *Polymer Nanofibers Assembled by Electrospinning.* Current Opinion in Colloid & Interface Science, 2003. **8**(1): p. 64–75.

CHAPTER 12

FIBER FORMATION DURING ELECTROSPINNING PROCESS: AN ENGINEERING INSIGHT

S. PORESKANDAR, F. RAEISI, SH. MAGHSOODLOU, and A. K. HAGHI

Department of Textile Engineering, Faculty of Engineering, University of Guilan, P.O. Box: 3756, Rasht, Iran

CONTENTS

ABSTRACT

Electrospinning is comparatively a simple method of producing nanofibers. Nanofibers produced by this method are widely utilized for varied applications like drug delivery and tissue scaffolding. For producing nanofibers in this method, an external electrostatic field should be acted on polymer solution/melt to produce a charged polymer jet. The most significant part of electrospinning is how to control the process. Therefore, we must know about the behavior of nanofiber jets during the electrospinning process. The aim of this chapter is to review various steps of fiber formation.

Symptoms	Definitions
V_c	Critical voltage
H	Distance between the capillary exit and the ground
L	Length of the capillary
R	Radius
γ	Surface tension of the liquid/ solution
nm	Nanometer

12.1 A BRIEF INSIGHT INTO ELECTROSPUN POLYMER NANOFIBERS

The polymer nanofibers have been associated with nano-scale diameters. Because of the unique properties like high surface area and high porosity, they prove to be useful in many important and varied applications such as tissue engineering scaffolds.[1,2]

Researchers and industries can produce them from a variation range of polymers.[3–5] Nowadays electrospinning of these materials has gained much attention mainly because of its being the cheapest and simplest method.[6–10]

In addition, the continuous growth in the number of publications and patents in this field has gained significance in recent years.[11–13] The result of it is shown in Figure12-1.

Electrospinning process requires an interaction between two forces shown in Figure 12-2. In this procedure, the polymer solution receives electrical charges from a high-voltage supply. These charges are carried by ions through the fluid. If the repulsive force between the charged ions overcomes the fluid surface tension, an electrified liquid jet could be formed and elongated toward the collector. With evaporation of the solvent, nanofibers jets are collected on the surface of the screen.[14,15] A schematic of electrospinning is shown in Figure 12-3.

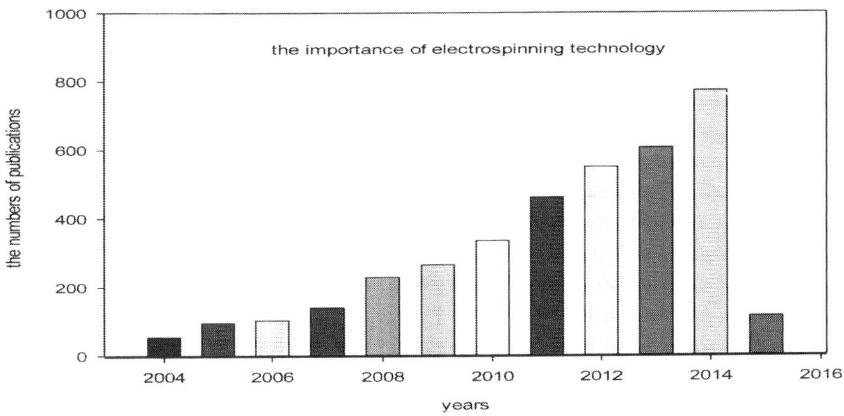

FIGURE 12-1 Numbers of publications about electrospinning.

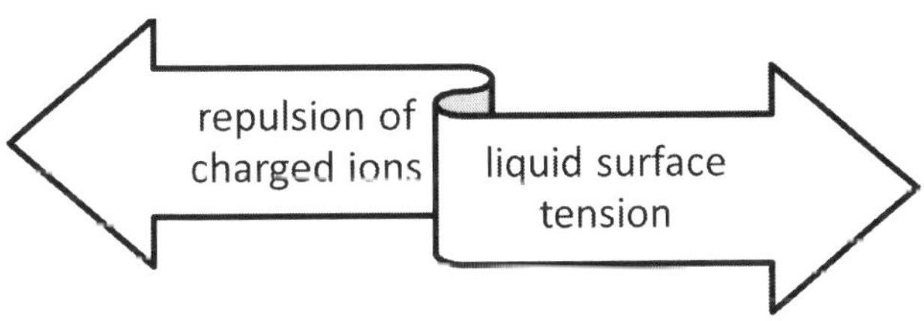

FIGURE 12-2 Forces affected electrospinning process.

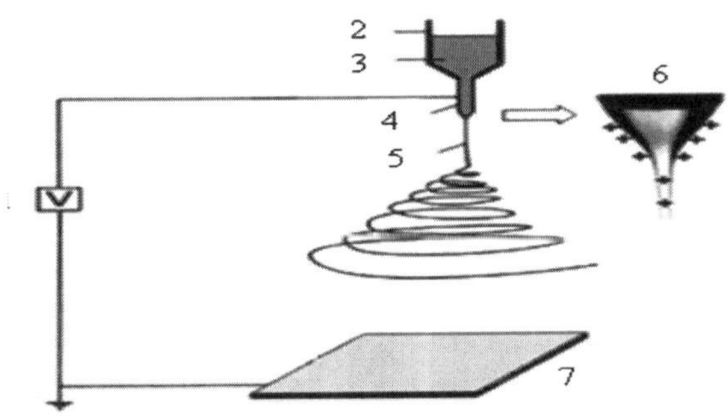

FIGURE 12-3 Sections of electrospinning process: (1) High voltage (2) Polymer solution (3) Syringe (4) Needle (5) Whipping instability (6) Taylor cone (7) Collector.

The most significant challenge in this process is to attain uniform nanofibers consistently and reproducibly.[11,16–18] Depending on several solution parameters, different results can be obtained using the same polymer and electrospinning setup.[19]

Factors that are studied to have a primary effect on the formation of uniform fibers are process parameters, environmental parameters, and solution parameters.[11,20–23] These factors are shown in the Figure 12-4. In addition, many researcher studies effect of these parameters on final fibers. A summary of the most important parameters are displayed in Table 12-1.

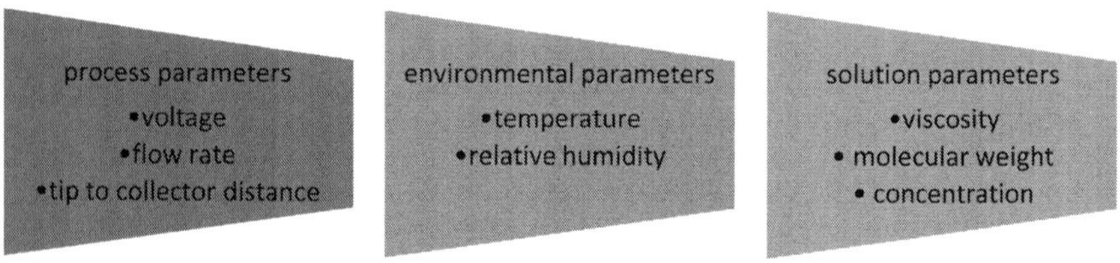

FIGURE 12-4 Parameters affect the morphology and size of electrospun nanofibers.

The mechanics of this process deserve a specific attention and necessary to predictive tools or a way for better understanding and optimization and controlling process.[32] Also, many researchers have studied the behavior of the jet during formation.[33] For this case, we must to know more about the parts of electrospinning process. In next section, we investigate this process in detail.

12.2 INVESTIGATED FORMATION PARTS OF ELECTROSPINNING PROCESS

As mention earlier, many parameters affect electrospun. Alike, this process is distinguished by four main sections.[34] Therefore, we can separate the process into four sections shown in Figure 12-5. A description of these stages follows in continues.

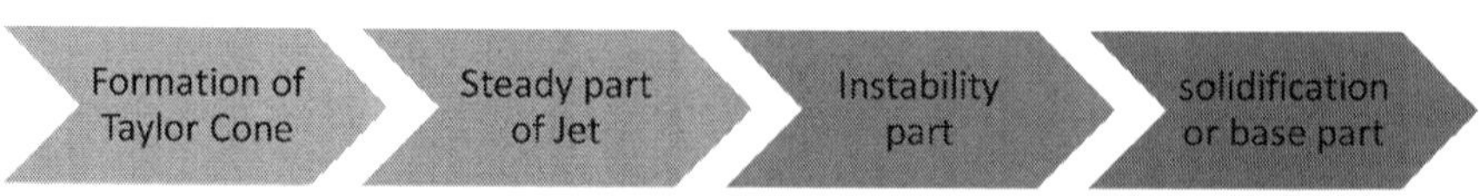

FIGURE 12-5 Different parts of electrospinning process.

TABLE 12-1 Effects of some parameters on electrospinning nanofibers in researchers' studies

	Parameters	Year	Name of Researcher	Effect	References
Process Parameters	Needle to collector distance	1999	Fong et al	Inversely proportional to bead formation density Inverse to the electric field strength	[24]
		2003	Gupta & Wilkes	Inversely proportional to bead formation density Inversely proportional to fiber diameter	[25]
		2004	Theron et al	Exponentially inverse to the volume charge density Inverse to the electric field strength	[26]
	Flow rate	2004	Theron et al	Directly proportional to the electric current Inversely related to surface charge density Inversely related to volume charge density	[26]
		2005	Sawicka et al	Directly proportional to the fiber diameter	[27]
	Voltage	2001	Deitzel et al	Direct effect on bead formation	[28]
		2003	Gupta & Wilkes	Inversely related to fiber diameter	[25]
		2004	Theron et al	Inversely proportional to surface charge density	[26]
		2004	Kessick et al	AC potential improved fiber uniformity	[29]
Solution Parameters	Concentration of polymer	2001	Deitzel et al	Power law relation to the fiber diameter	[28]
		2002	Demir et al	Cube of polymer concentration proportional to diameter	[30]
		2003	Gupta & Wilkes	Directly proportional to the fiber diameter	[25]
		2004	Hsu &Shivkumar	Parabolic-upper and lower limit relation to diameter	[31]
	Viscosity	2004	Hsu &Shivkumar	Parabolic relation to diameter and spinning ability	[31]
Environmental Parameters	Temperature	2002	Demir et al	Inversely proportional to viscosity Uniform fibers with less beading	[30]

12.2.1 FIRST STEP: FORMATION OF TAYLOR CONE

An electrospinning solution is usually an ionic solution that contains charged ions. The amounts of positively and negatively charged particles are equal; therefore, the solution is electrically neutral.[14]

When an electrical potential difference is given between needle and collector, a hemispherical surface of the polymeric droplet at the orifice of needle is gradually expanded. When potential comes into a critical value (equal 1), a flow of jet starts formation to drop. Therefore, Taylor's cone is formed. A schematic of these steps are shown in Figure 12-6.

$$V_C^2 = 4\frac{H^2}{L^2}\left(\ln\frac{2L}{R} - \frac{3}{2}\right)(0.117\pi\gamma R) \tag{12-1}$$

Many researchers like Rayleigh, Zeleny, and Taylor, have given their insight into the survey of the behavior of liquid jets. Taylor determined that an angle of 49.3° is required to balance the surface tension and the electrostatic force.[14]

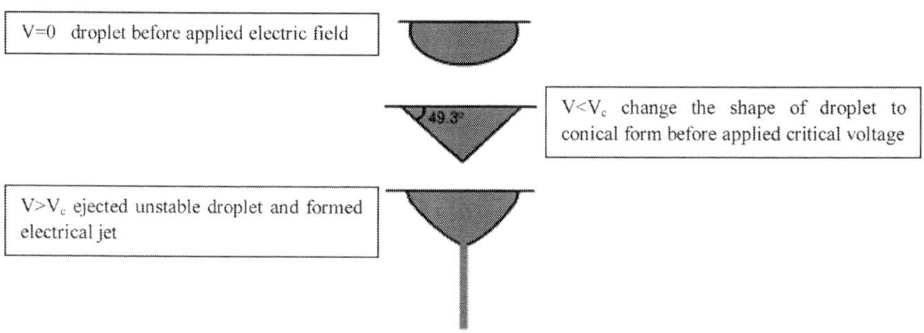

FIGURE 12-6 Changes in the polymer droplet with applied potential.

12.2.2 SECOND STEP: STEADY PART OF JET

As mention before, the jet is initiated from the droplet when the repelling forces of the surface charge overcome the surface tension and viscous forces of the droplet.[7] These repulsive forces between the jet segments will elongate the jet straight in the direction of its axis.[14]

A stable electrospinning jet travels from a polymer solution or melt to a collector. Electricity charges, commonly in the form of ions, tend to act in response to the electrical field that is associated with the potential. The electrical forces which stretch the fiber are resisted by the elongation viscosity of the jet.[12] Figure 12-7 shows stable part of the jet.

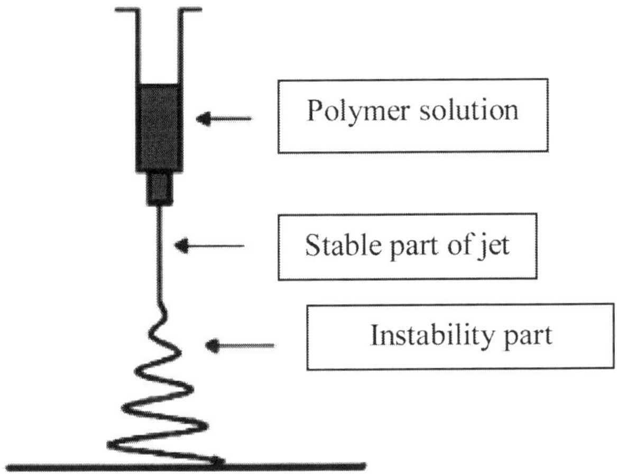

FIGURE 12-7 Instability in electrospinning.

The part of the jet leaving the tip was speeded up as the Coulomb forces; acting along the charges carried with the leading parts drew out the jet and extended it in its axis toward the collector.[33] During the elongation of the electrified liquid jet, the jet surface area increases dramatically.[6]

Therefore, for a thin, stable jet, it is significant to look for a balance of viscosity charge density and surface tension. Surface tension tries to lessen specific surface area, by changing jets into spheres. As the viscosity of a solution is increased, bead size increases, and the shape of the beads becomes more spindle shaped than spherical. As net charge density increases, the beads become a littler and more football shaped as well. Decreasing surface tension makes the beads disappear.[34]

Also, the straight, tapered part of the jet that went forth from the droplet has been mathematically modeled from many points of view. In future, we consider about modeling and simulating process.

12.2.3 THIRD STEP: INSTABILITY PART

After a small distance of stable traveling of the jet, it will start unstable behavior and separate into many fibers.[7,12,14] It is necessary to know that the charge on the fiber expands the jet in the radial directions and to extend it in the axial direction. This jet division occurs several times in rapid sequence and makes many small electrically charged fibers moving toward the collector.[12] In Figure 12-8, instability partof the jet is shown.

Recently, instability in electrospinning has received much attention. Many researchers like Reneker et al., Yarin et al., and Rutledge et al have studied jet instability. It is remarked that the jet followed a series of loops in which the loop diameters get larger.[14] Each instability grows at different rates.[34,35]

These instabilities vary and increase with distance, electrical field, and fiber diameter at different rates depending on the fluid parameters and performing conditions. Also, they influence the size and geometry of the deposited fibers.[36] These instabilities generate a looping and spiraling path and it contributes to the extreme elongation and the acceleration of the electrically charged liquid jet.[6]

12.2.3.1 VARIOUS MAIN SECTIONS OF INSTABILITY PART

As mention earlier, when the jet spirals toward the collector, higher order instabilities reveal themselves and result in spinning distance. These instabilities are separated into sections:[37]

- Bending instability (Figure 12-8a)
- Rayleigh instability (Figure 12-8b)
- Whipping instability (Figure 12-8c)

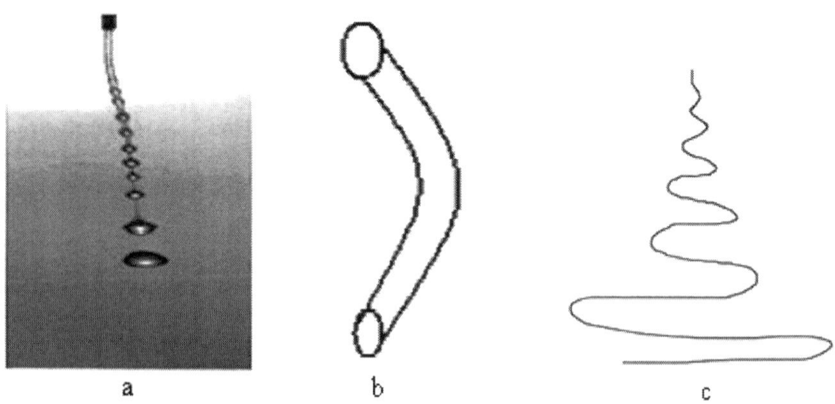

FIGURE 12-8 a) Rayleigh instability b) Bending instability c) Whipping instability.[34,38]

The polymer jet is influenced by these instabilities. These instabilities arise owing to the charge-charge repulsion between the excess charges present in the jet, which encourages the thinning and elongation of the jet. At high electric forces, the jet is dominated by bending (axisymmetric) and whipping instability (nonaxisymmetric), causing the jet to move around, and wave is produced in the jet. At higher electric fields and at enough charge density in the jet, the axisymmetric (i.e., Rayleigh and bending) instabilities are suppressed and the nonaxisymmetric instability is increased.[36] Axisymmetric conducting instabilities can be viewed as a direct competition between the surface charges with the surface tension of the fiber, while the fiber is moving.[34] The suits for the noted bending instability are explained and the technique modeled.[39] In the next part, these instabilities are summarized.

12.2.3.1.1 *Bending instability part*

When the jet moves straight toward the collector, bending instability develops into a series of loops expanding with time (Figure 12-9).[33,40–42]

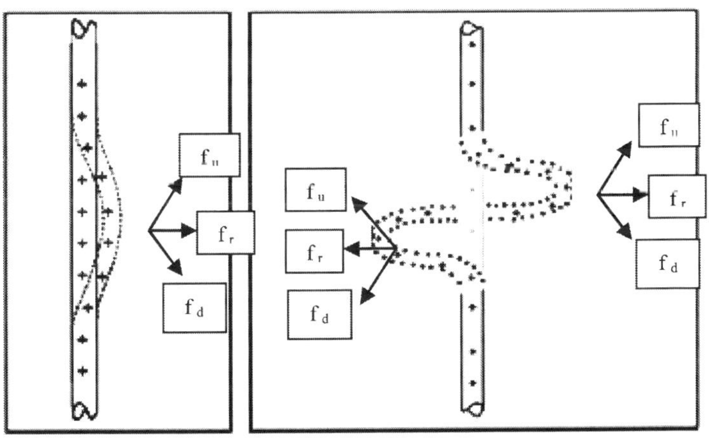

FIGURE 12-9 A part of the electrospun jet with act forces on this part.[33]

The bent part of the path was drawn out, and reduced in diameter. Typically, the electrical bending coil began at a particular distance from the orifice, and the diameter of the turns of the coil grew larger and led toward the collector. The growing bend developed into the first turn of a growing coil. The continuing electrical bending formed a spiral with many turns which expanded in diameter as the jet continued to expand in response to the Coulomb repulsion of the charge. Extending of each section of the electrical force caused by the charge transported by the jet continued. As the diameter of the jet decreased, the path of the jet again became unstable and a new, smaller diameter electrical bending instability developed. A succession of three or smaller diameter bending instabilities was often caught before the jet solidified. On the other hand, jets got fractal-like shapes, and their length increased enormously as their cross-sectional diameter decreased to a fraction of a micrometer. After several turns were made, a new electrical bending instability formed a smaller coil on a turn of the larger spiral. The turns of the smaller coil transformed into an even smaller coil and so forth until the elongation stopped, usually by solidification of the thin jet. Observations of the conical envelope surrounding the coils suggested the repulsive forces between the electrical charges on such a jet caused the loops to continue to expand, by cleaning up the multitude of small coils into the larger coils. The variability in the onset and the behavior of the series of bending instabilities can account for a significant part of the variation in diameter that is often noted in electrospun nanofibers.[33]

12.2.3.1.2 Rayleigh instability part

The first instability, also recognized as the Rayleigh instability is axisymmetric and occurs when the strength of electric field is low or when the viscosity of the solution is below the optimal value. Rayleigh instability is suppressed at high electric fields (higher charge densities) or when using higher concentrations of polymer in the solution [36]. It is almost dependent on the surface tension of the material.[34]

12.2.3.1.3 Whipping instability part

Whipping instability plays a key role in electrospinning for producing nanofibers.[9] It is the condition used for the bending instability of this rapidly moving jet.[14] In most operations of interest, the jet experiences a whipping instability, leading to bending and stretching of the jet, noted as loops of increasing size as the instability develops.[8]

When the electrostatic repulsion overcomes to viscoelasticity forces, the jet path changes to an expanding helix.[7] This instability is a result of small bends in the initial, uniformly charged straight fiber formation. As the fiber bends, the surface charges around the circumference of the jet are no longer uniform and a dipole moment is induced standing to the jet. These dipoles begin a localized torque, in response to each other and the applied electric field. This torque bends the jet. The bends expand as the jet moves further downstream, and then occurs the whipping instability that causes the decreased diameter of the fibers.[34] The whipping jet thins dramatically, while traveling the short distance between the electrodes.[8]

12.2.3.2 VARIOUS OTHER SECTIONS OF INSTABILITY PART

As comfortably as the electrical bending instability, other instabilities of different characteristics were observed:[33]

- Branching
- Formation of physical beads

If the excess charge density on the surface of the jet was high, waves were predicted to take shape along the surface of a cylindrical jet. These waves grew large enough to become unstable, and began branching which grew outward from the main jet. The familiar capillary instability that caused a cylindrical jet of liquid to collapse into separated droplets occurred when the surplus electrical charge carried by the jet was reduced. This structure was solidified to form beaded nanofibers. It is not unusual for beads or branches to occur on the coils produced by the bending instability, but it is rare for beads and branches to occur at the same time along the same section of the jet. Since beads occur when the charge per unit area is small, the branches occur when the charge per unit area is great. Empirical observation showed that modification of a single technique parameter sometimes produced predictable changes in the onset of particular instabilities. When an experimental apparatus was producing nanofiber, it was often possible to accommodate a single technique parameter to note changes in some features of the jet, such as the straight part at which the electrical bending began.[33]

12.2.3.2.1 Branching during electrospinning

Forming branches were viewed more often in more concentrated and more viscous solutions, and in electric fields higher than the minimum area required for producing a single jet. Many branches were observed to get large diameter of the jet, during the electrospinning of polycaprolactone at a concentration of 15% in acetone, in the straight path of the gate or in the first choice of bending instability. Bending and branching began after only a short distance from the tip. Branches that arise from a jet with a large diameter can become long and entangled. Small branches were viewed on fibers electrospun from molten polycaprolactone and on small jets of polyimide electrospun in a vacuum. Interesting branches were noted on jets of polyetherimide in a volatile solvent which tended to form a skin on the jet as it dried. The skin gave way into a ribbon.[33]

12.2.3.2.2 Beaded fibers formation

The capillary instability jet has been cleared since the times of Lord Rayleigh. He recognized that the surface energy of a particular volume of fluid in a cylindrical jet is more eminent than the same volume divided into droplets. The excess electrical charges carried by the jet create a strong elongation flow, and may either stabilize/destabilize the capillary instability, depending on the wavelength of the bead-forming instability. Stretching the entangled molecules in the strong elongation flow between the growing droplets produced a fluid, beads-on-string structure that solidified, leaving a beaded nanofiber. The beads were often spaced at recurring distances along the fiber, and sometimes in repeating patterns of large and small beads. When the concentration of the neutralizing counter ions from the corona was small, the electrospun nanofibers were smooth. As the concentration of the neutralizing ions increased, to a greater extent the charge on the jet was neutralized by the airborne ions. As the charge density on the surface of the jet decreased, the number of beads per unit length increased. The volume of polymer in the form of nanofibers, about the volume of polymer in beads, decreased. Increasing the viscoelasticity of the electrospinning solution or increasing the concentration of the dissolved salt in the polyethylene oxide solution stabilizes the jet against forming beads. Beads also formed as the partial pressure of the solvent in the surrounding gas approached saturation. Forming a bead is the effect of a surface-tension-driven instability which has a cylindrical jet of polymeric fluid to divide into thinner fibers and larger beads by a minuscule extrusion technique.[33]

12.2.4 FOURTH PART: SOLIDIFICATION

As the jet moves toward the collector, it continues to expand by going past through the loops. Jet solidification is based along the traveling distance of the fibers. The distance between the collector and the capillary tip has a direct effect on the jet solidification and fiber diameter. If the nozzle-to-collector distance is long enough or the whipping instability is high, there is more time for fibers to dry before being picked up.[14] The collection part is where the jet is broken off. The polymer fiber that stays after the solvent evaporates may be accumulated on a metal screen. For polymers dissolved in nonvolatile solvents, water, or other suitable liquids can be used to collect the jet, remove the solvent, and develop the polymer fiber. If the jet arrives with a high-speed at a stationary collector, the jet coil or fold. Since the jet is charged, fibers lying on the collector repel fibers that come afterward. The charge on the fibers can be changed by ions created in a corona perform and carried to the collection region by air currents. The charge may also be removed by charge migration through the fiber to the conducting substrate, although for dry fibers with low electrical conductivity, this charge migration may be slow.[12]

12.3 CONCLUDING REMARKS

Producing nanofibers by electrospinning is a simple and widely utilized technique for varied applications. The most significant part of electrospinning is how to control the process. Therefore, we must know about the behavior of every part of the process and control the instabilities in it. Also, the most important tools for controlling process are modeling and simulating. In future, we review them.

KEYWORDS

- **Electrospinning**
- **Nanofibers**
- **Nanofibers Formation**

REFERENCES

1. Gupta, D.; Jassal, M.; Agrawal, A. K. *Electrospinning of Poly(vinyl alcohol)- Based Boger Fluids to Understand the Role of Elasticity on Morphology of Nanofibers.* Industrial & Engineering Chemistry Research, 2015. **54**: p. 1547–1554.
2. Carroll, C. P., et al. *Nanofibers from Electrically driven Viscoelastic Jets: Modeling and Experiments.* Korea-Australia Rheology Journal, 2008. **20**: p. 153–164.
3. Fang, J.; Wang, X.; Lin, T. *Functional Applications of Electrospun Nanofibers.* Nanofibers—Production, Properties and Functional Applications, 2011: p. 287–326.
4. Reneker, D. H., et al. *Electrospinning of Nanofibers from Polymer Solutions and Melts.* Advances in Applied Mechanics, 2007. **41**: p. 43–346.
5. Karra, S. *Modeling Electrospinning Process and a Numerical Scheme Using Lattice Boltzmann Method to Simulate Viscoelastic Fluid Flows*, 2007, Texas A&M University.
6. Šimko, M.; Erhart, J,; Lukáš, D. *A Mathematical Model of External Electrostatic Field of a Special Collector for Electrospinning of Nanofibers.* Journal of Electrostatics, 2014. **72**(2): p. 161–165.
7. Brooks, H.; Tucker, N. *Electrospinning Predictions Using Artificial Neural Networks.* Polymer, 2015. **58**: p. 22–29.
8. Fridrikh, S. V., et al. *Controlling the Fiber Diameter during Electrospinning.* in *Physical Review Letters*, American Physical Society, 2003. **90:** p. 144502.
9. Zeng, Y., et al. *Numerical Simulation of Whipping Process in Electrospinning.* in *WSEAS International Conference. Proceedings. Mathematics and Computers in Science and Engineering.* 2009. World Scientific and Engineering Academy and Society.
10. Ciechańska, D. *Multifunctional Bacterial Cellulose/Chitosan Composite Materials for Medical Applications.* Fibres & Textiles in Eastern Europe, 2004. **12**(4): p. 69–72.
11. Lu, P.; Ding, B. *Applications of Electrospun Fibers.* Recent patents on nanotechnology, 2008. **2**(3): p. 169–182.
12. Reneker, D. H.; Chun, I. *Nanometre Diameter Fibres of Polymer, Produced by Electrospinning.* Nanotechnology, 1996. **7**(3): p. 216–223.
13. Haghi, A. K. *Electrospun Nanofiber Process Control.* Cellulose Chemistry & Technology, 2010. **44**(9): p. 343–352.
14. Ghochaghi, N. *Experimental Development of Advanced Air Filtration Media Based on Electrospun Polymer Fibers*, in *Mechnical and Nuclear Engineering* 2014, Virginia Commonwealth. p. 1–165.
15. Ziabari, M.; Mottaghitalab, V.; Haghi, A. K. *Evaluation of Electrospun Nanofiber Pore Structure Parameters.* Korean Journal of Chemical Engineering, 2008. **25**(4): p. 923–932.
16. Li, Z.; Wang, C. *Effects of Working Parameters on Electrospinning*, in *One-Dimensional nanostructures.* 2013, Springer. p. 15–28.
17. Bognitzki, M., et al. *Nanostructured Fibers via Electrospinning.* Advanced Materials, 2001. **13**(1): p. 70–72.
18. De. V, S., et al. *The Effect of Temperature and Humidity on Electrospinning.* Journal of materials science, 2009. **44**(5): p. 1357–1362.
19. Sill, T. J.; von. Recum, H. A. *Electrospinning: Applications in Drug Delivery and Tissue Engineering.* Biomaterials, 2008. **29**(13): p. 1989–2006.
20. Angammana, C. J. *A Study of the Effects of Solution and Process Parameters on the Electrospinning Process and Nanofibre Morphology*, 2011, University of Waterloo.

21. Bhardwaj, N.; Kundu, S. C. *Electrospinning: A Fascinating Fiber Fabrication Technique.* Biotechnology Advances, 2010. **28**(3): p. 325–347.
22. Rafiei, S., et al. *New Horizons in Modeling and Simulation of Electrospun Nanofibers: A Detailed Review.* Cellulose Chemistry and Technology, 2014. **48**(5–6): p. 401–424.
23. Tan, S. H., et al. *Systematic Parameter Study for Ultra-Fine Fiber Fabrication via Electrospinning Process.* Polymer, 2005. **46**(16): p. 6128–6134.
24. Fong, H.; Chun, I.; Reneker, D. H. *Beaded Nanofibers Formed during Electrospinning.* Polymer, 1999. **40**(16): p. 4585–4592.
25. Gupta, P.; Wilkes, G. L. *Some Investigations on the Fiber Formation by Utilizing a Side-by-Side Bicomponent Electrospinning Approach.* Polymer, 2003. **44**(20): p. 6353–6359.
26. Theron, S. A., Zussman, E.; Yarin, A. L. *Experimental Investigation of the Governing Parameters in the Electrospinning of Polymer Solutions.* Polymer, 2004. **45**(6): p. 2017–2030.
27. Sawicka, K.; Gouma, P.; Simon, S. *Electrospun Biocomposite Nanofibers for Urea Biosensing.* Sensors and Actuators B: Chemical, 2005. **108**(1): p. 585–588.
28. Deitzel, J. M., et al. *The Effect of Processing Variables on the Morphology of Electrospun Nanofibers and Textiles.* Polymer, 2001. **42**(1): p. 261–272.
29. Kessick, R.; Fenn, J.; Tepper, G. *The Use of AC Potentials in Electrospraying and Electrospinning Processes.* Polymer, 2004. **45**(9): p. 2981–2984.
30. Demir, M. M., et al. *Electrospinning of Polyurethane Fibers.* Polymer, 2002. **43**(11): p. 3303–3309.
31. Hsu, C. M.; Shivkumar, S. *Nano-Sized Beads and Porous Fiber Constructs of Poly(ε-Caprolactone) Produced by Electrospinning.* Journal of materials science, 2004. **39**(9): p. 3003–3013.
32. Yarin, A. L.; Koombhongse, S.; Reneker, D. H. *Bending Instability in Electrospinning of Nanofibers.* Journal of Applied Physics, 2001. **89**(5): p. 3018–3026.
33. Reneker, D.H. and A.L. Yarin, *Electrospinning Jets and Polymer Nanofibers.* Polymer, 2008. **49**(10): p. 2387-2425.
34. Zhang, S. *Mechanical and Physical Properties of Electrospun Nanofibers.* 2009: p. 1-83.
35. Wu, Y., et al. *Controlling Stability of the Electrospun Fiber by Magnetic Field.* Chaos, Solitons & Fractals, 2007. **32**(1): p. 5–7.
36. Baji, A., et al. *Electrospinning of Polymer Nanofibers: Effects on Oriented Morphology, Structures and Tensile Properties.* Composites Science and Technology, 2010. **70**(5): p. 703–718.
37. He, J. H.; Wu, Y.; Zuo, W.W. *Critical Length of Straight Jet in Electrospinning.* Polymer, 2005. **46**(26): p. 12637–12640.
38. Hohman, M. M., et al. *Electrospinning and Electrically Forced Jets. II. Applications.* Physics of Fluids (1994–present), 2001. **13**(8): p. 2221–2236.
39. Yarin, A. L.; Zussman, E. *Electrospinning of Nanofibers from Polymer Solutions.* XXI ICTAM.—Warsaw, Poland, 2004: p. 12–13.
40. Sawicka, K. M.; Gouma, P. *Electrospun Composite Nanofibers for Functional Applications.* Journal of Nanoparticle Research, 2006. **8**(6): p. 769–781.
41. Yousefzadeh, M., et al. *A Note on the 3D Structural Design of Electrospun Nanofibers.* Journal of Engineered Fabrics & Fibers (JEFF), 2012. **7**(2): p. 17–23.
42. Li, W. J., et al. *Electrospun Nanofibrous Structure: A Novel Scaffold for Tissue Engineering.* Journal of biomedical materials research, 2002. **60**(4): p. 613–621.

CHAPTER 13

CHARACTERISTICS OF FILM AND NONWOVEN FIBER MATERIALS PREPARED FROM POLYURETHANE AND STYRENE ACRYLONITRILE

S. G. KARPOVA, YU. A. NAUMOVA, L. P. LYUSOVA, and A. A. POPOV

CONTENTS

ABSTRACT

Structural and dynamic analysis, combining thermo-physical mea-surements by differential scanning calorimetry and measurements of segmental mobility by the EPR microprobe technique, is performed for films and nonwoven fiber materials prepared from polyurethane (PU) and a styrene–acrylonitrile copo-lymer (SAN), as well as mixed compositions thereof. The effect of tetrahydro-furan, ethyl acetate, and acetone solvents on the structure and molecular dynamics of films and matrices based on ultrathin PU and SAN fibers is examined. A weak effect of solvent on the molecular dynamics of chains in the PU film and nonwo-ven materials and a strong influence on the molecular mobility in SAN films and films of mixed compositions with a high SAN content are observed. For fibers, such influence is negligible. In PU and PU/SAN mixed formulations, mesomorphic structures are formed in both the film and nonwoven materials. The temperature dependence of rotational correlation time τ of the probe exhibits a kink at temperatures close to the melting point of the mesomorphic structures in the PU/SAN mixtures. All the studied dependences for both the films and fibers feature a kink at PU/SAN = 50/50, which is associated with phase inversion in the compositions. The probe measurements show the impact of an oxidant (ozone) on the amorphous phase in these polymers. Measuring the rotational dynamics of the probe before and after exposure to ozone of the film and ultrafine fiber materials showed that, for both the PU films and fibers, ozonation produces practically no effect on the molecular dynamics, while for PU/SAN compositions and pure SAN, τ changes significantly.

13.1 INTRODUCTION

Development and synthesis of new polymers is a long-term and expensive process. One effective way of solving the problem is to use the mixtures of existing polymers. By varying the composition of the mixture and types of constituent com-ponents, it is possible to prepare the required material.

Another aspect affecting the properties of polymers is the method for their processing. One of the most important ways of processing of polymers is to treat them in solutions. Structurization in solution is controlled by the nature of the solvent used, with the main reason for differences in the structure of the obtained materials being the influence of the thermodynamic quality of the solvent on the mode of interaction of polymer macromolecules with solvent molecules and, as a consequence, on the interaction between macromolecules themselves.[1–4]

The solvent plays a significantly more complicated role in solvent–polymer 1–polymer 2 systems. How the solvent affects the structure of films and nanoscale fibrous structures (fibers, filaments, networks, and porous fibrous matrices) is of considerable interest. Currently, ultrathin fibers and products thereof are exten-sively used in biomedicine, cell engineering, separation and filtration processes, in creating reinforced composites, in electronics, chemical analysis, sensory diag-nostics, and a number of other innovative areas.[5–13]

The choice of polymers is associated with a wide use of polyurethane (PU) in medicine and the light industry, for example, for producing fibers and non woven micro-and nanofibrous materials, adhesive compositions, varnishes, paints, protective agents, sealing materials, artificial leather, suede, filtering materials for respirators, and rubberized fabrics. In medicine, it is used to cultivate cells of various tissues and to manufacture bone prostheses, connections of nerve fibers, osmotic dialysis filters, artificial heart components, artificial blood vessels, catheters, andmatrices for drug delivery. At present, mixed compositions are widely used as filtering fiber materials for personal respiratory protection.

A high hemocompatibility and mechanical properties of these polymers put them above competition with respect to other polymeric implants. The structure and properties of PU were studied in refs [14–16]. In addition, mixed compositions of PU with chitosan,[17,18] polyhydroxybutyrate,[19] and styrene–acrylonitrile (SAN) copolymer[20] were investigated.

Thus, the selection of the components of mixtures of polymers and solvent for their processing in solutions makes it possible to widely change the set of physico-mechanical properties of the tested film and fibrous materials, leading to creation of products with valuable consumer qualities.

13.2 SUBJECTS AND METHODS

The objects of study were films prepared from Desmocoll 400 poly-urethane thermoplastic (Bayer, M_w = 1.0 × 105), SAN 350N styrene–acrylonitrile copolymer (Kumho, M_w = 1.0 × 105), and mixed compositions thereof at PU/SAN ratios (wt %) of 10/90, 30/70, 50/50, and 80/20 in the film material and 25/75, 50/50, 75/25 in the nonwoven fiber materials. The films were dissolved in organic solvents of different chemical classes: ethyl acetate (EA), tetrahydrofuran (THF), and acetone (A). The concentrations of the test solutions ranged within 0.1–2.0 g/100 mL. The films were prepared in Petri dishes by casting from 10 wt % polymer solutions, with subsequent evaporation of the solvent at constant temperature and humidity.

We also studied polymeric nonwoven materials with a fiber diameter of 1–10 μm prepared by electrospinning from Desmocoll 400, SAN, and mixtures thereof. The material has a mass per unit area of 20–70 g/m^2 and an aerodynamic resistance of 3–30 Pa at an air flow rate of 1 cm/s. At present, fibers are produced by means of an original technology based on electrospinning (ES), which enables to manufacture ultrathin fibers and nonwoven materials. The nonwoven materials tested in the present work were prepared by electrospinning, more specifically by the "nanospider" method, a patented technology of high-voltage capillary-free electrospinning of fibers.[6] Electrospinning process was carried out at a temperature of 20°C and a relative humidity of 60%. The distance to the receiving electrode was 0.2–0.3 m.

Thermal analysis of the test materials was performed on a DTAS-1300 thermal analyzer at temperatures from –90 to 140°C. The heating rate was 20 K/min. The temperature was measured to within ±0.5°C.

The molecular mobility in the samples was studied by the paramagnetic probe method. The probe was 2,2,6,6-tetramethylpiperidine-1-oxyl, a stable nitroxyl radical. The latter was introduced into the films and nonwoven materials with low and high SAN contents from the vapor phase at temperatures of respect-tively 40 and 70°C until a concentration of 10–3–10–4 mol/L was reached. The EPR spectra were recorded in the absence of saturation, which was verified by examining the dependence of the signal intensity on the microwave field power. The probe rotational correlation time τ was evaluated from EPR spectra by the formula[21]

$$\tau = {}_{\Delta}H_{+}\,(\sqrt{I_{+}/I_{-}} - 1) \times 6.65 \times 10^{-10}$$

where ${}_{\Delta}H$ is the width the spectral components located in the weak field; I_{+}/I_{-} is the ratio of intensities of the components in the weak and strong fields, respectively. The error in the determination of τ was ±5%.

13.3 RESULTS AND DISCUSSION

13.3.1 STUDY OF THE INFLUENCE OF THE SOLVENT ON THE CRYSTALLINE PHASE IN MIXED PU/SAN COMPOSITIONS

Differential scanning calorimetry (DSC) was used to study the structure of films and fibers of PU/SAN mixed compositions formed in THF solvent (Figure 13-1). It was found that the dependences of the fraction of mesomorphic structures and their melting points on the mixture composition are different for films and fibers.

The DSC data (Table 13-1) showed that the fractions of mesomorphic structures α in the PU films and fibers are similar, being characterized by 45–47 J/g (melting enthalpy). While a small increase in the SAN content significantly reduces the fraction of mesomorphic structures in the films (Figure 13-1a), an identical increase in the fibers produces an extremely small change (Figure 13-1b). At PU/SAN = 50/50, no mesomorphic structures are formed in films and only small fraction in fibers. At a higher SAN content in the mixed composition (over 50%), the fraction of mesomorphic structures in the films increases dramatically, while in the fibers, α remains almost unchanged.

Thus, for the films and fibers, the fraction α differently depends on the composition of the formulation, but their common feature is a kink in the dependences at 50% PU, which is associated, in our opinion, with a phase inversion. The different behavior of the dependence of α on the mixture composition for the films and fibers can be explained, in our opinion, by a difference in the morphology of the tested mixed compositions.

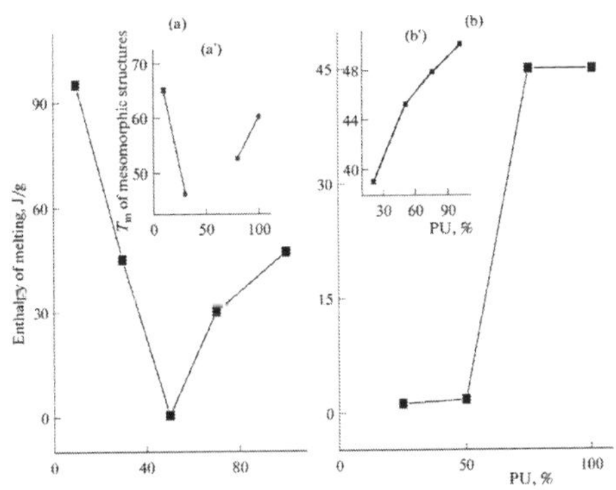

FIGURE 13-1 Dependence of the enthalpy of melting of mesomorphic structures in the (a) film and (b) nonwoven materials and of the melting temperature of mesomorphic structures in the (a') film and (b') nonwoven materials on the mixture composition.

The dependences of the melting temperature T_m on the composition for the films and fibers prepared from these polymers are also different. While the melting point of the films and fibers decreases with increasing SAN content (up to 50%), at higher SAN content in the mixture, T_m for the films increases, but, on the contrary, decreases sharply for the fibers. The observed patterns are also indicative of phase inversion in both the films and fibers at PU/SAN = 50/50. A comparison of the da-ta on T_m and α reveals that they change similarly. For example, as the SAN con-tent in the composition increases, so does the concentration of defects in the mesomorphic structures of PU in the films, which, in turn, reduces α and T_m. In the fib-ers, no appreciable change in α is observed (with the exception of 50/50 com-position), so T_m decreases slightly in mixed compositions containing up to 50% SAN. In PU/SAN = 50/50 compositions, phase inversion takes place, with a higher degree of entanglement of chains preventing the formation of mesomorphic struc-tures. As a result, no mesomorphic structures are formed in films, while their pro-portion in fibers is negligibly small. In mixed compositions, in which the SAN component forms a continuous phase, increasingly perfect PU mesomorphic struc-tures arise in films, which leads to an increase in α and T_m; however, in fibers (probably due to a high speed of spinning), they do not have time to form. Characteristically, the melting point of the films is higher than that of the fibers. For example, while T_m = 51.4°C for the PU/SAN = 30/70 films, T_m = 39°C for the fibers of the same composition; and T_m = 60.3 and 48.5°C, respectively, for the PU films and fibers (analogous data were obtained for other compositions), which is also indicative of more imperfect mesomorphic formations in the fibers compared with the films.

A characteristic feature is the fact that, while the glass transition temperature T_g for the SAN films is 80°C, it is 110°C for the fibers. In the mixed-composition films, no glass transition of SAN was observed in the temperature range covered (from –90 to 140°C), except for the PU/SAN = 50/50 composition, whereas the glass transition of the SAN occurs in the fibers of compositions PU/SAN = 25/75 and 50/50, indicative of a more rigid amorphous phase in fibers. The lack of grass transition in mixed formulations (except the above) is suggestive of a plasticizing effect of PU on the structure of mixtures and of a sufficiently high compatibility of these polymers.

Thus, the preparation of films from solution is slow enough, so equilibrium is established in the system, and addition of SAN (in small amounts) to PU redu-ces the fraction of PU mesomorphic structures due to the intermolecular interaction between PU and SAN molecules, as is shown in ref. [5].

TABLE 13-1 Effects of the Ratio of the Polymers (wt %) and the Type of Solvent on T_g and T_m (°C) as studied by DSC

Films

Solvent	PU		PU/SAN 80/20		PU/SAN 50/50		PU/SAN 30/70		PU/SAN 10/90		SAN
	T_m	T_g	T_m	T_g	T_m	T_g	T_m	T_g	T_m	T_g	T_g
THF	60.3	–35	57	–30	–	50	51.4	–	55	–30	80
EA	54	–	53	–20	48	–	65	56	80	77	85
A	54	–40	48	–	56	–	55	–	76	70	115

Fibers

Solvent	PU		PU/SAN 75/25		PU/SAN 50/50		PU/SAN 75/75		SAN
	T_m	T_g	T_m	T_g	T_m	T_g	T_m	T_g	T_g
THF	48.5	–40	48	–10	42	95 and –35	39	55	110

At higher (over 50%) SAN content, at which the PU forms a dispersed phase, PU inclusions with a dense surface composed of straightened chains arise. The fraction of PU mesomorphic structures in the film material increases at high SAN content (>50%) due to a lesser chain entanglement in the dispersed phase as compared to the dispersion phase. In the fibers, apparently due to a high rate of formation, the structure has no time to come to an equilibrium state, which makes the fraction of PU mesomorphic structures in mixed fiber compositions with high PU content extremely small (Figure 13-1a).

13.3.2 EFFECT OF SOLVENT ON THE DYNAMICS AND STRUCTURE OF THE AMORPHOUS PHASE IN FILMS AND FIBERS PREPARED FROM PU, SAN, AND MIXED COMPOSITIONS THEREOF

To examine the complex evolution of mobility of the probe, we used the model of binary distribution of segments in less dense and more dense intercrystalline regions, characterized, respectively, by a more rapid and less rapid rotation of TEMPO stable radicals. Studies of molecular dynamics in the films and fibers pre-pared from PU/SAN mixed compositions revealed that the amorphous phase (except PU, Figure 13-2) is heterogeneous. The heterogeneity of amorphous phase indicates the presence therein of structures with a sufficiently low I_+^1 and high I_+^2 mobility. In what follows, the time correlation was calculated only for the presence therein of structures with a sufficiently low and high mobility. In what follows, the time correlation was calculated only for the fast component.

Studies of the influence of the type of solvent showed that the solvent produces a sufficiently strong effect on the rotational mobility of the probe, and consequently, on the molecular mobility of the polymer chains (Figure 13-3). Figure 13-3a shows that the highest probe mobility is observed for the films formed in the THF solvent, while the lowest in acetone. The solvent does not significantly affect the molecular mobility in the fibers (Figure 13-3b). Introduction of SAN into the polymer matrix causes the growth of τ. Note that, with increasing content of SAN in the composition up to 50%, no significant change in the correlation time is observed: only at higher SAN content in the mixture, when the SAN component forms a continuous phase, a sharp rise in τ occurs, indicative of an increase in the rigidity of the amorphous regions. The observed dependence of τ on the composition of the films can be explained by the structure of the interfacial regions (Figure 13-4). To process and quantitatively interpret the data

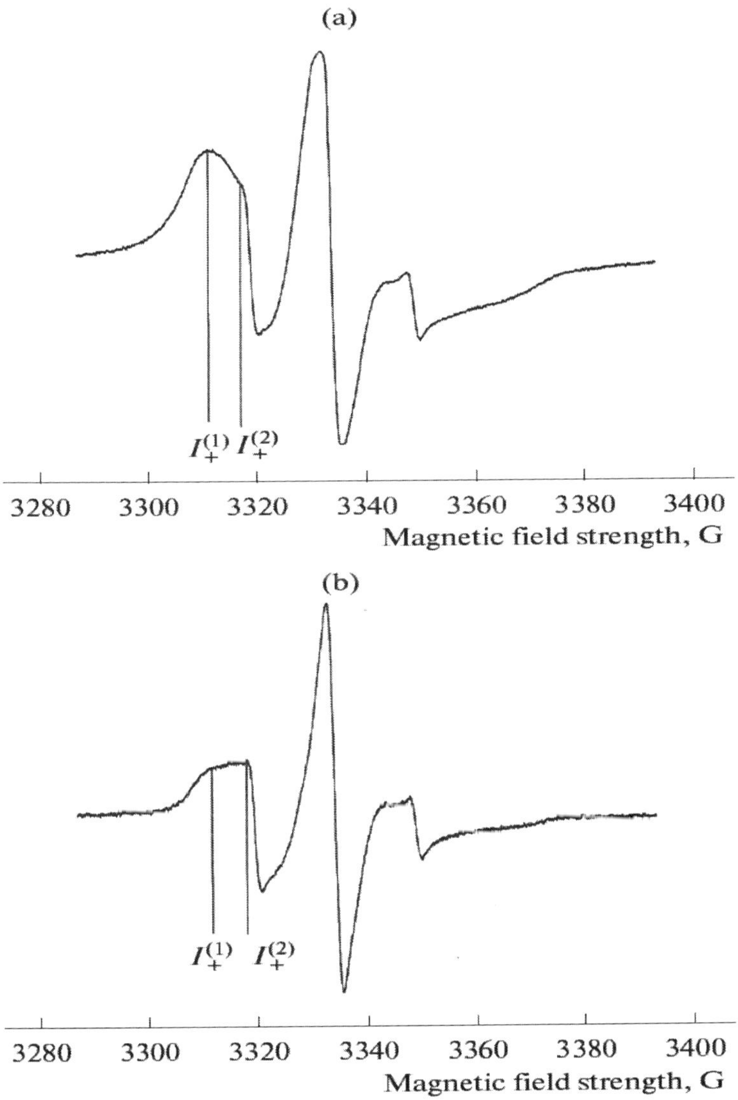

FIGURE 13-2 EPR spectra of the PU/SAN − 50/50 material formed in the THF solvent: (a) film material and (b) nonwoven material.

obtained by the EPR spectroscopy, we used the "Bruker," "Winer," and "Simfonia" software codes. These programs enabled to calculate the molecular mobility in the interfacial regions of the studied mixed compositions. Figure 13-4a shows that, in the films, most dense interfacial regions are formed in mixtures containing small amounts of SAN so that the transport of the radicals through this layer is hampered, which explains why the change in τ upon adding SAN (up to 50%) is small. The interfacial layer in compositions with a high content of SAN is characterized by a lower value of τ, indicating that the interfacial interlayer is looser, that is, more permeable for the radical. In the fibers, irrespective of the composition, the interfacial layer is characterized by a rather low τ, which indicates its high permeability for the radical and manifests itself through the composition dependence of τ (Figure 13-3b). For both the films and the fibers, the composition dependences of τ feature a kink at PU/SAN = 50/50, a result that confirms the above conclusion about phase inversion at this ratio of the contents of the polymers.

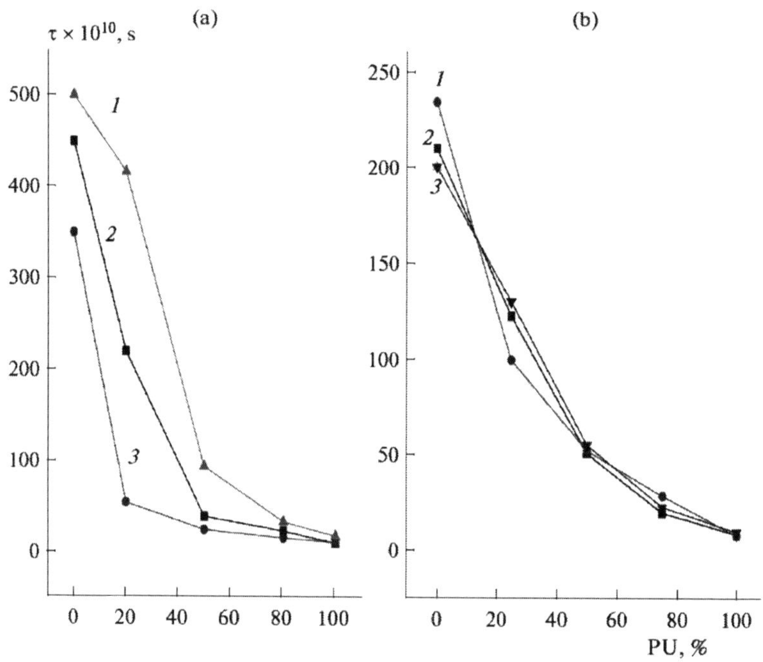

FIGURE 13-3 Dependence of τ on the composition of the formulation in the (*1*) A, (*2*), EA, and (*3*) THF solvents for the (a) film and (b) nonwoven materials.

It should be noted that the correlation times τ for PU samples formed in va-rious solvents, be it films or fibers, differ insignificantly, while for SAN films these differences are significant. Thus, the solvent changes the structure of the amorphous regions in the SAN component to a larger extent than it does in the PU component.

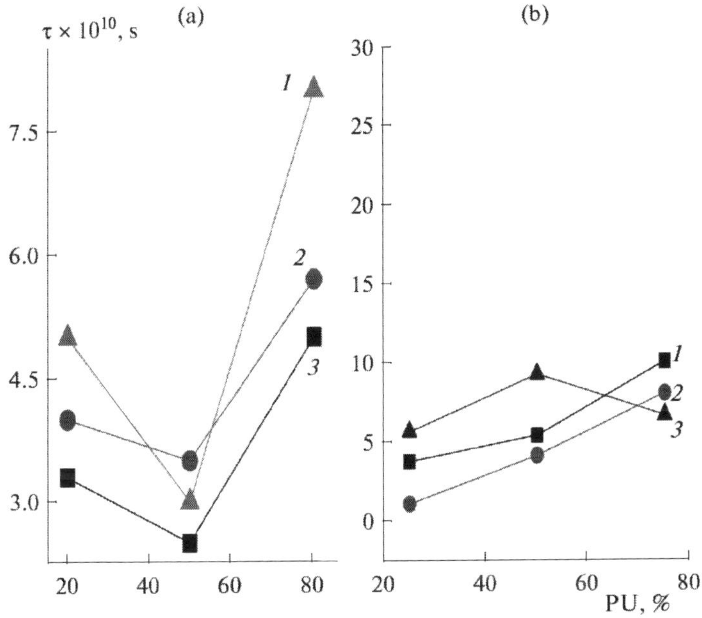

FIGURE 13-4 Dependence of τ for interphase layers in the (a) film and (b) nonwoven materials of the composition of the formulation in the (*1*) A, (*2*), EA, and (*3*) THF solvents.

According to ref. [5], for a given mixed composition, the thermodynamic quail-ties of the solvents change from "good" to "poor" in the order THF > EA > A. In "poor" solvents, the degree of interaction between the polymer and solvent is much higher than in "good" ones; therefore, solvents differing in quality create different polymeric structures in the solution, which persist after removal of the solvent. Differences in the interaction of the polymers in solutions lead to distinctions in the properties of the films and nonwoven fibrous materials prepared from solutions, since the structure formed in the solution in part survives.

In terms of thermodynamics, "good" and "poor" solvents differ only in their interaction with the polymer, resulting in a different interaction of macromolecules with each other. A higher degree of coiling of macromolecules in "poor" solvents favors an increase in the number of contacts not only between like but also bet-ween unlike macromolecules.[11] Thus, the solvent greatly influences the struc-ture and molecular mobility of the polymer it dissolves. In THF, the most "good" solvent, the structure of the polymer composition is close to the equilibrium. In passing to the "bad" solvent, the structure becomes increasingly nonequilibrium, the density of chains grows, and as a result, the rigidity of the macromolecules increases in the series of solvents THF, EA, and A (Figure 13-3a). For the fibers, the effect of the solvent on the amorphous structure of the polymers is much weaker (Figure 13-3b).

Figure 13-5a demonstrates how the fraction of macromolecules comprising the dense amorphous regions in the film compositions (evaluated from figure 13-5) increases in going from the "good" to the "poor" solvent. It can be seen that the increase of the PE content in the mixture causes a decrease in the fraction of regions with hindered motion, so that, for mixed compositions with low SAN content (less than 50%), only one spectrum is observed. In the fibers, these differences manifest themselves in a lesser extent (Figure 13-5b); however, a double spectrum is observed for mixed compositions with high PU content. Note that the kink in these dependences is also observed at PU/SAN = 50/50 (Figure 13-5b).

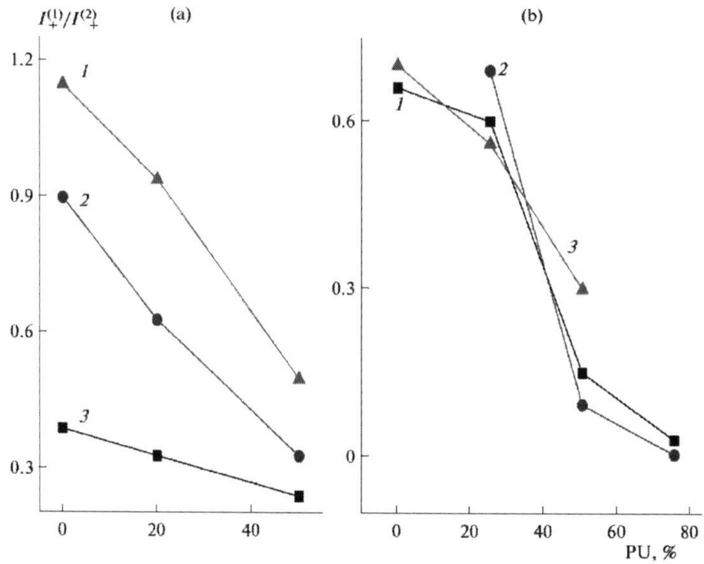

FIGURE 13-5 Dependence of $I_+^{(1)} / I_+^{(2)}$ on the composition of the formulation in the (*1*) A, (*2*), EA, and (*3*) THF solvents for the (a) film and (b) nonwoven materials.

The solvent determines the concentration of radicals in the studied polymers. That the density of the amorphous phase in PU, as well as in the compositions formed in the different solvents (Figure 13-6), is seen from the decrease in the concentration of the radical in passing from the "good" to the "poor" solvent, in both the films and the fibers prepared from the mixtures studied. However, an analysis of the concentration of the radical in the mixed compositions formed, for example, in the THF solvent shows that the observed changes in the films can be attributed to the growth of density of the samples in going from PU to SAN and to the decrease of the fraction of amorphous phase accessible to the radical, factors that cause a reduction in the concentration of the radical in the samples (Figure 13-6a).

Different dependences were observed for the PU/SAN fibers (Figure 13-6b). The composition of the fiber produced no significant effect on the concentration of the radical, as it happens in the films. Only for the PU/SAN = 50/50 samples, a jump in the dependence of the radical concentration on the composition was observed, indicative of the most loose packing of the chains in these fibers. Characteristically, that the extreme point was observed PU/SAN = 50/50 also supports the conclusion on phase inversion at these ratio of the contents of the polymers.

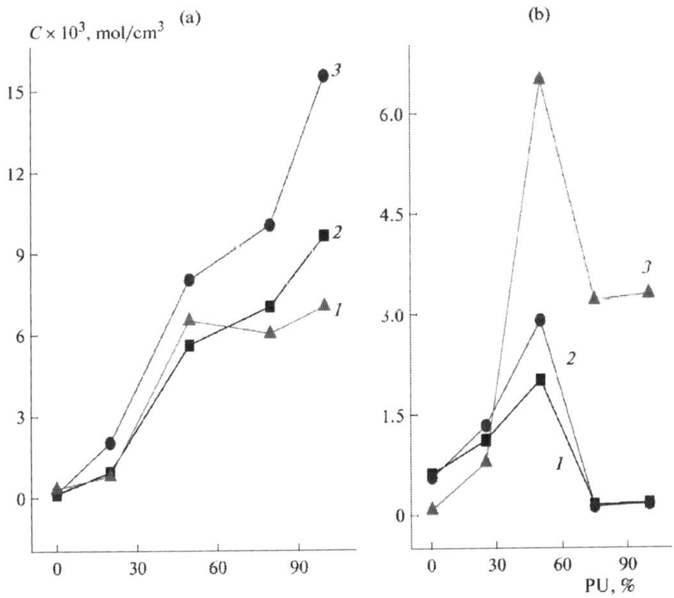

FIGURE 13-6 Dependence of the concentration of radical on the composition of the formulation in the (*1*) A, (*2*), EA, and (*3*) THF solvents for the (a) film and (b) nonwoven materials.

We also studied how the correlation time for films and fibers prepared from these mixtures depends on temperature. It was shown that these dependences have a kink at 45–55°C, which is apparently due to the "unfreezing" of mesomor-phic structures (Figure 13-1). A linear dependence was observed only for PU/SAN = 50/50 films; that is, for compositions without mesomorphic structures (in the fiber samples, their fraction is negligibly small).

The activation energy was found to change differently, depending on the composition of the polymer mixture, type of solvent, and method of preparation of the polymeric material (Figure 13-7). Calculation of the activation energy for the rotational mobility of the probe in the tested polymers showed that, for the PU films, the effect of solvent on the value of E_a is small; adding SAN (up to 50%) reduces the activation energy, which is apparently due to a decrease in the density of the mixed composition (Figure 13-7a). At higher SAN content in the mixture, changes in E_a were not so conspicuous for any of the solvents. Note that the activation energy increases in passing from the "good" to the "poor" solvent.

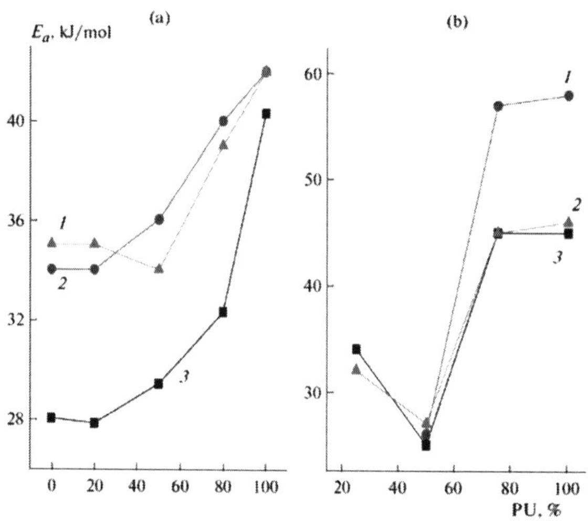

FIGURE 13-7 Dependence of E_a on the composition of the formulation in the (*1*) A, (*2*), EA, and (*3*) THF solvents for the (a) film and (b) nonwoven materials.

In case of the fiber samples, the activation energy behaves differently (Figure 13-7b). Like α (Figure 13-1b), E_a for the mixtures with <50% PU practically does not change, at 50% PU, E_a decreases sharply, whereas for the compositions with higher SAN content, the activation energy increases slightly. Note that the values of E_a for the fibrous polymers formed in the different solvents (except for acetone) are similar, while for the films, these differences are significant (Figure 13-7a). It is important that, for all the mixed compositions (except for 50/50), the temperature dependences of the correlation time exhibit a kink within 45–55°C, temperatures close to the melting point of the mesomorphic structures. Therefore, we explain the observed kinks in the temperature dependences of τ by the unfreezing of mesomor-phic structures. That the activation energy E_a increases at temperatures above 50°C can be attributed to the fact that the activation energy for mesomorphic struc-tures is higher.

Using the microprobe method, we demonstrated that the oxidant (ozone) influences the amorphous phase of the above polymers. Studies have shown that regardless of the method of preparation, the polymeric compositions with high SAN content were oxidized. Polyurethane and mixtures thereof with small amount of SAN were more resistant to the impact of ozone (Figure 13-8).

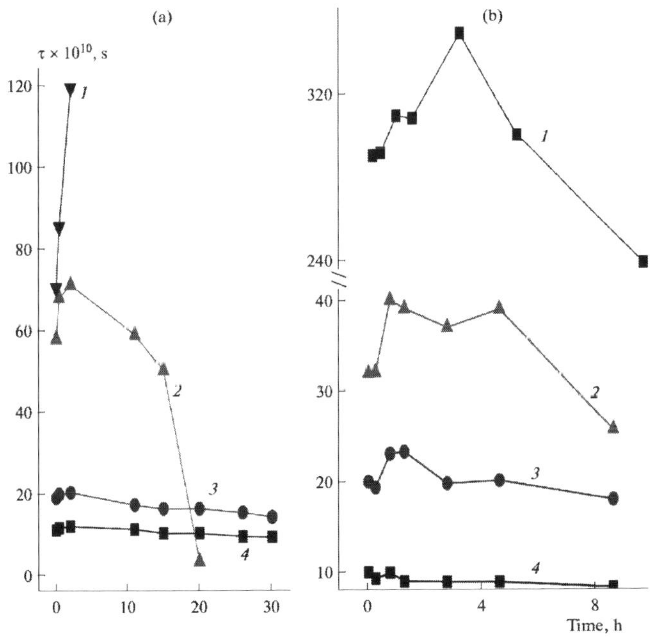

FIGURE 13-8 Dependence of τ on the duration of treatment in ozone of the (a) film [PU/SAN = (*1*) 30/70, (*2*) 50/50, (*3*) 80/20, and (*4*) 100/0] and (b) nonwoven [PU/SAN = (*1*) 25/75, (*2*)50/50, (*3*) 75/25, and (*4*) 100/0] materials.

For the PU and PU/SAN = 80/20 samples, the correlation time changed only slightly after ozonation of the films and fibers formed in THF. In the mixed formulations with high SAN content, τ for both the films and fibers increased after a short exposure to ozone (up to 2 h). Further oxidation with ozone decreased τ, with these changes manifesting themselves most clearly for the PU/SAN = 25/75 composition (PU/SAN = 30/70 for the film sample). Oxidation processes are known to involve the physical and chemical crosslinking of macromolecules, as well as the degradation of chains. At the initial stage of the oxidation, crosslinking processes prevail, leading to a decrease in molecular mobility and, consequently, to an increase in τ. At longer exposures to ozone, the processes of degradation of chains begin to dominate, causing a decrease in τ. Thus, we can conclude that SAN is most strongly susceptible to ozone oxidation. Similar patterns were obtained for the samples formed in the A and EA media.

A comprehensive study of film and nonwoven materials made it possible to analyze changes in the molecular mobility of the polymer molecules in early stages of their interaction with oxidizing aggressive media. We also examined how the ratio changes with the duration of treatment in ozone of the fibers formed in THF. Figure 13-9 shows that, in the initial stage of exposure to ozone (up to 2 h), this quantity increases, but later reduces. The growth of indicates an increase in the fraction of dense amorphous domains, which is apparently associated with the physical crosslinking of macromolecules during the ozonation, so the decrease of this parameter indicates the destruction of these regions. Like the correlation time, the parameter symbatically, which confirms the conclusion on the dominant role of crosslinks between chains at the initial stage of ozone oxidation.

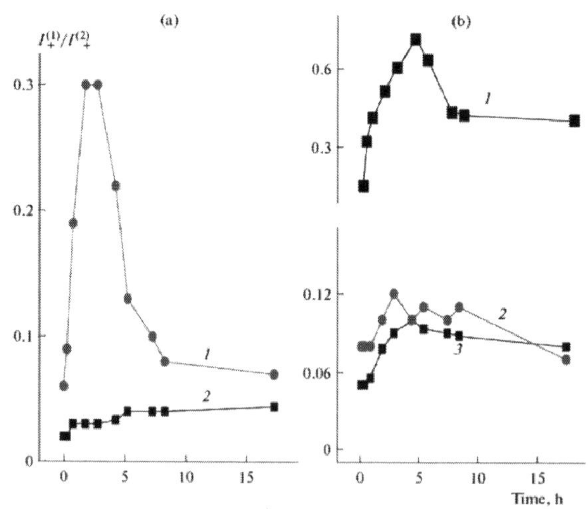

FIGURE 13-9 Dependence of I_+^1/I_+^2 on the duration of exposure to ozone of the (a) film [PU/SAN = (*1*) 30/70 and (*2*) 50/50] and (b) nonwoven [PU/SAN = (*1*) 25/75, (*2*) 50/50, and (*3*) 75/25] materials.

13.4 CONCLUSIONS

Thus, we have shown that the formation of the structure of a polymer in solution is determined by the nature of the solvent, with the main reason for differences in structure being the thermodynamic affinity between the solvent and solute.

The strongest influence of the type of solvent on the structure and molecular dynamics was observed for SAN and mixed compositions with a high content of the latter.

It was shown that mesomorphic structures are formed in both the film and nonwoven materials.

At the PU/SAN = 50/50 ratio, the dependences of τ, E_a, the concentration of the radical, and the fraction of mesomorphic structures on the composition of the film and nonwoven materials exhibit kinks because of phase inversion.

Using the probe method, we revealed the influence of an oxidant (ozone) on the amorphous phase in the film and nonwoven materials. It was shown that, irrespective of the mode of formation of compositions, those with a high SAN content are more readily oxidized. Polyurethane and mixtures thereof with small amounts of SAN are more resistant to the effect of ozone.

While for the film material, the type of solvent greatly affects the molecular mobility, for the nonwoven fabric, this effect is significantly smoothed.

It has also been shown that, while in the films, the concentration of the radial decreases by almost an order of magnitude in substituting SAN for PU, in the nonwoven material, these changes are small.

KEYWORDS

- **Effects of Temperature**
- **Electrospinning Method**
- **Correlation**
- **EPR Microprobe**
- **TEMPO Stable Radical**
- **Time Oxidation by Ozone**
- **Ultrathin Fibers**

REFERENCES

1. Tager, A. A. *Physical Chemistry of Polymers* (Scientific World, Moscow, 2007) [in Rus.].
2. Krokhina, L. S. Macromolecular compounds. **18**, 663 (1976).
3. Pendyala, V. N. S.; Xavier, S. F. Polymer **38**, 3565 (1997).
4. Filatov, Yu. N.; Kapustin, I. A.; Filatov, I. Yu.; Yakushkin, M. S. Life Safety. No. 54, 6 (2011).
5. Boksha, M. Yu. Thesis PhD (Lomonosov Moscow State Univ. Fine Chem. Technol., Moscow, 2010).
6. Pocius, A. *Adhesion and Adhesives Technology* (Hanser, Munchen, 2012).
7. Ol'khov, A. A.; Staroverova, O. V.; Filatov, Yu. N. et al., Gazette Kazan Technological University, **16** (8), 157 (2013).
8. Kapustin, I. A.; Filatov, I. Yu.; Filatov, Yu. N. Chemical Fiber, No. 5, 37 (2012).
9. Filatov, Y.; Budyka, A.; Kirichenko, V. *Electrospinning of Micro- and Nanofibers: Fundamentals in Separation and Filtration Processes* (Begell House, New York, 2007).
10. Budyka, A. K.; Borisov, N. B. *Fiber Filters for Control of Air Pollution* (IzdAT, Moscow, 2008) [in Rus.].
11. Filatov, Yu. N.; Filatov, I. Yu.; Nebratenko, M. Yu. RF Patent No. 2357785, Byull. Izobret. No. 16 (2009).
12. Naumova, Yu. A.; Lyusova, L. R.; Karpova, S. G. et al., Rubber, No. 4, 48 (2013).
13. Greiner, A.; Wendoff, J. H. Adv. Polym. Sci. **219**, 107 (2008).
14. Shtilman, M. I. *Polymers of Medical_Biological Purposes* (Akadembook, Moscow, 2006) [in Rus.].
15. Karpova, S. G.; Iordanskii, A. L.; Klenina, N. S.; Popov, A. A.; Lomakin, S. M.; Shilkina, N. G.; Rebrov, A. V. Russ. J. Phys. Chem. B **7**, 225 (2013).
16. Ikejima, T.; Inoue, Y. Carbohydr. Polym. **41**, 351 (2000).
17. Karpova, S. G.; Iordanskii, A. L.; Chvalun, S. N.; Shcherbina, M. A.; Lomakin, S. M.; Shilkina, N. G.; Rogovina, S. Z.; Markin, V. S.; Popov, A. A.; Berlin, A. A. Reports Phys. Chem. **446**, 176 (2012).
18. Karpova, S. G.; Iordanskii, A. L.; Popov, A. A.; Shilkina, N. G.; Lomakin, S. M.; Shcherbin, M. A.; Chvalun, S. N.; Berlin, A. A. Russ. J. Phys. Chem. B **6**, 72 (2012).
19. Ratzsch, M.; Haudel, G.; Pompe, G.; Meyer, E. J. Macromol. Sci. Chem. A **27**, 1631 (1990).
20. Karpova, S. G.; Naumova, Yu. A.; Lukanina, Yu. K.; Lyusova, L. R.; Popov, A. A. Polymer Sci., Ser. A **56**, 472 (2014).
21. Vasserman, A. M.; Kovarskii, A. L. *Spin Labels and Probes in Physical Chemistry of Polymers* (Science, Moscow, 1986) [in Rus.].

PART III
Chemical Engineering Science

CHAPTER 14

GENERALIZED KINETIC OF BIODEGRADATION

G. E. ZAIKOV, K. Z. GUMARGALIEVA*, I. G. KALININA*, M. I. ARTSIS, and L. A. ZIMINA

N. M. Emanuel Institute of Biochemical Physics, Russian Academy of Sciences 4 Kosygin str., Moscow 119334, Russia
E-mail: chembio@sky.chph.ras.ru

**N.N. Semenov Institute of Chemical Physics, Russian Academy of Sciences 4 Kosygin str., Moscow 119991, Russia*

CONTENTS

ABSTRACT

The information about generalized kinetic data which can describe biodegradation of polymeric materials (kinetics of biomass growth and methods of investigation of its formal mechanism of biodegradation, microorganism adhesion) is presented in this paper.

14.1 INTRODUCTION

In the seventies-eighties academician N.M. Emanuel has drawn attention of the scientific community to the necessity of assessing the characteristics of time-dependent macroscopic processes (should it be the growth of tumor cells or aging of polymeric materials under action of the ambient medium) using the so-called generalized kinetic curves requiring a great body of experimental data. In particular, this approach was for the first time developed by N.M. Emanuel in his fundamental monograph[1] for quantitative description of the tumor-cell growth.

Biodegradation of polymers occurs when living organisms are in contact with these materials; it changes their operational properties.[2,3] In general, the following processes take place in biologically damaged polymers:

- Adsorption on the material surface of microorganisms or species from tissues of living organisms
- Material degradation due to either specific action (living organisms use polymeric materials as the culture medium) or under action of the metabolism products

In the first process, the chemical structure of the polymer is not changed as a rule, the material serves as a support on which microorganism colonies adhere and grow (bio-overgrowing), or collagenic capsules are formed with the implanted material. Microorganism adhesion is the initial stage of the material overgrowth, which predetermines the further behavior of bio-overgrowth and bio-damage of polymeric materials. Assessment of the biomass on the polymer surface at the bio-overgrowth stage is of great interest because the biomass amount influences the "surface" operational properties, such as optical, adhesion, and so on.

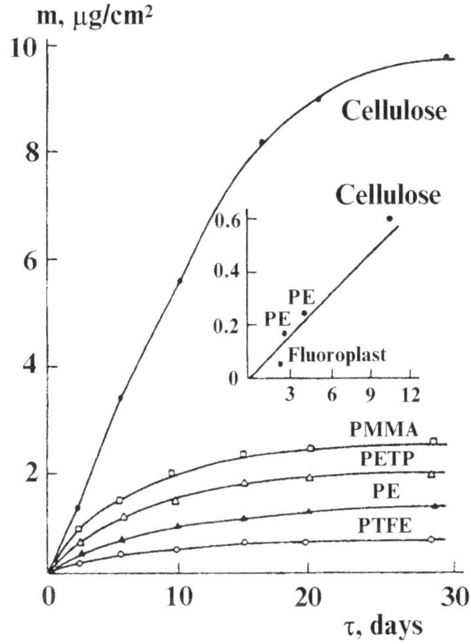

FIGURE 14-1 The kinetics of growth of microscopic fungi (the standard GOST set) on the surface of different polymers at $\phi = 98\%$ and T = 30°C (PMMA is polymethylmethacrylate, PETP is polyethyleneterephthalate, PE is polyethylene, PTFE is polytetrafluoroethylene). The inset shows the rate constant for the biomass growth as a function of the amount of water adsorbed by the polymer.

The second process leads directly to polymer aging under action of chemically active species. In this case, the "bulk" operational properties, such as mechanical, dielectric, and so on, will change.

In this paper we have employed the method of generalized kinetic curves to assess the growth of microscopic fungi on the surface of polymeric material, to study the subsequent process of material biodegradation directly by the biological medium of organism or by the products of living activity of growing microorganisms (metabolites), and to estimate the kinetic parameters of microorganism adhesion on the polymer surface.

14.2 KINETICS OF BIOMASS GROWTH: METHODS AND RESULTS

The growth and development of microscopic fungi directly on solid polymeric surfaces are conventionally assessed from the data on the growth of the diameter of certain colonies or of the colonies of some microscopic fungi using the five-grade scale and standard methods. This is because measurements of biomass in amounts of a few μg/cm^2 at the initial growth stage present serious experimental difficulties.

To determine the kinetic parameters of biomass build-up, we have used a sensitive radio-isotopic method;[4,5] polymeric plates contaminated by a suspension of microscopic fungi in water or Chapek-Dox culture medium with a concentration of 10^6 cell/ml were held in the atmosphere of tritium water vapor. Tritium is accumulated in the biomass proportionally to the growing biomass containing on average 85 wt.% of water.[6,7] The biomass value was measured as a difference between the weights of the contaminated and reference polymer samples. The radiation intensity was measured on a Mark-388 liquid-phase scintillation counter.

Spread of microorganisms was assessed from changes in the amount of dry biomass per unit surface of a sample.

Figure 14-1 shows the kinetics of biomass build-up on various polymeric surfaces. The biomass variations on all the polymers studied are similar and are satisfactorily described by an empirical equation of the type of

$$\frac{m}{m_\infty} = 1 - \exp(-kt) \tag{14-1}$$

TABLE 14-1 The initial rate of the biomass buildup V_{init}, and the limiting biomass value m_∞ on the surface of various polymericmaterials where **m** is the instantaneous biomass, $\mathbf{m}_\infty$ is the equilibrium biomass, and **k** is the effective rate constant for biomass growth.

Material studied	Treating by a mixture of spores in the Chapek-Dox medium			Treating by a mixture of spores in water		
	m_∞, μg/cm^2	V_{init}, μg/cm^2·day	$k_{eff}\cdot 10^6$, s^{-1}	m_∞, μg/cm^2	V_{init}, μg/cm^2·day	$k_{eff}\cdot 10^6$, s^{-1}
Cellophane	10.5 1	0.60 ± 0.05	1.0	5.80 ±0.60	0.40 ± 0.080	1.5
Polyethylene terephthalate	2.4 ± 0.15	0.16 ± 0.02	1.2	0.27 ± 0.02	0.027 ± 0.003	0.9
Polyethylene	1.5 ± 0.20	0.27 ± 0.02	0.9	0.27 ± 0.02	0.01 ± 0.001	1.3
Polytetrafluoroethylene	1.1 ± 0.1	0.04 ± 0.01	0.7	0.17 ± 0.02	0.01 ± 0.001	0.8
Polymethylmethacrylate	–	–	–	0.91 ± 0.50	–	–
Polyimide	–	–	–	0.05 ± 0.01	–	–

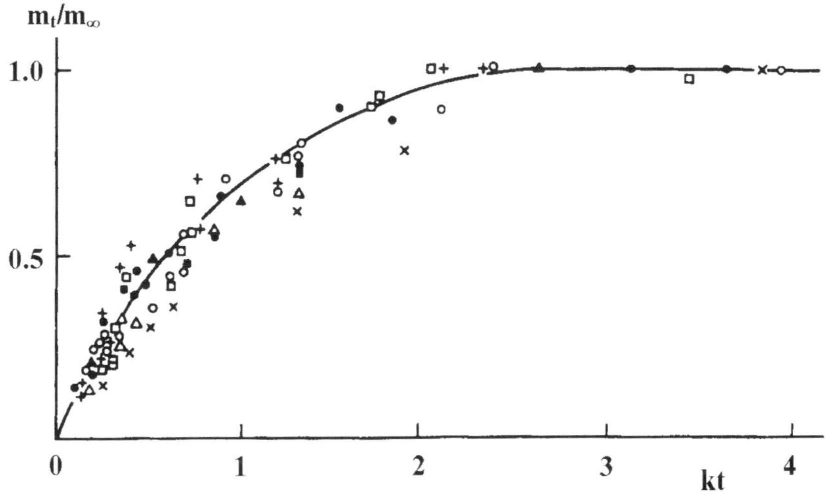

FIGURE 14-2 The generalized kinetic curve of the growth of *Aspergillus niger* fungus on the surface of various polymers at a humidity of 90% and T = 30°C. The fungus layer is applied from the aqueous suspension:

•–PE, o–PMMA, ×–PETP, □–cellulose, ■–PTFE; from the Chapek-Don medium: +–PE, Δ–PETP, ▲–PMMA.

Table 14-1 lists the values of $\mathbf{m}_\infty$, the initial rate of biomass growth $\mathbf{V}_{init}$, and $\mathbf{k}_{eff}$ determined from the logarithmic plot Eq. (14-1) for two sorts of spore suspensions–in water and in the Chapek-Don culture medium. The values of $\mathbf{m}_\infty$ and $\mathbf{V}_{init}$ for the suspensions in the culture medium is twice as higher as that in water, while the $\mathbf{k}_{eff}$ for these two media are nearly equal. The higher (as compared to other polymers) bio-overgrowth on cellulosic polymers is noteworthy. This is associated with their hydrophily. Indeed, the $\mathbf{k}_{eff}$ and $\mathbf{m}_\infty$ values are greater, the more soluble polymer in water (see inset in Figure 14-1), which enables the biomass growth to be predicted for polymers with known water solubilities in the polymer. The generalized kinetic curve plotted in the $\frac{m}{m_\infty} - kt$ coordinates is presented in Figure 14-2. It follows from this figure that all the biomass values fit a single curve whatever the nature of polymers, culture media, and microorganism types.

Similar dependences are also obtained for inhibited biomass growth in the presence of biostabilizers referred to as biocides.[8]

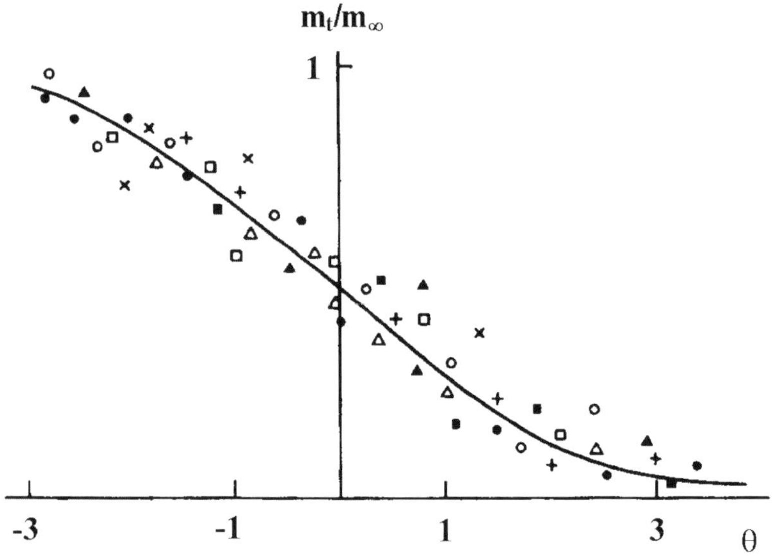

FIGURE 14-3 Generalized kinetics of inhibited biomass growth in the presence of biocides: o–Ionol, •–Phlamal, □–Merthiolate, Δ–$CuSO_2$, ■–nitsedin, +–oxyphenyl, ▲–alkylbenzyl-dimethylammonium chloride (ABDM), ×–*N-para*-tolylmaleimide.

The generalized kinetic plot for the build-up of *Aspergillus niger* microscopic fungus in the presence of various biocides is displayed in Figure 14-3, where the relative biomass amount is described by the following empirical equation:

$$\frac{m}{m_\infty} = \left(1 + e^{\theta}\right)^{-1} \tag{14-2}$$

where $\theta = \ln \mathbf{a} + \ln \mathbf{b}$; **a** and **b** are constants specifying biocide affinity.

14.3 BIODEGRADATION

As mentioned earlier, biomass buildup or long-time contact of a material with a biological medium is attended by degradation processes.

Bioaging of polymers is most frequently appraised from variations of the mechanical properties of the samples, for example, breaking stress σ. For the majority of polymers the correlation between σ and the reciprocal average-number polymerization degree $\left(\overline{P}_n\right)$ is linear:

$$\sigma = A - \frac{P}{\bar{P}_0}. \tag{14-3}$$

For polymeric implanted materials degrading, according to the law of random processes, we can derive a simple formula relating σ to the implantation time:

$$\frac{\sigma}{\sigma_0} = 1 - \frac{k}{\sigma_0}t. \tag{14-4}$$

Denoting the time of complete strength loss by

$$\tau = \frac{\sigma_0}{k} \tag{14-5}$$

yields

$$\frac{\sigma}{\sigma_0} = 1 - t \cdot \tau \tag{14-6}$$

Figure 14-4 displays the generalized plot in the coordinates of Eq. (14-6) for bioaging of various medical polymers in the subcutaneous tissue of rabbits.[9,10] Thus, upon estimating the value of **k** from the initial portions of the kinetic curves, one can calculate by Eq. (14-5) the time of complete loss of the strength of the material implanted in a living organism.

The relative change in the breaking stress of elementary fibers of a canvas tissue upon incubation in it of a suspension of *Aspergillus niger* spores ($\mathbf{C}_0$ = 10^6 spore/ml) is shown in Figure 14-5. The value of σ drops linearly with time due to material degradation on the fiber surface:

$$\frac{\sigma}{\sigma_0} = 1 - \frac{k_{\text{eff(surf)}}}{l_0 \rho} \tag{14-7}$$

where ρ is the density, and **l** is the film thickness or fiber diameter.

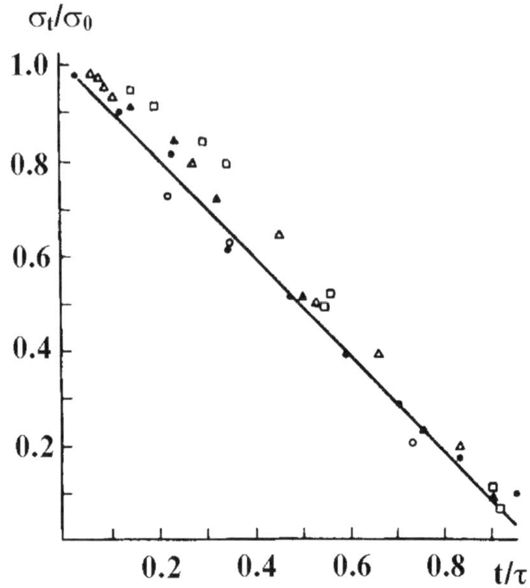

FIGURE 14-4 Generalized dependence of the relative change in the broking stress of various polymers implanted into the subcutaneous tissue of rabbits: •–PE, Δ–polyglycolide, ▲–polyglactine, □–polyamide, –polypropylene.

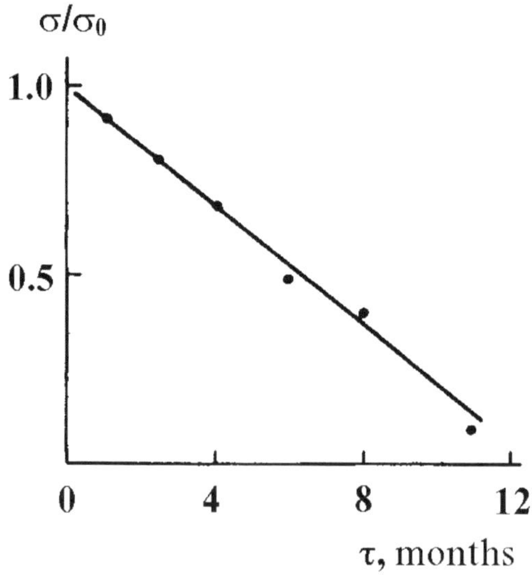

FIGURE 14-5 Relative values of the breaking stress for fibers of a woven cellulose-based cloth as a function of time during which microscopic *Aspergillus niger* fungi at a concentration of 10^6 cell/ml act on the sample.

Biodegradation of any particular polymer in the medium of a living organism is quantitatively specified by effective rate constants for cleavage of different bonds under the effect of hydrolyzing factors. Previously, in refs. [10, 11] we have shown that polyethylene terephthalate undergoes bulk degradation only in an acidic medium of the organism (e.g., when inflammation processes take place in it) with a rate constant of $1.3 \cdot 10^{-6}$ s^{-1}, whereas polycaproamide degrades via the mixed mechanism (both on the surface and in the bulk) with rate constants of $0.15 \cdot 10^{-8}$ s^{-1} and 10^{-1} s^{-1}, respectively.

In the final run, each particular polymer must be characterized by a set of effective degradation rate constants, as listed in Table 14-2.

TABLE 14-2 Effective rate constants for degradation of biomedical polymers in various media of living organisms*.

Polymer	k_{acid}, s^{-1}	k_{sal}, s^{-1}	k_{H_2O}, s^{-1}	k_{enz}, s^{-1}
Polycarpoamide	–	$0.15 \cdot 10^{-8}$	10^{-10}	–
Polyglicolide	–	$1.30 \cdot 10^{-4}$	$0.8 \cdot 10^{-3}$	$5.00 \cdot 10^{-4}$
Polyglactine	–	$5.00 \cdot 10^{-4}$	$0.9 \cdot 10^{-3}$	$0.13 \cdot 10^{-4}$
PETP	$1.3 \cdot 10^{-6}$	–	–	–
Polyethylene	–	–		$1.2 \cdot 10^{-9}$ [11]
Cellulose	–	–		$0.5 \cdot 10^{-6}$

* The data obtained by the authors.

14.4 KINETICS OF MICROORGANISM ADHESION

Biodegradation, as a result of microorganism interaction with the polymer surface, starts with the adhesion process. The adhered cells act as aggressive bioagents by virtue of exoenzymes or other low-molecular species evolved by the organism. Therefore, the quantitative adhesion parameters govern the rates of bio-overgrowing (biomass build-up) and biodegradation.[12]

The adhesion strength ($\mathbf{F}_{adh}$) as measured by the centrifugal break-off technique is a parameter amenable to quantitative analysis. To estimate the value of $\mathbf{F}_{adh}$, suspension of microorganism cells (10^6 spore/ml) was applied to the polymer film surface and incubated for a period of time under various external conditions (temperature 7°C and humidity ϕ, %). Then, the films were fastened on metal plates and centrifuged in the field of forces directed normally to the surface. The number of cells (γ) torn from the surface in a given field of forces and collected in the centrifuge cup filled with distilled water were counted under microscope. Adhesion was quantified by the force of spore break-off of the polymer surface:

$$F = \left(\frac{1}{675}\right)\pi^3 r^3 \omega^3 R\left(\rho_{\text{cell}} - \rho_{\text{liq}}\right) \tag{14-8}$$

where **r** is the spore radius, ω is the angular rotation velocity, **R** is the distance from the rotor axis, ρ_{cell} is the spore (cell) density, and ρ_{liq} is the density of a liquid in which the spores are torn off.

TABLE 14-3 Parameters of *Aspergillus niger* spore adhesion to the surface of various polymeric materials.

Material	k, h^{-1}	F^{50}, dyn/cell	γ_∞, %
Polyethylene	0.06	$3.3\cdot10^{-4}$	55
Epoxy resin	0.08	$6.6\cdot10^{-4}$	70
Polymethylmethacrylate	0.36	$1.0\cdot10^{-3}$	80
Cellophane	0.36	$1.6\cdot10^{-3}$	85
Acetylcellulose	0.41	$2.5\cdot10^{-3}$	90
Polyethyleneterephthalate	0.41	$2.5\cdot10^{-3}$	90

Typical dependences of γ_F on the time of microorganism spore holding on the surface are **S**-shaped adhesion curves specified by two parameters: the equilibrium adhesion number under given conditions, γ_∞, and the rate constant of formation of the adhesion forces, $\mathbf{k}_{eff}$, which is determined from the expression describing the experimental curves:

$$\ln\frac{\gamma}{\gamma_\infty} = -k_{\text{ads}}t \tag{14-9}$$

The effect of the material type on adhesion of *Aspergillus niger* spores is shown in Figure 14-6, which displays the time histories of adhesion under certain thermal and humidity conditions for various polymers. The adhesion curves level off in 24 h. The calculated rate constants for formation of the adhesion forces γ_∞, and the adhesion forces themselves are listed in Table 14-3.

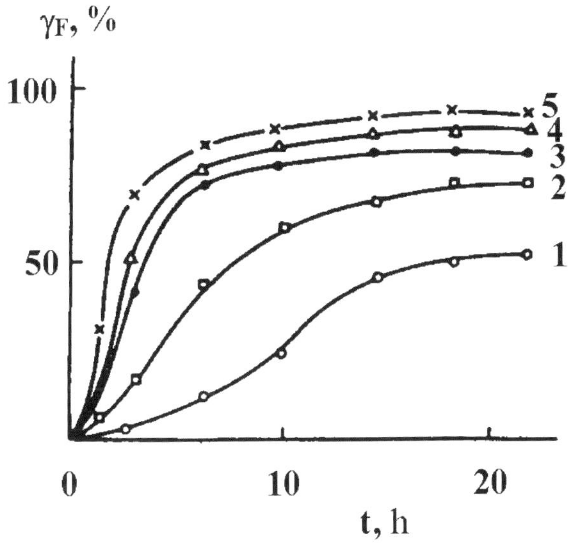

FIGURE 14-6 Kinetics of adhesion of *Aspergillus niger* cells at 22°C and ϕ = 98% to the surface of various polymers: (1) PE; (2) epoxy-polymer; (3) PMMA; (4) acetylcellulose; (5) cellophane.

Polymers can be arranged in the order of increasing spore adhesion from hydrophobic polyethylene to polyethylene-terephthalate.

From the experimental data on the numbers of *Aspergillus niger* spore adhesion at a constant humidity at various temperatures, we calculated the parameters for two polymers with extreme values of water solubility: polyethylene and cellophane (Table 14-4).

TABLE 14-4 Adhesion parameters of the interaction for polyethylene and cellophane at various temperatures and a relative humidity ϕ = 30%.

T, °C	Polyethylene			Cellophane		
	γ_∞, %	k, h^{-1}	F^{50}, dyn/cell	γ_∞, %	k, h^{-1}	F^{50}, dyn/cell
10	85	0.06	$5.2 \cdot 10^{-4}$	90	0.36	$2.6 \cdot 10^{-3}$
22	55	0.06	$3.3 \cdot 10^{-4}$	85	0.36	$1.6 \cdot 10^{-3}$
38	50	0.06	$1.3 \cdot 10^{-4}$	–	–	–

For example, the biostability value inferred from this expression for cellophane (humidity ϕ = 98%) is $0.5 \cdot 10^6 s \cdot cm^2/(dyn \cdot \mu g)$.

The temperature dependence of adhesion is not profound for each polymer, which is supported by the invariable rate constants for the formation of adhesion forces (Table 14-4). In addition, the adhesion force increases with decreasing temperature.

From the kinetic curves for adhesion of *Aspergillus niger* spores to the polyethylene surface at a constant temperature and various humidity of the medium, we calculated the rate constants; it follows from the analysis of these calculations that the **k** values change appreciably, whereas γ_∞ values differ insignificantly (Table 14-5).

TABLE 14-5 Adhesion parameters for *Aspergillus niger* spores on polyethylene at various air humidity and a constant temperature T = 10°C.

ϕ, %	k, h^{-1}	γ_∞, %	F^{50}, dyn/cell
0	0.08	70	$3.0 \cdot 10^{-3}$
30	0.66	85	$5.2 \cdot 10^{-4}$
100	0.56	100	$1.9 \cdot 10^{-3}$

The adhesion parameters were calculated (Table 14-7) from the kinetic curves measured for adhesion of various microscopic fungi to the polyethylene surface shown in Figure 14-7.

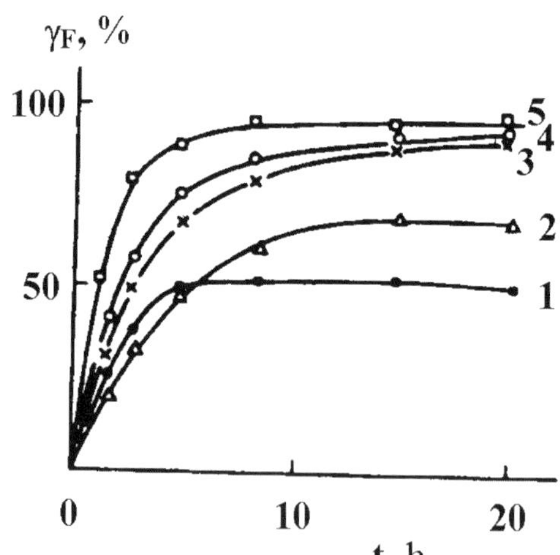

FIGURE 14-7 Kinetics of adhesion of various microscopic fungi to the polyethylene surface: (1) *Aspergillus niger*; (2) *Penicillium cyclopuim*; (3) *Paec. varioti*; (4) *Penicillium chrysogenum*; (5) *Aspergillus terreus*.

The final goal of investigations into the quantitative parameters of adhesion leading to bio-overgrowth is evaluation of the kinetic parameters for polymer biodamage. The experimental data are included in the summarizing table, from which one can infer that the biomass adhesion and growth are related by a linear dependence; as to the degradation process, it calls for additional studies in model media to determine the degradation type. The adhesion number γ_∞ for various fungi ranges between 98 and 55.

The experimental values of the kinetic parameters characterizing interactions of microorganism cells (spores) with a polymer surface make it possible to forecast biological stability of the material. Generally, biological stability (**B**) depends on the adhesion force, biomass amount, and effective rate constants for cleavage of the accessible bonds:

$$B \propto \frac{1}{F_{\text{ads}}} \frac{1}{\Delta m_\infty} \frac{1}{k_{\text{eff}}}$$

TABLE 14-6 Adhesion parameters for microscopic fungi on the polyethylene surface at ϕ = 98% and T = 22°C.

Fungus type	γ_∞, %	**k, h^{-1}**	**r, μm**	**F^{50}, dyn/cell**
Aspergillus niger	80	0.23	5.0 ± 0.5	$0.26 \cdot 10^{-1}$
Aspergillus terreus	94	1.96	1.0 ± 0.05	$0.31 \cdot 10^{-3}$
Paec. Varioti	98	0.30	5.5 ± 0.7	$0.76 \cdot 10^{-1}$
Penicillium Chrisogenum	94	0.40	1.5 ± 0.06	$0.45 \cdot 10^{-3}$
Penicillium cyclopium	55	0.50	2.6 ± 0.4	$0.10 \cdot 10^{-1}$

Note: **r** is the cell (spore) radius of microscopic fungi.

Thus, it is shown that investigations into the kinetics of such macroscopic processes as adhesion, bio-overgrowth, and bio-degradation (Table 14-7) allow the mechanism of complex processes responsible for biostability and biodegradation of polymeric materials to be modelled.

TABLE 14-7 Parameters of polymer biodegradation

Material	Adhesion parameters				Growth			
	k_{eff}^{adh}, s^{-1}	F^{50}, dyn/cell	γ_∞, %	m_∞, μg/cm²	k_{acid}^{eff}, s^{-1}	k_{sal}^{eff}, s^{-1}	k_{H_2O}, s^{-1}	k_{enz}, s^{-1}
Polyethylene (LPPE)	$1.6 \cdot 10^{-5}$	$3.3 \cdot 10^{-4}$	55 ± 5	1.5 ± 0.2	–	–	–	$1.2 \cdot 10^{-9}$
Epoxy polymer	$2.2 \cdot 10^{-5}$	$6.6 \cdot 10^{-4}$	70 ± 5	4.5 ± 0.7	–	–	–	$1.7 \cdot 10^{-7}$
Polymethyl-methacrylate	$2.1 \cdot 10^{-5}$	$1.1 \cdot 10^{-3}$	80 ± 5	3.8 ± 0.7	Was not studied	Was not studied	Was not studied	Was not studied
Polyethylene Terephthalate	$1.1 \cdot 10^{-4}$	$2.5 \cdot 10^{-3}$	90 ± 5	2.4 ± 0.5	$-1.3 \cdot 10^{-6}$	–	–	–
Cellulose	$1.0 \cdot 10^{-4}$	$1.6 \cdot 10^{-3}$	85 ± 5	10.5 ± 1.0	–	–	–	$0.5 \cdot 10^{-6}$
Polyimide	$1.7 \cdot 10^{-4}$	$1.9 \cdot 10^{-3}$	82 ± 5	3.9 ± 0.5	–	$0.2 \cdot 10^{-8}$	10^{-10}	–

Note to Table 14-7: k_{eff}^{adh} is the rate constant for adhesion of microscopic *Aspergillus niger* fungus; **F** is the force of a single cell adhesion to the polymer surface; γ_∞ is the limiting adhesion number; $\mathbf{m}_\infty$ is the biomass per unit surface area; k_{acid}^{eff} is the mass rate constant for degradation in an acidic medium; k_{sal}^{eff} is the mass rate constant for degradation under the effect of phosphate ions; k_{enz}^{eff} is the rate constant of degradation induced by enzymes.[13–21]

KEYWORDS

- **Adhesion**
- **Biodegradation**
- **Biomass Growth**
- **Kinetics**
- **Methods of Investigations**
- **Polymeric Materials**

REFERENCES

1. Emanuel, N. M. *Kinetics of Experimental Tumor Processes (in Rus.)* , Science Publisher (Nauka), Moscow, 1977, 450 pp.
2. *Urgent Problems of Biodamage (in Rus.),* Ed. by Emanuel, N. M. Scientific Council of USSR Academy of Sciences on Biodamage, Science Publisher (Nauka), Moscow, 1983, 520 pp.
3. Andreyuk, E. I.; Bilai, V. I.; Koval', E. Z.; Kozlova, I. A. *Microbe Corrosion and Its Initiators (in Rus.),* Science Publisher (Naukova Dumka), Kiev, 1980, 480 pp.
4. Kinley, Mc.; Kelton, R.; *Verh. Int. Ver. Theor. and Limnol.*, 1981, **21**, 1348 pp..
5. Mazur, P. P. *Biological Testing of Timber by Virtue of Radioactive Isotope Technique (in Rus.),* State Building Publisher (Gosstroiizdat), Moscow, 1959, pp. 19 and 79.
6. Paton, A. M. *Intern. Biodeter. Bull.*, 1982, **18**(1), 3 pp.
7. Zhdanova, N. N.; Vasilevskaya, A. I. *Experimental Ecology of Pungi m Nature and Tests (in Rus.),* Science Publisher (Naukova Dumka), Kiev, 1982, 117 pp.
8. Gumargalieva, K. Z.; Kalinina, I. G.; Zaikov, G. E. *Polymer Degrad. and Stab.*, 1999, **45**, 5–17 pp.
9. Gumargalieva, K. Z.; Moiseev, Yu. V.; Daurova, T. T. *Biomaterials*, 1950, **1**, 213–218 pp.
10. Moiseev, Yu. V.; Daurova, T. T.; Voronkova, O. S., et al., *J. Polym. Sci., Polym. Symp.*, 1979, **66**, 269–278 pp.
11. Moiseev, Yu. V.; Zaikov, G. E. *Chemical Stability of Polymers in Aggressive Media (in Rus.),* Chemistry Publishing House (Khimiya), Moscow, 1979, 252 pp.
12. Marshall, K. C. *Microbial Adhesion and Aggregation*, Springer Publisher (Verlag), Berlin, 1984.
13. Zaikov, G. E. Biochemical Physics, Nova Science Publ., New York, 2007, 388 pp.
14. "Modeling and Prediction of Polymer Nanocomposite Properties." Ed. by Mittal, V. Wiley-VCH Verlag GmbH & Co. KGaA, 2013, Kozlov, G. V.; Yanovsky, Yu. G.; Zaikov, G. E. pp. 39-62, "Modern Experimental and Theoretical Analysis Methods of Particulate-Filled Nanocomposites Structure."
15. "Organic Chemistry, Biochemistry, Biotechnology and Renewable Resources. Research and Development, Volume 1, Today and Tomorrow." Ed. by Zaikov, G.; Pudel, F.; Spychalski, G. Nova Science Publ., 2013, 212 pp.

16. "Organic Chemistry, Biochemistry, Biotechnology and Renewable Resources. Research and Development, Volume 2, Tomorrow and Perspectives." Ed. by Zaikov, G. E.; Stoyanov, O. V.; Pekhtesheva, E. L. Nova Science Publishers, New York, 2013, 398 pp.
17. "New Steps in Physical Chemistry, Chemical Physics and Biochemical Physics." Ed. by Zaikov, G. E.; Pearce, E. M.; Kirshenbaum, G. Nova Sci. Publ., 2013, 330 pp.
18. "Applied Methodologies in Polymer Research and Technology." Ed. by Hamrang, A.; Balkose, D.; Zaikov, G. E.; Haghi, A. K. Apple Academic Press, 2015, 245 pp.
19. "Polymers and Polymeric Composites: Properties, Optimization and Application," Zaikov, G. E.; Bazylak, L. I.; Haghi, A. K. Apple Academic Press, Toronto-New Jersey, 2014, 278 pp.
20. "High Performance Elastomer Materials: An Engineering Approach." Ed. by Bielinski, D. M.; Kozlowski, R.; Zaikov, G. E. Apple Academic Press, Toronto-New Jersey, 2015, 268 pp.
21. "Ecological Consequences of Increasing Crop Productivity: Plant Breeding and Biotic Diversity." Ed. by Opalko, A. I.; Weisfeld, L. I.; Bekuzarova, S. A.; Bome, N. A.; Zaikov, G. E. Apple Academic Press, Toronto-New Jersey, 2015, 478 pp.

CHAPTER 15

REACTION OF TELOMERIZATION OF ETHYLENE AND TRICHLORACETIC ACID ETHYL ESTER

NODAR CHKHUBIANISHVILI and LALI KRISTESASHVILI

Georgian Technical University
E-mail: rusikoch@yahoo.com

CONTENTS

ABSTRACT

The interest toward radical telomerization is caused by the fact that it enables to obtain comparably low-molecular long-chain organic substances with various functional groups that are hard to obtain with ordinary methods and that can be widely applied in practice. The reaction of telomerization of ethylene and trichloracetic acid ethyl ester[1] is especially interesting. The reaction proceeds in the following way:

$$nCH_2 = CH_2 + CCl_3COOEt \xrightarrow{\text{Initiator}} Cl\,(CH_2CH_2)nCCl_2COOEt$$

Telomers, received from ethylene and trichloracetic acid esters, are utilized to obtain the psychotropic drug Gamalon, ε-Capron (ε-hexanoic acid), and biological active substances Lysin and methion.

15.1 RESULTS AND DISCUSSIONS

15.1.1 TELOMERIZATION OF ETHYLENE AND TRICHLOROROACETIC ACID

In order to study this reaction, a series of autoclave experiments were conducted: we had an experiment with the equipment which is presented in Figure 15-1. It was a 200cm^3 volume stainless steel autoclave (1) which was equipped with electromagnetic stirrer (2) and a pocket of thermocouple. We added solution of initiator to telogen. During each experiment 103.5gr (75ml) trichloracetic acid ethyl ester was loaded into the autoclave under the conditions of high pressure and closed valves (4, 5, 11). The autoclave was cooled at up to -20–25°C and the vacuum in the system was created through the valve (6) connected to a vacuum line. The residual pressure was 5–10mmHg. Then we closed the valve (6), opened valves (4, 11), and delivered ethylene to evacuate air. Ethylene was delivered from the balloon (7). The autoclave was warmed up to room temperature, electromagnetic stirrer was turned on and telogen was saturated with ethylene to defined pressure of manometer (9). Once pressure decay was stopped, the autoclave was placed into the thermostat (8) and heated to the reaction temperature. Liquid silicone was used as a heat transmitter. The stirring in the autoclave was done with nitrogen sparging. The temperature was regulated via contact thermometer connected to the network through electromagnetic switch. The control over telomerization was implemented according to the pressure decay on the manometer (9). The reaction was conducted until the drop of ethylene pressure was finished. Calculation of the amount of ethylene loaded into the autoclave was determined by ethylene amounts reacted (based on the data of telomers composition) and returned after the reaction or by solubility of ethylene.

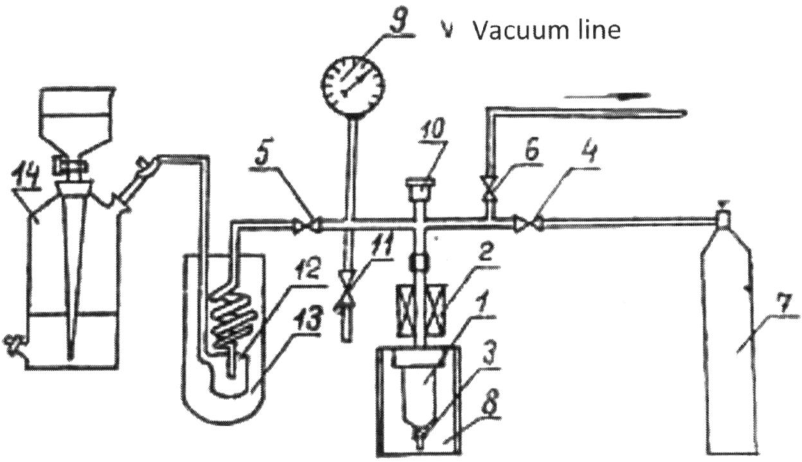

FIGURE 15-1 The scheme of the equipment to carry out the reaction of telomerization in the intermittently operational autoclave.

For security reasons, the equipment was outfitted with blasting plate (10) designed for 200 atmosphere pressure and emergency valve (11).

After finishing the reaction, autoclave was cooled down to room temperature. Throttling of nonreacted ethylene was done in fixer (12). It was cooled down to -15–20°C in Dewar vessel and was collected in gasometer (14). The fixer was installed to hold the reaction mass taken away during throttling. After throttling of air, reaction products were removed, nonreacted initial ester was removed by distillation and the mix of telomeres was distilled on the rectifying column of 7–10 theoretical plate effectiveness.

Samples were analyzed on air-liquid chromatograph, on heat-conducting detector.

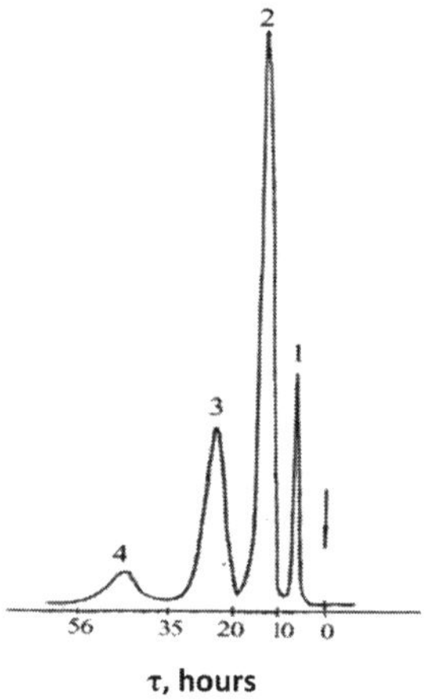

FIGURE 15-2 Chromatogram of telomer mix 1–n_1; 2–n_2; 3–n_3; 4–n_4.

It was established that the product chromatogram had four peaks (see Figure 15-2):

1 – corresponded to telomer n_1, 1,1,3-thrichlorerbor acid ethyl ester
2 – corresponded to telomer n_2, 1,1,5-thrichlorcapron acid ethyl ester
3 – corresponded to telomer n_3, 1,1,7-thrichlorcapril acid ethyl ester
4 – corresponded to telomer n_4, high molecule esther

Besides chromatographic analysis, identity of telomers was established by boiling temperature, density, refraction, infrared spectroscopy, and element analysis.

15.1.2 THERMODYNAMICS OF THE REACTION OF TELOMERIZATION OF ETHYLENE AND TRICHLORACETIC ACID ETHYL ESTER

The telomerization reaction is an exothermic process. The approximate estimation of thermal effect of the reaction is possible through the energy of chemical bonds. The total thermal effect of telomerization reactions is mostly determined by the energies of breaking double C=C bonds and generation of single C-C bonds.

There are no data available in the literature about thermodynamic characteristics of the reaction of telomerization of ethylene and trichloracetic acid ethyl ester. That was why we estimated standard thermal effect of the reaction, coefficients of heat capacity equation $c^p = \varphi(T)$, and entropy for initial and final products of the reaction based on molecular structure with group revision method.[2] Results are reported in Table 15-1. Figure 15-3 shows dependence of ΔF and LgK_p values on temperature. Approximate values of temperature allowing the reaction to be conducted thermodynamically are determined graphically.

TABLE 15-1 Standard thermal effects and entropy increment of reactions

#	Reaction	ΔH_{298} kcal/mol	ΔS_{298} kcal/mol
A	$CCl_3COOC_2H_5 + C_2H_4 \rightarrow Cl(CH_2CH_2)CCl_2COOC_2H_5$	-27.7	-31.1
B	$CCl_3COOC_2H_5 + 2C_2H_4 \rightarrow Cl(CH_2CH_2)_2CCl_2COOC_2H_5$	-50.8	-65.5
C	$CCl_3COOC_2H_5 + 3C_2H_4 \rightarrow Cl(CH_2CH_2)_3CCl_2COOC_2H_5$	-73.9	-100.1

As long as the reaction of telomerization takes place in liquid but not in air phase, calculations of thermal effects of reaction should take into account the heat of phase transition, the heat of condensation of trichloracetic acid ethyl ester and telomers, and the heat of dilution of ethylene into the telogen. The heat of dilution of telomers in telogens can be ignored. Combination of reaction equation and phase transformation equation leads to the following:

1. For 1,1,3 – trichlor butyric acid ethyl ester (telomer n_1):

$$\Delta H'_{c_6} = \Delta H_{c_6} - \Delta H_{telog.dilut.} + \Delta H_{telog.evapor.} - \Delta H_{c_6 evapor.}$$

2. For 1,1,5 – trichlor hexanoic acid ethyl ester:

$$\Delta H'_{c_8} = \Delta H_{c_8} - 2\Delta H_{ester.dilut.} + \Delta H_{telog.evapor.} - \Delta H_{c_8 evapor.}$$

3. For 1,1,7 – trichlor hexyl acetic acid ethyl ester

$$\Delta H'_{c_{10}} = \Delta H_{c_{10}} - 3\Delta H_{ester.dilut.} + \Delta H_{telog.evapor.} - \Delta H_{c_{10} evapor.}$$

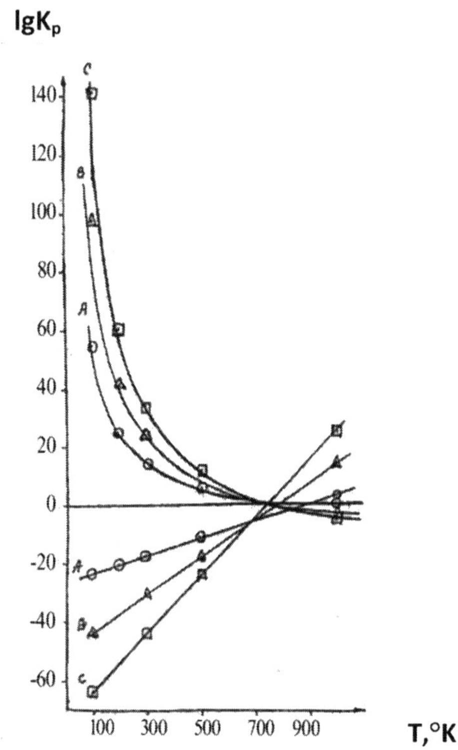

FIGURE 15-3 The dependence of equilibrium constant logarithm and ΔF isobaric potential on temperature for A, B, C telomerization reactions.

The heat of evaporation of trichloracetic acid and telomers is calculated with respective empirical formula through extrapolation of boiling temperature data:

$$\Delta H_{\text{evapor}} \approx (8.75 + 4.575 \ \lg T_{\text{boil.}})^{*} T_{\text{boil.}}$$

In Figure 15-3, the heat of ethylene dilution in trichloracetic acid ethyl ester is ~ 3.5kcal/mol=14.5kJ/mol.

$$\Delta H^{'}_{c_6} = 27.7 + 3.5 + 9.5 - 10.5 = 30.2 \text{ kcal / mol} = 126.5 \text{ kJ / mol}$$

$$\Delta H^{'}_{c_8} = 50.8 + 23.5 + 9.5 - 11.4 = 55.9 \text{ kcal / mol} = 234.2 \text{ kJ / mol}$$

$$\Delta H^{'}_{c_{10}} = 73.9 + 3.35 + 9.5 - 11.9 = 82.0 \text{ kcal / mol} = 343.6 \text{ kJ / mol}$$

The data show that thermal effect of these reactions in liquid phase is slightly different from that in air phase. Therefore, thermal effect of telomerization of ethylene and trichloracetic acid ethyl ester is on average 28.5 kcal (119.3 kJ/mol) on 1 mol ethylene, which is close to polymerization heat of ethylene (30 kcal/mol).

15.1.3 SOLUBILITY OF ETHYLENE IN TRICHLORACETIC ACID ETHYL ESTER

To study the kinetics of telomerization reactions, having knowledge of the condition of air and liquid phases is of big importance as it determines liquid and air phase composition, dependence of the reaction speed on components' concentration.

The solubility of ethylene was studied on special equipment (Figure 15-4).

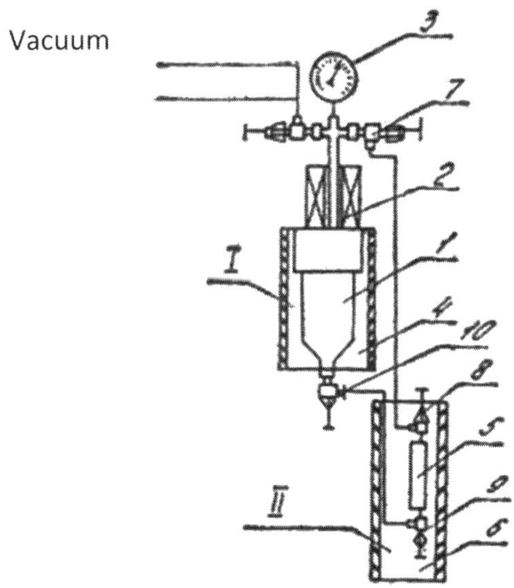

FIGURE 15-4 The scheme of the equipment to study ethylene solubility.

Creation of vacuum in the equipment was done under the condition of opened valves of both systems. We closed valves (10) and (7) and loaded trichloracetic acid ethyl ester into the autoclave. The thermostat of the autoclave was cooled down to -15–20°C. Then we opened valve (7) and created vacuum in the system. Subsequently, we closed the vacuum line and supplied the systems with ethylene. Required temperature was reached in both thermostats and the stirrer was turned on. We turned the equipment off from ethylene delivery line and observed the termination of the saturation of trichloracetic acid ethyl ester with ethylene by ending the pressure decay. After that, we closed the stirrer, opened the valve (10) and the systems remained under similar temperature during one hour. Then the valves (7–10) were closed and the second system was separated from the first. The solubility of ethylene was studied under different pressure and temperature.

Air solubility depends on the nature of air and solvent, temperature, and pressure. The association between air solubility and partial pressure is expressed by Henry's Law:

$$N_2^{'} = kP_2,$$

Where: $N_2^{'}$ – a molar fraction of liquid soluble air
P_2 – partial pressure of a given air above solution
k – Henry's law constant (atmospheric/molar fraction)

Ethylene solubility under different pressure and temperature is shown in Figure 15-5. The figure shows that ethylene solubility is directly proportional to ethylene pressure and responds to Henry's law. Out of experimental data, we calculated Henry's law constant at 25, 90, 100, 120°C and it was respectively 0.428, 1.195, 1,395 and 1. 725.

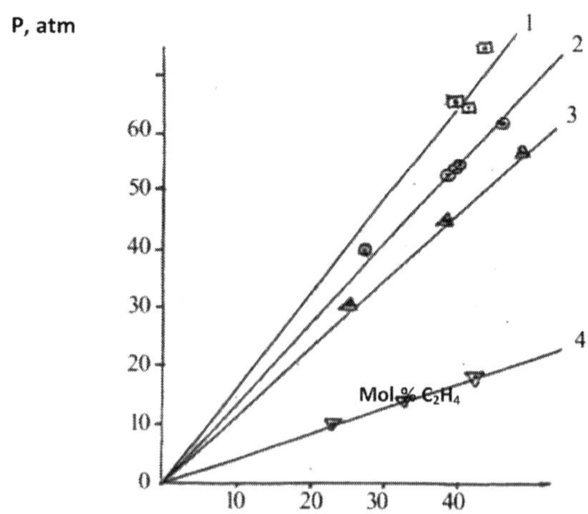

FIGURE 15-5 Ethylene solubility in trichloracetic ethyl ester at various pressure and temperature 1–120°C; 2–100°C; 3–90°C; 4–25°C.

Based on these data we found the heat of ethylene dilution in trichloracetic acid ethyl ester. For this reason, we designed the curve representing association between logarithm of Henry's law constant and inverse of absolute temperature. This association is shown in Figure 15-6, which is drawn as a straight line and tangent of its angle is a thermal heat of ethylene.

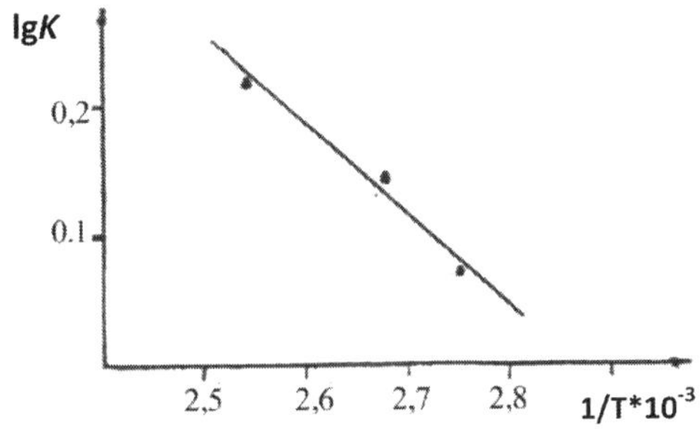

FIGURE 15-6 Dependence of logarithm of Henry's law constant on inverse of absolute temperature in the system—ethylene trichloracetic acid ester.

The diagram and Arrhenius equation help determine ethylene dilution heat, which is approximately 3.5kcal/mol.

Figure 15-7 shows the dependence of molar volumes of studied solutions on pressure and temperature. It is clear that the change of molar volumes of solutions is directly proportional to pressure in the area of temperature and pressure of the research interest.

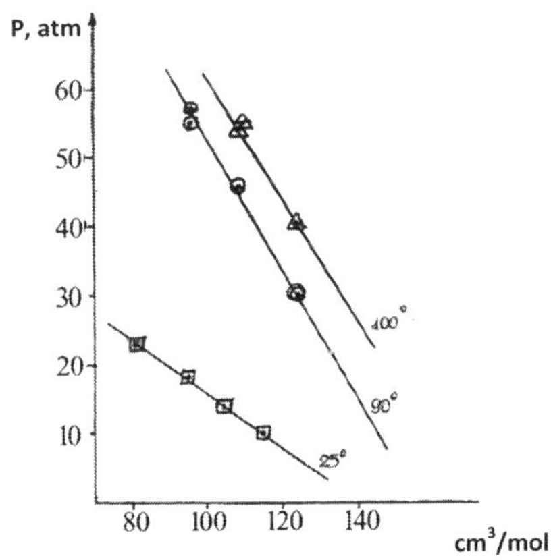

FIGURE 15-7 Dependence of molar volumes of ethylene solution on pressure at 25, 90, and 100°C temperature.

The goal of conducted series of autoclave experiments of telomerization of Ethylene and trichloracetic ethyl ester was to study the following issues:

1. Effect of temperature on the yield and composition of telomers
2. Effect of concentration of initiator on the yield and composition of telomers
3. Effect of mixture of initial reagents on the composition of telomers

15.1.4 EFFECT OF TEMPERATURE

One of the important factors for telomerization reactions is reaction temperature. It determines the velocity of dissociation of the initiator and reaction potential of reacting components and affects the chain transmission constants. The effect of temperature on ethylene conversion was studied under the condition of loading equal quantities of reagents. Figure 15-8 depicts the dependence of ethylene conversion on temperature.

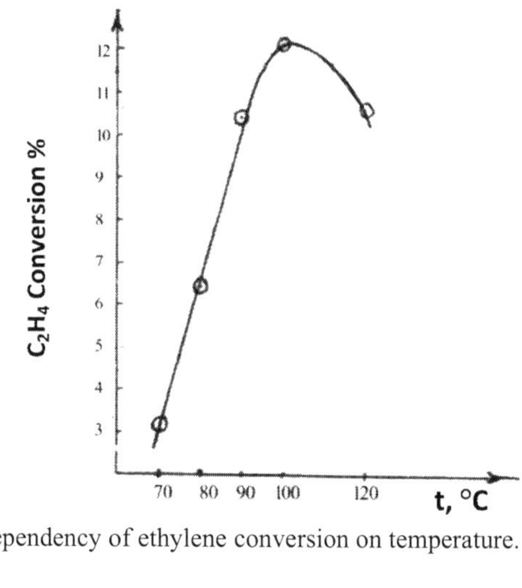

FIGURE 15-8 Dependency of ethylene conversion on temperature.

The highest conversion of ethylene is achieved at 100°C (12.3mol%). Conversion is reduced through decreased temperature, which can be attributable to the incomplete dissociation of initiator. Temperature increase above 100°C apparently considerably raises the recombination velocity of initiator's free radicals.

Figure 15-9 describes the dependence of pressure decay on time at various temperatures. As far as the velocity of telomerization reaction is limited by the velocity of dissociation of initiator, the data of pressure decay can be used to determine the velocity of telomerization. As shown in Figure 15-9, the duration of reaction at 70°C is 10.5 hours, while the reactions at 100 and 120°C end in ~ 4.5 and 1.5 hours, respectively.

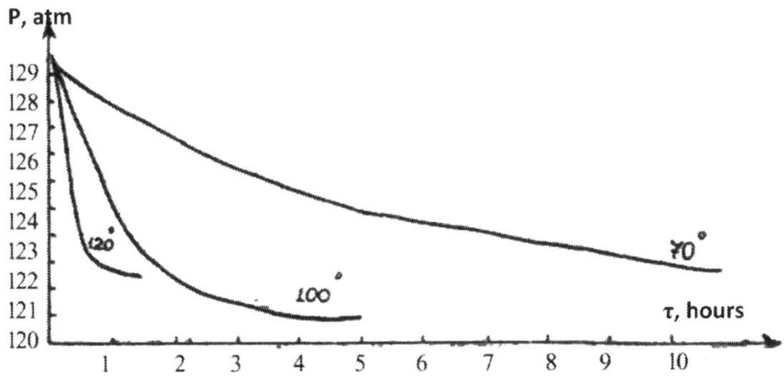

FIGURE 15-9 Pressure drop dependence on time at various temperatures.

15.1.5 EFFECT OF INITIATOR

According to the kinetic equation of telomerization, the velocity of telomer yield is proportional to initiator concentration in power ½ (for squared discontinuity and disproportionation) or 1 (for allylic discontinuity). The study of the effect of quantity of initiator on telomer compositions and yields was conducted in the following telogen concentration intervals: BP(Benzoyl Peroxide) $11 \cdot 10^{-3}$–$82 \cdot 10^{-3}$ mol/liter and AIBN (*azo-bis-izobutironitrile)* $24 \cdot 10^{-3}$–$163 \cdot 10^{-3}$ mol/liter.

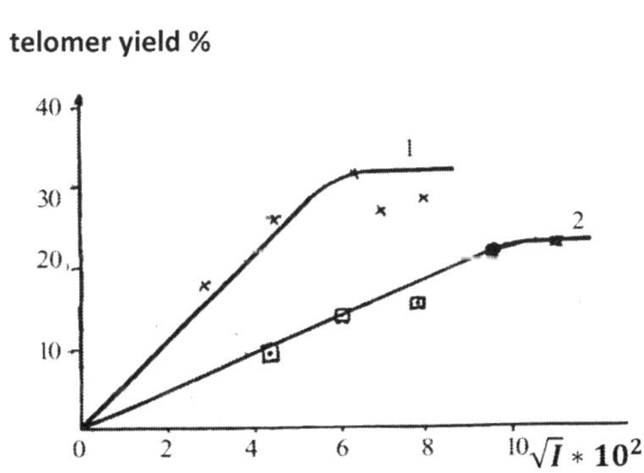

FIGURE 15-10 Effect of initiator on telomer yields 1–Benzoyl peroxide t=120°C, τ =2.5 hours 2–AIBN t=100°C, τ =1.5 hours.

As we see from Figure 15-10, with the increase of concentrations of initiators BP and AIBN up to some limit, the linear dependence between telomer yield and square root of initiator concentration takes place. It is found that the linear dependence takes place up to $27.5 \cdot 10^{-3}$ mol/liter for BP and up to $121.9 \cdot 10^{-3}$ mol/liter for AIBN. Above these concentrations, linear dependence is no longer in place.

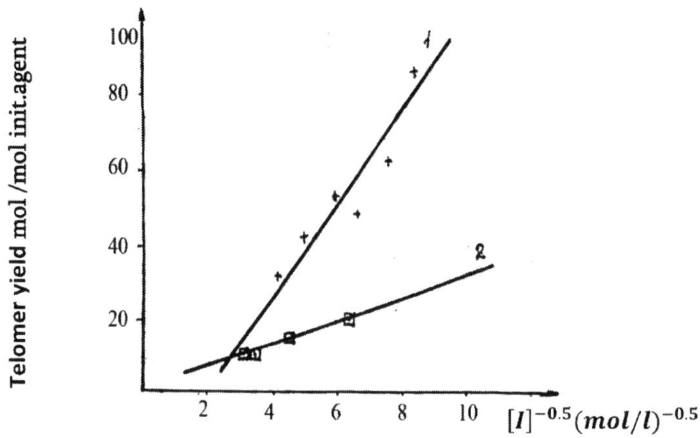

FIGURE 15-11 Effectiveness of initiator.

1–Benzoyl peroxide, t=120°C, τ =2.5 hours

2–AIBN, t=100°C, τ =1.5 hours

Based on Figure 15-11, for BP with average concentration of $27.5 \cdot 10^{-3}$mol/liter telomers are received with ~49 mol yield and for AIBN with similar concentration telomers are received with ~18 mol yield per mole of initiator. Therefore, BP is 2.7 times as effective as AIBN. Thus, the length of kinetic chain for BP is ~100 and for AIBN it is ~40. Derived from experimental data, chain transmission constants are calculated at 100°C: $c_1 = 94 \cdot 10^{-3} \pm 12.6 \cdot 10^{-3}$; $c_2 = 1.76 \pm 0.24$; $c_3 = 3.26 \pm 0.42$.

15.1.6 EFFECT OF ETHYLENE MOLAR CONCENTRATION ON THE COMPOSITION OF TELOMERS

Figure 15-12 describes the dependence of telomer composition change on molar composition of ethylene in the mixture at 100°C.

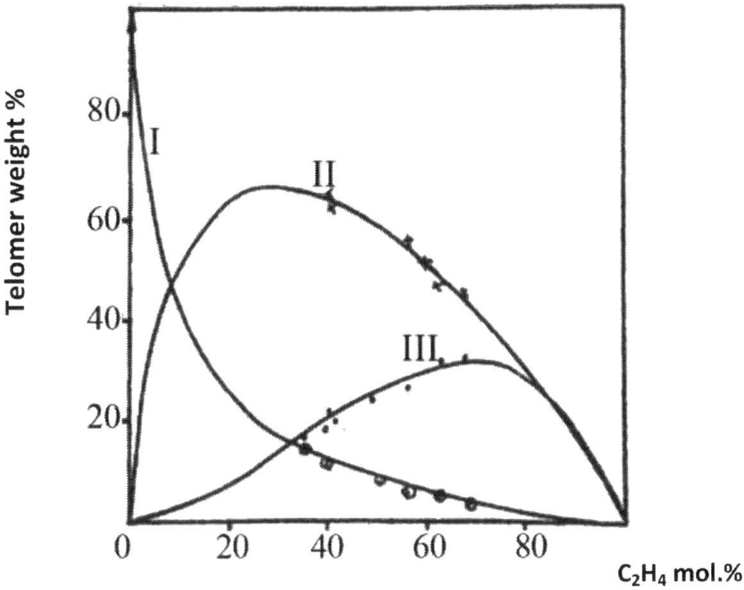

FIGURE 15-12 I–telomer n=1; II–Telomer n=2; III–telomer n=3.

As we see in the diagram, when ethylene concentration goes to zero in the original mixture, telomer n1 composition reaches 100%. In case of significant increase of ethylene concentration in initial mixture, the quantity of lower order telomers approaches zero and upper order telomers ($n>3$) will be the major products of telomerization reaction.

The curves n=2 and n=3 have maximums. Therefore, the adequate composition of ethylene in the mixture to receive n=2 (70 weight %) is 45 mol % and to receive telomer n=3 (33 weight %) is 80 mol %.

According to the data from this study, it is recommended to conduct telomerization reaction of ethylene and trichloracetic acid ethyl ester under the following conditions:

- Initiator---------------------------- BP (0.5 mol/l telogen)
- Telogen/Oliphin ratio ----------- 1:0.8
- Temperature---------------------- 100°C
- Reaction time-------------------- 1 hour

KEYWORDS

- **Ethylene**
- **Ethyl Ester**
- **Mechanism**
- **Reaction**
- **Reactivity**
- **Telomerization**
- **Trichloracetic Acid**

REFERENCES

1. Nodar, C. Thermodynamics of telomerization reaction of ethylene and trichloracetic acid ethyl. Georgian Chemical Journal, 2009, vol. 9, no. 2, pp.130–132. (in Georgian).
2. Karapetyants, M. Kh. Chemical thermodynamics, 4th ed., Moscow, Khimiya Publication, 2013, 584 p. ISBN 987-5-397-03700-6. (in Russian).
3. Chkhubianishvili, N.; Kristesashvili, L. Study of ethylene dilution in trichloracetic acid ethyl ester. The works of Georgian Technical University, 2002, no. 6 (445), pp. 58–60. (in Georgian).

CHAPTER 16

SYNTHESIS AND SPECTRAL-FLUORESCENT STUDY OF PROTEIN COATINGS ON MAGNETIC NANOPARTICLES USING CARBOCYANINE DYES

P. G. PRONKIN, A. V. BYCHKOVA, O. N. SOROKINA, A. S. TATIKOLOV, A. L. KOVARSKII, and M. A. ROSENFELD

Emanuel Institute of Biochemical Physics, Russian Academy of Sciences, Moscow, Russia; E-mail: pronkinp@gmail.com

CONTENTS

ABSTRACT

Formation of the biocompatible coatings on the surface of magnetic nanoparticles (MNPs) opens up prospects for their use in medicine. In this work, magnetite nanoparticles have been obtained with stable coatings formed from bovine and human serum albumins (BSA and HSA) by free radical cross-linking (Fenton reaction). We have studied the interaction of anionic carbocyanine dyes with BSA and HSA composed of the coatings on magnetic nanoparticles. Electronic excitation energy transfer (EEET) between molecules of carbocyanine dyes has been studied both in samples of HSA solutions and in systems containing BSA/HSA coated MNPs. The effect of HSA denaturation on the spectral and fluorescent properties of the dyes and EEET has been studied, and the data on donor fluorescence quenching by the acceptor have been obtained. It has been found that ~90% of albumin molecules forming the MNPs coatings retain their capability of binding with a fluorescent dye. Albumins were shown to retain their functional properties in the cross-linked coating, thereby providing evidence for biocompatibility of the coatings obtained by free radical oxidative protein modification. Carbocyanine dyes are suggested as spectral-fluorescent probes for assessment of the functional properties of the albumin molecules forming the MNPs coatings.

16.1 AIM AND BACKGROUND

Magnetic nanoparticles have many applications in different areas of biology and medicine such as hyperthermia, magnetic resonance imaging, immunoassay, cell and molecular separation, a smart delivery of drugs to target cells.[1–7] Magnetic nanosystems (MNSs) in biomedicine consist of one or more magnetic cores and biological or synthetic molecules which serve as a basis for multifunctional coatings on MNPs surface.

The creation of functional coatings capable of targeting the particles to biological objects and possessing therapeutic effects makes possible efficient application of MNPs for solving biomedical problems.[6] Natural macromolecules (including proteins) are promising as materials for creation of coatings on MNPs due to their biocompatibility, an ability to protect magnetic cores from the influence of biological liquids, an ability to prevent agglomeration of MNSs in dispersion, and their possible activity as therapeutic products. [8–10] The creation of stable protein coatings with retention of native properties of molecules is still an important biomedical problem because of disadvantages of the commonly used methods such as formation of a polydisperse ensemble of particles, nonselective linking of proteins leading to cross-linking of macromolecules in solution, and desorption of coatings. Serum albumin is one of the most important protein components of blood plasma. The preservation of the native structure and function of albumin upon its immobilization on the surface of nanoparticles provides them with biocompatibility and hinders unwanted adsorption of other proteins.[11]

The aim of this work was to create stable protein coatings (consisting of HSA and BSA) on the surface of individual MNPs using the free radical approach[12] and estimate functional properties and activity of proteins in the MNPs using method of spectral and fluorescent probes. Creation of coatings consisting of HSA and BSA is the first step to obtain functional multicomponent coatings on MNPs. Our objective was also to show the possibility of using fluorescent dye probes (including their use in EEET) to assess the extent of HSA denaturation in solution and in the systems containing protein-coated magnetic nanoparticles.

16.2 INTRODUCTION

At present, bifunctional cross-linking agents (glutaraldehyde, carbodiimide) have been widely used for the creation of protein coatings on nanoparticle surfaces.[13] However, the use of these compounds leads to the formation of polydisperse ensembles of particles as a result of cross-linking of protein molecules belonging to different nanoparticles or, on the contrary, to the desorption of protein molecules from MNPs surfaces as a result of incomplete coating cross-linking[14] Nonselective cross-linking of a surface layer composed of adsorbed molecules and the involvement of all molecules that are present in a reaction system and accessible for a cross-linking agent in the covalent bonding can also take place. The free radical method proposed in patent[12] for the creation of stable protein coatings on MNPs surfaces makes it possible to form coatings on individual nanoparticles. This method is based on the ability of proteins to undergo free radical oxidative modification with the formation of intermolecular covalent bonds.[15] Free radicals are generated on MNPs surface through the Fenton reaction, thereby involving proteins adsorbed on the surface in the free radical cross-linking. The efficiency of the formation of coatings on MNPs by the free radical mechanism has been proven by the work.[16]

Various approaches have been developed to study the processes of protein adsorption on MNPs and the functional properties of proteins with the use of infrared (IR) spectroscopy, UV spectrophotometry, erythrocyte sedimentation rate (ESR) spectroscopy of spin labels, and other techniques.[16] These approaches are of limited use as regard to studying the functional properties of proteins adsorbed on MNPs. Note that adsorption of proteins on the surface of nanoparticles and occurrence of free radical processes in such systems can lead to conformational changes in macromolecules and a loss of their functions (protein denaturation).[17–19] In this work, spectral and fluorescence probes technique was used to assess retaining the natural functional properties of albumins in systems containing magnetite nanoparticles with protein coatings. It is known that denaturation of albumins changes their functional properties and, hence, can affect the noncovalent interaction of the albumins with probe molecules. Anionic meso-substituted polymethine (cyanine) dyes were used as spectral and fluorescence probes since their photophysical and photochemical properties depend on the properties of the medium.[20,21] This is due to the presence of a flexible polymethine chain in their molecules.[21] The application of polymethine dyes as spectral and fluorescent probes is based on the suppression of nonradiative deactivation processes (including cis–trans photoisomerization about C–C bonds of the polymethine chain[22]) in a complex of the dyes with biomacromolecules, which leads to growing fluorescence.[20]

We have studied the noncovalent interaction of thiacarbocyanine dyes 3,3'-di-(γ-sulfopropyl)-9-methylcarbocyanine betaine (D1) and 3,3'-di-(γ-sulfopropyl)-4,5,4',5'-dibenzo-9-ethylthiacarbocyanine betaine (D2) with BSA and HSA in the samples of MNPs. The structural formulas of the dyes are presented in the following:

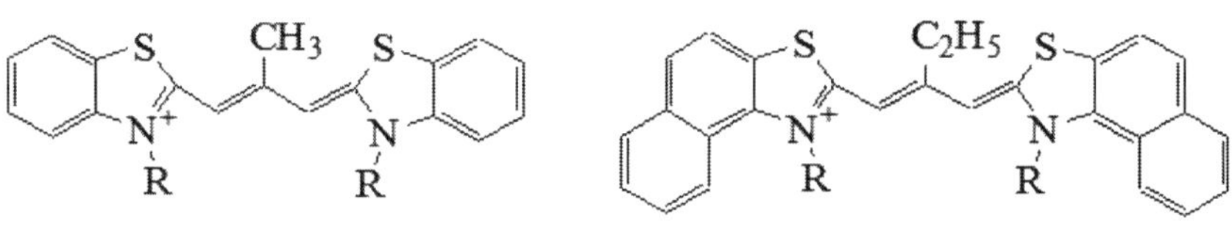

D1, R = $(CH_2)_3SO_3^-$ D2, R = $(CH_2)_3SO_3^-$.

The interaction of these dyes with different serum albumins was studied earlier;[23,24] these dyes were shown to form stable complexes with high binding constants and with spatial fixation of molecular fragments; the interaction with the proteins is accompanied by a shift of the cis–trans equilibrium and a steep rise in the fluorescence quantum yield.

EEET) between molecules of carbocyanine dyes was studied in systems containing protein-coated magnetic nanoparticles. The dye D1 was used as an electronic excitation energy donor, and the dye D2 was used as an acceptor. The results obtained on the BSA/HSA coated MNPs were compared with the data for HSA control solutions. The effect of HSA denaturation on the spectral and fluorescent properties of the dyes and EEET was studied, and the data on donor fluorescence quenching by the acceptor were obtained.

16.3 EXPERIMENTAL PART

Magnetic nanoparticles were synthesized by the method of coprecipitation of di- and trivalent iron salts in aqueous medium in the presence of ammonium hydroxide and the subsequent electrostatic stabilization in the 0.1 M phosphate–citrate buffer (0.05 M NaCl) with pH 4.0. To create the coatings, the new method of free radical linking of proteins on the surface of nanoparticles was used.[6,12,16]

Upon synthesis of MNPs and obtaining coatings, iron(II) sulfate heptahydrate, iron(III) chloride hexahydrate (reagent grade), ammonium hydroxide (high purity grade), sodium dihydrogen phosphate dihydrate (analytical grade), disodium hydrogen phosphate dodecahydrate (reagent grade), citric acid monohydrate (high purity grade), hydrogen peroxide (3%, medical drug, Russia), HSA, BSA, and human fibrinogen (Sigma-Aldrich, USA) were used.

Synthesis of MNPs and the creation of coatings is described in detail in the following.

The particle size was evaluated by the method of dynamic light scattering (DLS). The DLS measurements were performed on a Zetasizer Nano-S (Malvern, England) instrument without additional dilution of the samples. The protein unbound in the coatings was removed from the solution using magnetic separation.[16] The stability of coatings on MNPs was controlled by the competitive substitution of fibrinogen for albumin.[16]

The dyes given by Prof. B.I. Shapiro (NIIKHIMFOTOPROEKT) were used. For protein denaturation, crystalline urea (reagent grade) was used. Dye solutions were prepared immediately before measurements.

Absorption spectra of the dyes were measured with an SF2000 spectrophotometer (Russia), while fluorescence measurements were performed with a Fluorat-02-Panorama spectrofluorimeter (Russia) in standard 1cm quartz cells. Fluorescence and fluorescence excitation spectra were corrected for the reference channel signal and the sample transmission signal. During the experiments, dyes concentrations in solutions were varied in a range of 1–5 × 10^{-6} mol/L; the dyes were introduced into the systems from concentrated stock solutions. Upon fluorescent titration, equal volumes of MNPs and protein-containing samples were placed into a cell with a dye solution. In experimental conditions, the total albumin concentration in cells was lower than 2 mg/ml. Distilled water and 0.1 M phosphate–citrate buffer (0.05 M NaCl) with pH 4.0 were used as solvents. The effective degrees of the dye–albumin interaction in different samples were comparatively estimated in terms of effective protein concentration (c_{eff}, %), which is equal to the fluorescence intensity ratio between probe–protein complexes in MNPs containing and MNPs free (reference) samples.[25]

Eq. (16-1) (Hill's equation) has been used to evaluate the efficiency of dye-protein interaction:

$$\lg\theta = m\lg[Q] - m\lg K_d \tag{16-1}$$

where $\theta=(I_0-I)/I$ is the part of protein binding sites occupied by the dye (I_0 and I are the fluorescence intensities in the absence and in the presence of albumin); $[Q]$ is the concentration of unbound ligand; K_d is the effective dissociation constant of the complex (mol L^{-1}); m is the Hill coefficient characterizing cooperative binding ($m>0$ is cooperative binding, $m<0$ is noncooperative binding).[26,27]

The efficiency (r) of EEET between molecules of the dyes (donor and acceptor) was determined from the spectral and fluorescent data using Eq. (16-2):

$$r = \frac{I_{ex.D} / I_{ex.A}}{Abs_D / Abs_A} \tag{16-2}$$

where $I_{ex.D}$, $I_{ex.A}$, Abs_D, and Abs_A are the intensities the donor (D) and acceptor (A) bands in the fluorescence excitation (I_{ex}) and absorption (Abs) spectra, respectively.[28]

The critical radius of energy transfer (R_0) was determined from the Förster theory:[28]

$$R_0 = 0.2108(\kappa^2 \Phi_{f0} n^{-4} \int F(\lambda)\varepsilon(\lambda)\lambda^4 d\lambda)^{1/6} \tag{16-3}$$

where κ^2 is the orientation factor, which depends on mutual orientation of the transition moments of the donor and the acceptor; Φ_{f0} is the fluorescence quantum yield for the donor in the absence of the acceptor; n is the refractive index of the medium; $F(\lambda)$ is the normalized fluorescence spectrum of the donor, $\varepsilon(\lambda)$ is the molar absorption coefficient of the acceptor, L mol^{-1} cm^{-1}; and λ is the wavelength, nm. In the calculations of R_0 by Eq. (16-3), the value of the orientation factor $\kappa^2 = 2/3$ has been used, which corresponds to the random orientation of donor and acceptor molecules owing to rotational diffusion.

The distance (R) between the donor and the acceptor is determined from Eq. (16-4):[28]

$$r = R_0^6 \,/\, (R_0^6 + R^6) \tag{16-4}$$

The quenching of donor fluorescence upon introduction of an acceptor into the system is described by the Eq. (16-5) (Stern–Volmer equation):[28]

$$I_0 \,/\, I = 1 + K_{SV}[Q] \tag{16-5}$$

where I_0 and I are the fluorescence intensities (quantum yields) in the absence and in the presence of the quencher, respectively; $[Q]$ is the quencher concentration; and K_{SV} is the Stern–Volmer quenching constant.

All experiments were performed at room temperature of 22 ± 2°C.

16.4 RESULTS AND DISCUSSION

16.4.1 *SYNTHESIS OF MAGNETIC NANOPARTICLES AND THE CREATION OF COATINGS*

MNPs were synthesized by precipitation of bi- and trivalent iron salts at 10°C from a mixed aqueous solution in the presence of ammonium hydroxide in accordance with the following scheme:

$$FeSO_4 + 2FeCl_3 + 8NH_4OH \rightarrow Fe_3O_4\downarrow + (NH_4)_2SO_4 + 6NH_4Cl + 4H_2O$$

Iron salts, $FeSO_4 \cdot 7H_2O$ (2.68 g) and $FeCl_3 \cdot 6H_2O$ (5.72 g), were dissolved in distilled water (200 mL) (with molecular ratio Fe^{2+}:Fe^{3+} = 1:2), and the solution was filtered. The cooled solution of the salts was added to a cooled 12.5% NH_4OH solution (40 mL) under continuous stirring on a magnetic stirrer. After the magnetite was formed, the reaction mixture was allowed to stand on a magnet until the sediment was formed; supernatant was decanted and replaced by distilled water. The procedure was repeated until the neutral pH value was attained. An electrostatically stabilized MNPs dispersion was obtained by adding 60 mL of 0.1 M phosphate–citrate buffer (0.05M NaCl) with pH 4.0 to the magnetic sediment, while the system was dispersed on an ultrasound generator MEF 314 (Russia). Then, the dispersion was centrifuged to remove coarse particles. The concentration of MNPs in the dispersion was estimated by drying its known volume and determining the mass of a residue with allowance for the mass of compounds composing the buffer. The concentration of MNPs in the dispersion after centrifugation was 9 mg/mL. The average size of particles evaluated by DLS was 24–28 nm (Figure 16-1a).

BSA and HSA coatings cross-linked via the free radical mechanism were prepared using a 10 mg/mL protein solution (4.45 mL; 65 wt. % BSA and 35 wt. % HSA) in a 0.05M phosphate buffer (pH 6.5), a 3% hydrogen peroxide solution (50 μL), and a diluted magnetite hydrosol (0.55 mL; sample S1). Hydrogen peroxide reacts with the surface of magnetite nanoparticles and ferrous ions to give hydroxyl radicals (Fenton reaction):

$$Fe^{2+} + H_2O_2 \rightarrow Fe^{3+} + OH\bullet + OH^-.$$

In contrast to S1, sample S2 did not contain hydrogen peroxide. Samples S1 and S2 were incubated for 24 h at room temperature and placed into the permanent magnetic field to separate nanoparticles. The concentration of MNPs in the samples was 0.35 mg/mL.

The stability of the MNPs coatings was proven by the absence of nanoparticle aggregation upon the addition of a fibrinogen (FG) solution to the samples (see Figure 16-1b). A 3 mg/mL FG solution was added to samples S1 and S2 in a volume ratio of 1:4.[16] Histograms of the volume distribution of particle sizes do not show the formation of large clusters of particles with a size larger than 1000 nm. This indicates that fibrinogen does not displace the albumins from both cross-linked and adsorption coatings. The average particle size of the samples S1 and S2 after magnetic separation was 44 nm (Figure 16-1c, d).

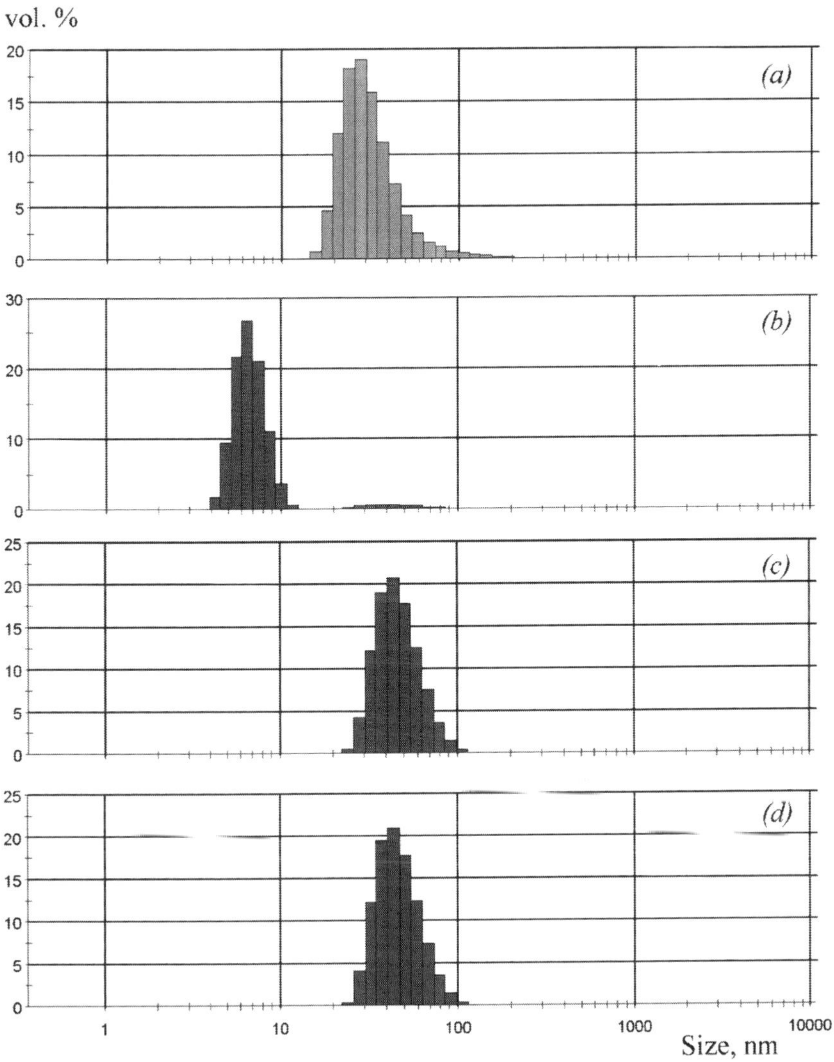

FIGURE 16-1 Histograms of the volume distribution of particle sizes (nm): uncoated MNPs with a concentration of 0.2 mg/ml (a), the sample of BSA/HSA coated MNPs after addition of fibrinogen (b), the sediment of BSA/HSA mixture and MNPs (c), the sediment of BSA/HSA mixture, H_2O_2 and MNPs (d) dissolved in the buffer. The sediments were obtained by separation using permanent magnetic field.

16.4.2 INTERACTION OF DYES D1 AND D2 WITH ALBUMINS IN SOLUTION

To study the systems containing MNPs with BSA/HSA coatings, the carbocyanine dye D1 was used. As noted earlier, *meso*-substituted thiacarbocyanine dyes are characterized by the equilibrium between the *cis*- and *trans*-isomers, which depend on the properties of the medium.[22,24] Namely, these dyes in polar solvents occur mainly as the nonfluorescent *cis*-isomers, whereas the equilibrium in low-polarity solvents is shifted toward the *trans*-isomers.

Dye D1 in water as a polar solvent should occur mainly in the *cis*-form.[24] Upon introduction of increasing concentrations of HSA or BSA into an aqueous solution of dye D1, a drop in the intensity of the initial absorption band of the monomeric D1 (λ_{abs} = 542 nm) and an appearance and a growth of the long-wavelength band (λ_{abs} = 562 and 552 nm in case of HSA and BSA respectively) belonging to the dye bound to albumin are observed (Figure 16-2).

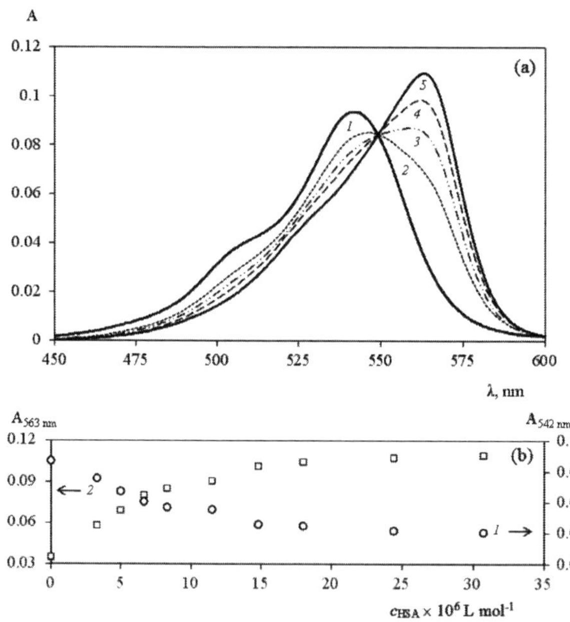

FIGURE 16-2 Absorption spectra of D1 in water solution upon introduction of increasing HSA concentrations. (a): c_{HSA} = 0 (*1*), 3.32 × 10^{-6} (*2*), 6.62 × 10^{-6} (*3*), 1.15 × 10^{-5} (*4*), and 3.07 × 10^{-5} (*5*) mol L^{-1}; (b) and absorbance changes of D1 at λ_{abs} = 542 nm (*1*, right axis) and at λ_{abs} = 563 (*2*, left axis) upon introduction of HSA.

Furthermore, the fluorescence excitation spectrum of D1 is shifted to the long-wavelength side with respect to the absorption spectrum. This is explained by the fact that fluorescence of D1 in aqueous media results from the presence of a low concentration of the *trans*-isomer, which possesses longer-wavelength absorption than the *cis*-isomer. Therefore, we may suppose that dye D1 bound to albumins also contains the *cis*- and *trans*-isomers. Dye D1 quite efficiently (with constants K_{eq} on the order of 1.2 $\times 10^5$ L mol^{-1}) interacts with both HSA and BSA, with a steep growth of fluorescence and with partial *cis–trans* conversion of the dye (Figure 16-3).[24]

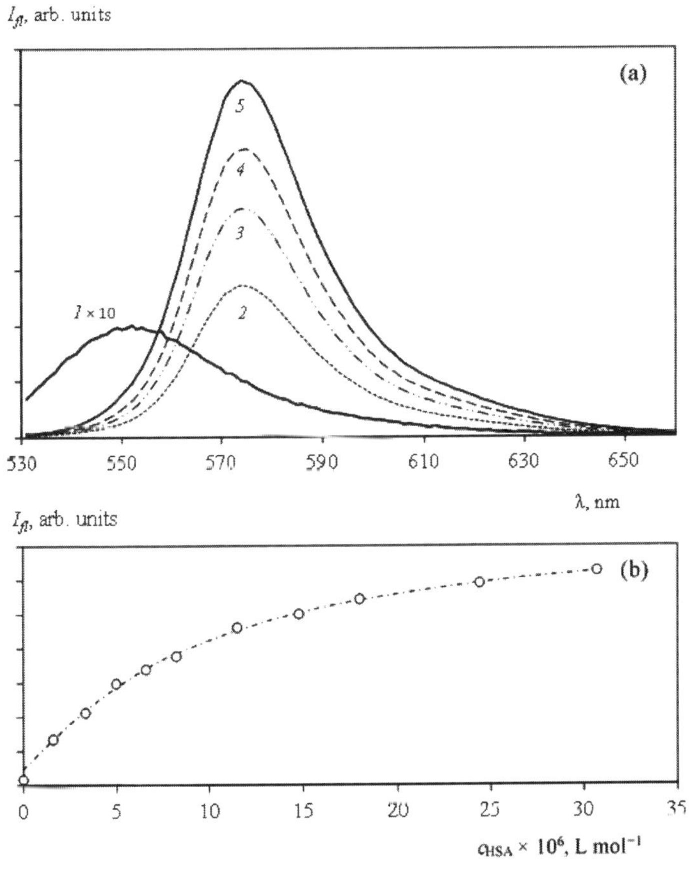

FIGURE 16-3 Changes in the fluorescence spectra (λ_{ex} = 525 nm) of D1 upon introduction of increasing HSA concentrations. (a): c_{HSA} = 0 (*1*), 3.32 × 10^{-6} (*2*), 6.62 × 10^{-6} (*3*), 1.15 × 10^{-5} (*4*), and 3.07 × 10^{-5} (*5*) mol L^{-1} (in water solution); growth of fluorescence intensity of D1 upon introduction of HSA (b); dashed line was obtained from Eq. (16-1) with K_{eq} = 1.2 × 10^5 L mol^{-1} and n=1.1).

In aqueous solution, D2 occurs mainly as the practically nonfluorescent *cis*-dimer (λ_{abs} = 535 nm,) and can form stable J-aggregates.[23] J-aggregates of D2 are characterized by narrow peaks with λ_{abs} = 645 nm and λ_{fl} = 660 nm. In the presence of HSA, a noncovalent complex of the *trans*-monomer of D2 (λ_{abs} = 612 nm) is formed. Dye D2 interacts with BSA considerably weaker than HSA. In the presence of BSA, only small portion of D2 molecules undergoes *cis–trans* conversion to form the fluorescing *trans*-isomer.[23]

Effect of albumin denaturation on the spectral and fluorescent properties of D1 and D2.

In this work, we studied the dependence of the spectral and fluorescent properties of the dyes on the extent of HSA denaturation. To denature HSA, portions of crystalline urea were added to the solution, with a urea concentration in the samples attaining 6 mol L^{-1}. For qualitative control of HSA denaturation, the DLS method was used (experiments without addition of dyes). An increase in the urea concentration in the sample to 4.5 mol L^{-1} resulted in an increase by 4.5 nm (ΔD) in the size of the particles of the microfraction of the largest contribution to the particle size distribution; at the urea concentration of 6 mol L^{-1} ΔD = 7.5 nm (Figure 16-4[*1*]), which approximately agrees with published data.[29]

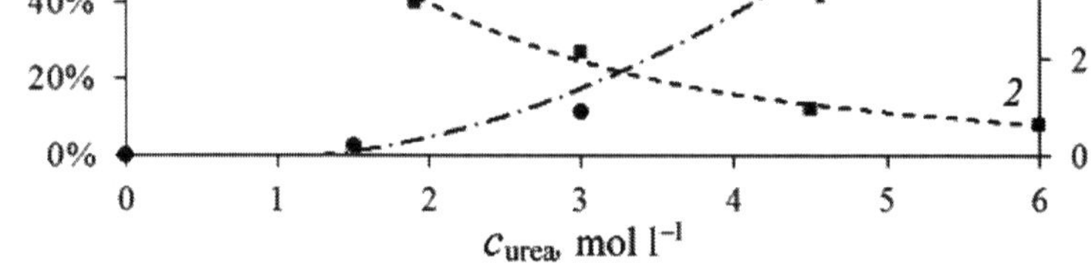

FIGURE 16-4 Dependence of the change of the size of HSA particles (ΔD, nm) on the urea concentration in the sample upon denaturation, which was obtained by the DLS method on the basis of size distribution of HSA particles (*1*, right axis); values of the effective protein concentration (c_{eff}, %) for dyes D1 (*2*) and D2 (*3*) in a complex with HSA (c_{HSA} = 3.8 × 10^{-5} mol L^{-1}) in the presence of urea in the sample (left axis).

In the presence of urea, the fluorescence intensity of dye D1 significantly decreases (Figure 16-4 [*2*]): c_{eff} = 27% in HSA solution at a urea concentration of 3 mol L^{-1} and c_{eff} = 8% at a urea concentration of 6 mol L^{-1}; that is, the D1–HSA complex decomposes almost completely. For D2, the decrease in the fluorescence intensity is small (c_{eff} = 74% at a urea concentration of 6 mol L^{-1}; see Figure 16-4 [3]). At a urea concentration of 6 mol L^{-1}, there are no substantial changes in the absorption spectra of D2 bound to HSA (insignificant long-wavelength shift of the maximum and some broadening of the band are observed in the absorption spectrum).

It is known that serum albumin denaturation is a multistep process; first of all, domain I of the protein is denatured.[29,30] At the urea concentration of 3 mol L^{-1}, 65–75% of HSA molecules in the sample are denatured in domain I. At the urea concentration of 3 mol L^{-1}, ~8% of HSA molecules are denatured in domain II, whereas ~65% of HSA molecules are denatured in this domain at 6 mol L^{-1}.[29,30] We may suppose that dye D1, whose complex with HSA decomposes in the presence of 3 mol L^{-1} of urea, forms a complex in domain I of the protein molecule. Dye D2, having higher values of c_{eff}, should probably bind with parts of the HSA molecule more tolerant to denaturation.

16.4.3 ESTIMATION OF THE TOTAL EFFECT OF THE DYE-ALBUMIN INTERACTION IN SYSTEMS WITH MNPS

Because D2 is much less effective as a fluorescent probe for BSA than for HSA, dye D1 was chosen to assess the state of the proteins in the MNPs systems with BSA/HSA coatings. The estimation of BSA/HSA interaction with D1 in the samples was performed in terms of the effective protein concentration (c_{eff}).[25] Fluorescent titration of dye D1 with BSA/HSA coated MNPs showed that albumine adsorption on MNPs and the reaction with H_2O_2 have a weak effect on the interaction of D1 with the proteins in these systems. Introduction of solutions of S1, S2 into a cell with D1 leads to a growth of the fluorescence intensity (complex of the *trans*-monomer). In case of uncoated MNPs solution, fluorescence did not grow. The effective concentration of the complex of D1 (c_{eff}) is ~90% for S1 and S2 (100% in the reference BSA/HSA solution with the same albumins concentrations).

In the systems S2 (coated by an adsorption layer of the albumins not subjected to free radical oxidative modification) and S1 (with the linked BSA/HSA coating), magnetic separation reduces the concentration of the albumins capable of binding a dye probe in a solution (at 15 ± 5% and 11 ± 5%, respectively). As a result of magnetic separation, both MNPs with adsorbed BSA/HSA coating and those with BSA/HSA coating linked by the free radical method are precipitated. However, in comparison with the results of magnetic separation process observed by the authors,[31] coatings in S2 systems were desorbed to a much lesser extent. This indicates its stability.

16.4.4 ELECTRONIC EXCITATION ENERGY TRANSFER IN THE DYE PAIR D1 AND D2

In the absence of interaction with the protein, EEET between the dyes does not occur in a homogeneous solution because of a long distance between donor and acceptor molecules (500 Å in solution at dye concentrations of ~2.0×10^{-6} mol L^{-1}). In the sample containing 1 mg mL^{-1} HSA ($c_{HSA} = 1.49 \times 10^{-5}$ mol L^{-1}) in the phosphate buffer, the band of D1 ($\lambda_{max\ fl} \sim 565$ nm) is observed in the fluorescence excitation spectra of D1 ($\lambda_{fl} = 612$ nm), which is due to EEET (Figure 16-5, curve *2'*).

EEET in the dye pair D1 and D2 was also studied in systems containing BSA/HSA coated MNPs. The total albumin concentration in the test samples was 1 mg mL^{-1} ($c_{BSA} = 1.00 \times 10^{-5}$ mol L^{-1} and $c_{HSA} = 5.2 \times 10^{-6}$ mol L^{-1}). The concentrations of D1 and D2 were 1.61×10^{-6} and 1.11×10^{-6} mol L^{-1}, respectively. The EEET efficiency r in this system was compared with that for the reference protein solution with the same HSA and BSA concentrations. The values of r were found to be similar, but they were lower than for the dyes bound to HSA: $r = 0.23$ for the BSA/HSA coated MNPs, whereas $r = 0.22$ for the reference BSA/HSA solution. In the sample containing 1 mg mL^{-1}, HSA $r = 0.37$ (at $c_{D2} = 1.95 \times 10^{-6}$ mol L^{-1}). The cause of the lower EEET efficiency for the BSA/HSA mixture is apparently the weaker interaction of dye D2 with BSA as compared with HSA.

An increase in the D2 concentration in the sample increases the efficiency of EEET, which is due to a growth of the proportion of acceptor molecules bound in the complex with HSA ($r = 0.54$ at $c_{D2} = 2.9 \times 10^{-6}$ mol L^{-1}).

For BSA/HSA coated MNPs (S1 and S2) and the reference BSA/HSA solution, similar values of the critical EEET radius ($R_0 = 52$ Å) and the distance between the donor and the acceptor in the complex ($R = 64$ Å) were obtained, which agrees satisfactorily with the characteristic dimensions of the albumin molecule in solution.[32]

Addition of urea to the solution containing HSA and dyes D1 and D2 prevents detection of EEET. In the fluorescence spectra of the D1–D2 system in an HSA solution (with a urea concentration of 6 mol L^{-1}), two bands are observed corresponding to fluorescence of dyes D1 and D2. In the fluorescence excitation spectra of the system in the range of $\lambda = 640$–660 nm (spectra not shown), only one band is present corresponding to fluorescence excitation of the acceptor dye (the contribution of the D band is absent), which indicates the lack of EEET.

16.4.5 QUENCHING OF DONOR FLUORESCENCE BY THE ACCEPTOR IN THE PRESENCE OF HSA

For all samples, we have studied quenching of donor fluorescence by the acceptor bound in a complex with HSA. An increase in the concentration of D2 leads to a noticeable drop in the intensity of the D1 fluorescence bands (Figure 16-5b, curves *1–4*). For a solution of HSA (1.49×10^{-5} mol L^{-1}) in the phosphate buffer, the Stern–Volmer quenching constant (K_{SV}) is 2.48×10^5 L mol^{-1} (Figure 16-5 inset, curve *1*). For the systems containing a mixture of the albumins BSA and HSA ($c_{BSA} = 1.00 \times 10^{-5}$ mol L^{-1}, $c_{HSA} = 5.2 \times 10^{-6}$ mol L^{-1}), the values of the quenching constants were found to be lower by a factor of 1.5–2. In particular, $K_{SV} = 1.29 \times 10^5$ L mol^{-1} for the MNPs sample S1 with the linked BSA/HSA coating (Figure 16-5 inset, curve *3*). For the MNPs sample S2 (coated by an adsorption layer of the albumins not subjected to free radical oxidative modification), $K_{SV} = 1.79 \times 10^5$ L mol^{-1} (Figure 16-5 inset, curve *2*). For the reference BSA/HSA solution, $K_{SV} = 1.64 \times 10^5$ L mol^{-1} (Figure. 16-5 inset, curve *2*).

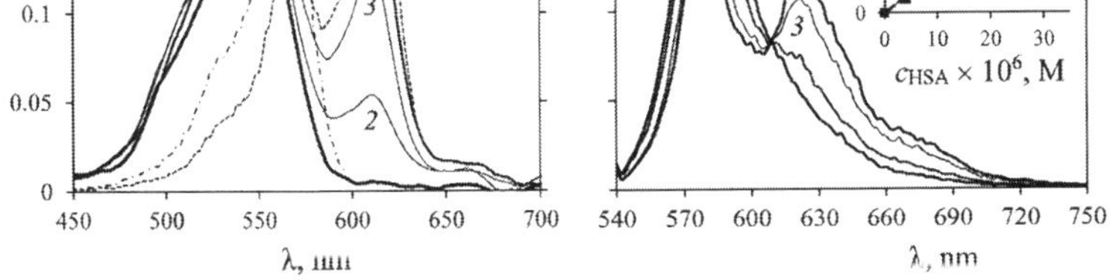

FIGURE 16-5 Absorption and fluorescence excitation spectra of dyes D1 and D2 (a; $c_{D1} = 2.0 \times 10^{-6}$ mol L^{-1}, $c_{HSA} = 1.49 \times 10^{-5}$ mol L^{-1}), fluorescence spectra of dyes D1 and D2 (b; $\lambda_{ex} = 530$ nm) at $c_{D2} = 0$ (*1* and *1`*), $6.0 \times 10^{-7}$ (*2* and *2`*), 1.56×10^{-6} (*3*), and 2.3×10^{-6} (*4*) mol L^{-1}. Inset: experimental data on fluorescence quenching of D1 by the acceptor D2 in the presence of HSA (*1*), for the BSA/HSA reference solution (*2*), for MNPs S1 (*3*) and S2 (*4*) in the coordinates of the Stern–Volmer plot (I_0/I-1 against the volume concentration of A) and the calculation using Eq. (16-5).

The high values of Stern–Volmer quenching constants are consistent with the static character of the quenching process, which occurs between donor and acceptor molecules bound to a biopolymer. The similarity of the values of the K_{SV} for the S1 and S2 samples, in our opinion, confirms that albumins retain their properties on the MNPs surface.

16.5 CONCLUSIONS

Magnetite MNPs have been synthesized and coatings have been formed from BSA and HSA on the surface of nanoparticles. A free radical procedure previously developed by the authors for protein cross-linking has been used for fixing the coatings on magnetite nanoparticles, this procedure being based on the ability of proteins to form intermolecular covalent bonds under the action of free radicals. The properties of the obtained coatings and their stability have been studied with the help of dynamic light scattering (DLS) and spectral-fluorescent probes. It has been shown that it is reasonable to use anionic carbocyanine dyes for qualitative and quantitative assessment of the properties of proteins composing the coatings when studying the adsorption and free radical cross-linking of BSA/HSA on MNPs. The developed and experimentally substantiated approaches to the analysis of an adsorption layer of biomacromolecules on the surface of magnetic nanoparticles based on the spectra of fluorescent probe dyes have led us to draw initial conclusions on conformational transformations of a protein on MNPs surfaces from the accessibility of its binding sites to a dye.

EEET between molecules of anionic polymethine dyes, which form complexes with HSA, was studied. The effect of denaturation on the spectral and fluorescent properties of these dyes in the complexes was examined. It has been shown that denaturation of the protein weakens its complexation with dye D1, whereas HSA molecules incorporated into the linked coating on MNPs retain their ability to bind the fluorescent probe, which indicates the lack of protein denaturation in the coating.

The similarity of the values of the quenching constants obtained for the MNPs samples with the BSA/HSA coating and those for protein solutions additionally confirms retaining of the structural and functional properties of the albumin adsorbed and linked on the surface of MNPs. This may be indicative of the preservation of their native functional properties after cross-linking, and, hence, the biocompatibility of the BSA/HSA coatings obtained on MNPs surfaces and the possibility of using albumin coatings for targeted drug delivery.

ACKNOWLEDGMENTS

The authors would like to thank Prof. B.I. Shapiro (Research Center NIIKHIMFOTOPROEKT) for providing the dyes.

This work has been supported by the Russian Foundation for Basic Research, project nos. 13-03-00863 and 14-03-31196-mol_a.

KEYWORDS

- **Bovine Serum Albumin**
- **Carbocyanine Dyes**
- **Human Serum Albumin**
- **Magnetic Nanoparticles**
- **Spectral-Fluorescent Probes**

REFERENCES

1 Tomitaka, A.; Yamada, T.; Takemura, Magnetic nanoparticle hyperthermia using Pluronic-coated Fe_3O_4 nanoparticles: An in vitro study. *J. Nano mater.* Vol. 2012, №4, (2012), ID 480626.
2 Rümenapp, C.; Gleich, B.; Haase, A. Magnetic nanoparticles in magnetic resonance imaging and diagnostics. *Pharm. Res.* Vol. 29, (2012), pp. 1165–1179.
3 Shahbazi-Gahrouei, D.; Abdolahi, M.; Zarkesh-Esfahani, S. H.; Laurent, S.; Sermeus, C.; Gruettner, C. Functionalized magnetic nanoparticles for the detection and quantitative analysis of cell surface antigen. *Biomed Res. Int.* Vol. 2013, (2013), ID 349408.
4 Gupta, A. K.; Gupta, M. Synthesis and surface engineering of iron oxide nanoparticles for biomedical applications, *Biomaterials*. Vol. 26, № 18, (2005), pp. 3995–4021.
5 Laurent, S.; Forge, D.; Port, M.; Roch, A.; Robic, C.; Vander Elst, L.; and Muller, R. N. Magnetic iron oxide nanoparticles: Synthesis, stabilization, vectorization, physicochemical characterizations, and biological applications. *Chem. Rev.* Vol. 108, № 6, (2008), pp. 2064–2110.
6 Bychkova, A. V.; Sorokina, O. N.; Rosenfeld, M. A.; Kovarski, A. L. Multifunctional biocompatible coatings on magnetic nanoparticles. *Russ. Chem. Rev.* Vol. 81, (2012), pp.1026–1050.
7 Yang, H. W.; Hua, M. Y.; Liu, H. L.; Huang, C. Y.; Wei, K. C. Potential of magnetic nanoparticles for targeted drug delivery. *Nanotechnol. Sci. Appl.* Vol. 5, (2012), pp. 73–86.
8 Koneracka, M.; Kopcansky, P.; Antalik, M.; Timko, M.; Ramchand, C. N.; Lobo, D.; Mehta, R. V.; Upadhyay. R. V. Immobilization of proteins and enzymes to fine magnetic particles. *J. Magn. Magn. Mater.* Vol. 201, (1999), pp. 427–430.

9 Huang, J.; Wang, L.; Lin, R.; Wang, A. Y.; Yang, L.; Kuang, M.; Qian, W.; Mao, H. Casein-coated iron oxide nanoparticles for high MRI contrast enhancement and efficient cell targeting. *ACS Appl. Mater. Interfaces.* Vol. 5, № 11, (2013), pp. 4632–4639.
10 Mahdavi, M.; Ahmad, M. B.; Haron, M. J.; Namvar, F.; Nadi, B.; Rahman, M. Z.; Amin, J. Synthesis, surface modification and characterisation of biocompatible magnetic iron oxide nanoparticles for biomedical applications. *Molecules.* Vol. 18, (2013), pp. 7533–7548.
11 Ballet, T.; Boulange, L.; Brechet, Y.; Bruckert, F.; Weidenhaupt, M. Protein conformational changes induced by adsorption onto material surfaces: An important issue for biomedical applications of material science. *Bull. Polish Acad. Sci. Tech. Sci.* Vol. 58, №2, (2010), pp. 303–315.
12 Rozenfel'd, M. A.; Bychkova, A. V.; Sorokina, O. N.; Kovarskii, A. L.; Leonova, V. B.; Lomakin, S. M.; Makarov, G. G. RF Patent 2.484.178 (2013).
13 Horak, D.; Svobodova, Z.; Autebert, J.; Coudert, B.; Plichta, Z.; Kralovec, K.; Bılkova, Z.; Viovy, J. L. Albumin-coated monodisperse magnetic poly(glycidyl methacrylate) microspheres with immobilized antibodies: Application to the capture of epithelial cancer cells. *J. Biomed. Mater. Res. A.* Vol. 101, № .1, (2013), pp. 23–32.
14 Peng, Z. G.; Hidajat, K.; Uddin. M. S. Adsorption of bovine serum albumin on nanosized magnetic particles. *J. Colloid Interface Sci.* Vol. 271, № 2, (2004), pp. 277–283.
15 Stadtman, E. R.; Levine. R. L. Free radical-mediated oxidation of free amino acids and amino acid residues in proteins. *Amino Acids.* Vol. 25, (2003), pp. 207–218.
16 Bychkova, A. V.; Rosenfeld, M. A.; Leonova, V. B.; Sorokina, O. N.; Lomakin, S. M.; Kovarskii, A. L. Free-radical cross-linking of serum albumin molecules on the surface of magnetite nanoparticles in aqueous dispersion. *Colloid J.* Vol. 75, (2013), pp. 7–13.
17 Rosenfeld, M. A. Leonova, V. B.; Konstantinova, M. L.; Razumovskii, S. D. Self-assembly of fibrin monomers and fibrinogen aggregation during ozone oxidation. *Biochemistry-Moscow.* Vol. 74, (2009), pp. 41–46.
18 Rosenfeld, M. A.; Shchegolikhin, A. N.; Bychkova, A. V.; Leonova, V. B.; Kostanova, E. A.; Biryukova, M. I.; Razumovskii,S. D.; Konstantinova, M. L. Free-radical oxidation of plasma fibrin-stabilizing factor. *Doklady* (*"Reports", in Rus.*) *Biochemistry and Biophysics.* Vol. 446, (2012), pp. 213–216.
19 Pretorius, E.; Bester, J.; Vermeulen, N.; Lipinski, B. Oxidation inhibits iron-induced blood coagulation. *Curr. Drug Targets.* Vol. 14, № 1, (2013), pp. 13–19.
20 Tatikolov, A. S. Polymethine dyes as spectral-fluorescent probes for biomacromolecules, *J. Photochem. Photobiol. C: Photochem. Rev.* Vol. 13, (2012), pp. 55–90.
21 Ishchenko, A. A. Stroenie i spektral'no-lyuminestsentnye svoistva polimetinovykh krasitelei (Structure and Spectral–Luminescent Properties of Polymethine Dyes). Kiev: *Naukova Dumka ("Scientific Thought", in* Ukr.*).* (1994), 232 p.
22 West, W.; Pearce, S.; Grum, F. Stereoisomerism in cyanine dyes *meso*-substituted thiacarbocyanines. *J. Phys. Chem.* Vol. 71, №. 5, (1967), pp. 1316–1326.
23 Tatikolov, A. S.; Costa, S. M. B. Complexation of polymethine dyes with human serum albumin: A spectroscopic study. *Biophys. Chem.* Vol. 107, (2004), 33–49.
24 Kashin, A. S.; Tatikolov, A. S. Spectral and fluorescent study of the interaction of anionic cyanine dyes with serum albumins. *High Energy Chem.* Vol. 43, №. 6, (2009), pp. 480–488.
25 Azizova, O. A.; Aseichev, A. V.; Vekman, E. M.; Moskvina, S. N.; Skotnikova, O. I..; Smolina, N. V.; Gryzunov, Yu. A.; Dobretsov, G. E. Investigation of changes of albumin

transport function using the fluorescent probe K-35 under the action of oxidants influence of hypochlorite. *Bull. Exp. Biol. Med.* Vol. 152, (2011), pp. 653–657.

26 Hill, A. V. The heat produced in contracture and muscular tone. *Journal of Physiology*. Vol. 40, (1910), pp. iv–vii.

27 Goutellea, S.; Maurinc, M.; Rougierb, F.; Barbautb, X.; Bourguignona, L.; Ducherb, M.; Mairea, P. The Hill equation: A review of its capabilities in pharmacological modelling. *Biochem. & Biophy.* Vol. 22, (2008), pp. 633–648.

28 Lakowicz, J. R. Principles of Fluorescence Spectroscopy. New York: *Plenum Press,* (1983), 496 p.

29 Leggio, C.; Galantini, L.; Konarev, P. V.; Pavel, N. V. Urea-induced denaturation process on defatted human serum albumin and in the presence of palmitic acid. *J. Phys. Chem. B*. Vol. 113, (2009), pp. 12590–12602.

31 Pronkin, P. G.; Bychkova, A. V.; Sorokina, O. N.; Kovarskii, A. L.; Rosenfeld, M. A.; Tatikolov, A. S. Study of protein coatings on magnetic nanoparticles by the method of spectral and fluorescent probes. *High Energy Chem.* Vol. 47, №. 5, (2013), pp. 268–270.

32 Kiselev, M. A.; Gryzunov, Iu. A.; Dobretsov, G. E.; Komarova, M. N. The size of human serum albumin molecules in solution. *Biophysics (Moscow)*. Vol. 46, № 3, (2001), pp. 402–405.

CHAPTER 17

STRUCTURE OF MULTILAYER THERMAL SHRINK FILMS FOR PACKAGING

R. M. GARIPOV, V. N. SEROVA, A. I. ZAGIDULLIN, A. I. KHASANOV, and A. A EFREMOVA

Kazan National Research Technological University, Kazan, Russia

CONTENTS

ABSTRACT

The paper describes the characteristics of polyolefins used in film production. It shows the results of the elastic mechanical properties of films based on a mixture of polyolefins and analyzes the results of tests determining the elastic and mechanical properties of the resulting formulations. This work is devoted to the study of structure and operational characteristics of nine-layer shrink barrier films on basis of the polyolefin plastomer, polyamide, and copolymer of ethylene vinyl alcohol. Using a digital optical microscope obtained microphotographs of cross-sections of films, we studied the resistance of films to light and thermal-oxidative aging.

17.1 INTRODUCTION

In the last few years, the packaging industry has developed immensely. Therefore, the manufacturers have started to show their interest in barrier materials with various product-protecting properties. Nowadays, the dominating packaging material is the multilayer films.[1–3] The properties of the so-called "plastic" packaging can be improved by changing the general thickness of the polymer material, the thickness of the separate layers, and by changing the layer order. It also counts for the most important attribute of the food packaging, which is the oxygen permeability. The O_2 permeability of the single-layer polyethylene packaging films lays within 500–1000 cm^3/m^2 in 24 hours, whereas the PA ones have the O_2 permeability of 30-100 cm^3/m^2 in 24 hours. However, the O_2 permeability of the multilayer films made of these polymers is only 10–50 cm^3/m^2 in 24 hours.[4,5] According to the requirements, the O_2 permeability of the plastic food packaging must be within 1–100 cm^3/m^2 in 24 hours. It has also been proven with the results of our research on the barrier properties of single- and multilayer polymer films.[6,7]

Due to the high level of competition in the packaging industry, requirements for the plastic packaging become even stricter. The modern packaging not only has to be barrier resistant, strong, bright, and sanitary but also has to be light resistant and thermally stable. Therefore, the purpose of this work is to study the structure and resistance to light and thermal-oxidative aging of multilayer thermal shrink film for food packaging. These 9 layers films were produced on the extrusion line 650/9 by "GAP

S.r.l.", Italy. The barrier properties of these films were insured by the use of polyamide (PA) and ethylene copolymer with vinyl alcohol (EVOH). The polyolefin plastomer (POP) served as a weld layer. The separate film layers were connected with adhesives (Tie). The film structures were as follows:

- POP/POP/Tie/PA/Tie/PA/Tie/POP/POP is a 50 mμ medium barrier film for cheese packaging;
- PA/Tie/Tie/PA/EVOH/PA/Tie/Tie/POP is a 110 mμ high barrier film for meat packaging.

17.2 EXPERIMENTAL PROCEDURE

The structure of the films was studied with the digital optical microscope «Keyence VH-Z500R» in polarized transmitted light.

The light transmission coefficient of the film samples was registered with the spectrophotometer SF-46. We used the spectrodensitometer Spectrodens by Techkon, Germany to measure the optical density and register the spectral curves. In order to study their resistance to light, we put the film samples under the integral light of high-pressure mercury lamp DRT-240, which is a powerful source of UV-light from 20 cm.

The infrared specters of polymer films were registered, according to the attenuation total reflection method, on the infrared Fourier spectrometer "InfraLUM-08," the spectral range of which is 780–400 cm^{-1}.

The films' strength attributes were measured on the Testwinner 992 dynamometer, Germany, which also included the load-elongation curves. That said the elongation speed counted 50 mm/min. The registration of thermogravimetric analysis curves was held out with the Perkin Elmer STA6000 device at the 30–400°C temperature range at the temperature increase speed of 5°C/min.

17.3 RESULTS AND DISCUSSION

Figure 17-1 shows the microphotographs of the films' cross sections that were taken with the digital microscope at 1000x magnification.

a **b**

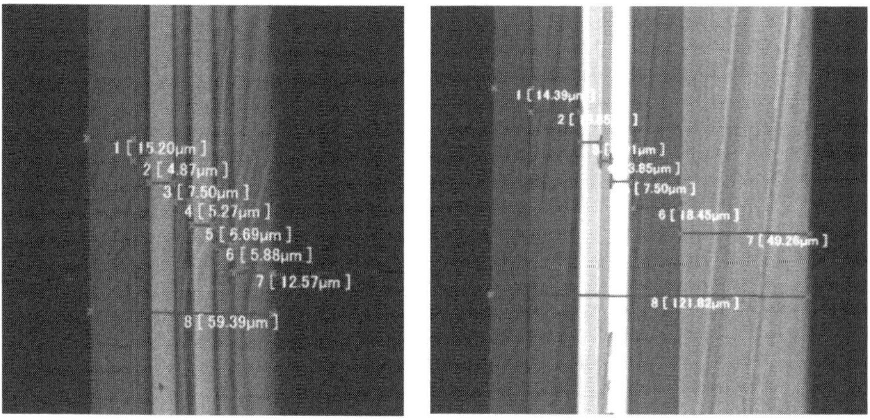

a – medium-barrier film; b – high-barrier film

FIGURE 17-1 Cross-sectional micrograph of the films at 1000 times magnification in transmitted light. a – medium-barrier film; b – high-barrier film.

The results were analyzed in Table 17-1.

According to the microphotographs, both films have only seven layers. However, based on the thickness comparison in Table 17-1, thickness of the side layers of the medium-barrier film corresponds to the thickness sum A+B and H+I of the formulation layers. With the high-barrier film, thickness of the second layer corresponds to the layers sum B+C and the sixth layer thickness to the G+H sum, accordingly. The reason for it is that these layers are made of the same materials and as a result have the same transmission coefficient, and the optical microscope does not separate them.

TABLE 17-1 The layers thickness of shrink-barrier films

Layer	Material	The thickness of the layer the recipe, mc	The thickness of the layer produced film, mc
		Medium-barrier film	
A	POP	5	15,2
B	POP	5,5	
C	Tie	5	4,87
D	PA	7	7,5
E	Tie	5	5,27
F	PA	7	6,69
G	Tie	5	5,88

TABLE 17-1 *(Continued)*

Layer	Material	The thickness of the layer the recipe, mc	The thickness of the layer produced film, mc
H	POP	5	12,57
I	POP	5,5	
Total:		50	57,98
		High-barrier film	
A	PA	14,3	14,39
B	Tie	9,35	18,85
C	Tie	8,8	
D	Pa	8,25	7,91
E	EVOH	4,4	3,85
F	PA	8,25	7,5
G	Tie	7,15	18,45
H	Tie	7,7	
I	POP	41,8	49,26
Total:		110	120,21

Figure 17-2 illustrates the spectral transmittance (τ) of the films in the wavelength interval λ = 240-–680 nm.

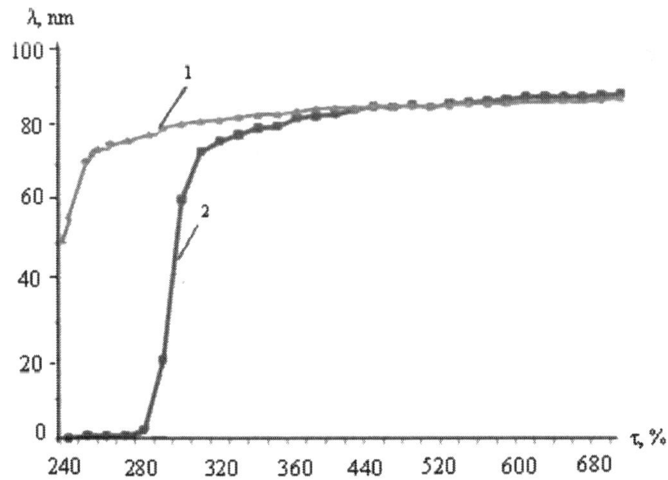

FIGURE 17-2 Spectral transmittance of the films. 1 – medium-barrier film; 2 – high-barrier film.

It is obvious that the light transmittance of the polymer films depend on their content as well as on the thickness. Figure 17-2 clearly shows that the 50 mμ medium-barrier film has better transmittance in the UV area of the spectrum and the blue area of the visible range ($\lambda = 400–450$ nm). The cutoff edge of the spectral curve of the 110 mμ high-barrier film is clearly shifted to the area of longer waves, the reason of which is a big amount of chromophoric groups in the film.[8]

The light (photochemical) aging of the film samples in the process of UV irradiation was shown in the fact that they considerably lost their transparency, especially in the $\lambda = 320–400$ nm area. The reason for this is that they absorbed products formed due to the photo-oxidative degradation of the polymers.[9] In the photochemical process of aging, the increase in optical density of the films named in the wavelength interval was observed. In this case, we also observed a decrease in strength characteristics of the film samples. Thus, we took the relative changes of the transmittance coefficient ($\Delta\tau$), optical density (ΔD), critical stress ($\Delta\sigma$), and extension elongation ($\Delta\varepsilon$) as light resistance criteria:

$$\Delta\tau = (\tau_o - \tau_t)\bullet 100/\tau_o,\ \Delta D = (D_t - D_o)\bullet 100/D_o,\ \Delta\sigma = (\sigma_o - \sigma_t)\bullet 100/\sigma_o,$$
$$\Delta\varepsilon = (\varepsilon_o - \varepsilon_t)\bullet 100/\varepsilon_o$$

where τ_o, D_o, σ_o, and ε_o are the transmittance coefficient (at constant wavelength), critical stress, and extension elongation before the UV irradiation, respectively; τ_t, D_t, σt, and ε_t are the same attribute after the UV irradiation for t hours.

The dependence of τ and D on the duration of UV irradiation (t) in hours—$\tau = f(t)$ and $D = f(t)$ shown in Figures 17-3 and 17-4—reflects the light ageing kinetics of the films. They were obtained at fixed values of the wavelength λ.

As we can see, if we increase the UV irradiation duration, the rate of ongoing photochemical processes gradually decreases. The values of the initial speed of decrease of the light transmittance (W_τ) and the initial speed of increase of the films' optical density (W_D) were calculated according to the initial (straight-line) areas of the kinetic curves $\tau = f(t)$ and $D = f(t)$. All of the light-aging characteristics of the film samples are given in the table along with the results of research with the TGA method. Having compared the values of the optical characteristics W_τ, $\Delta\tau$, W_D, and ΔD, we may state that the 50 mμ medium-barrier film has the highest light resistance.

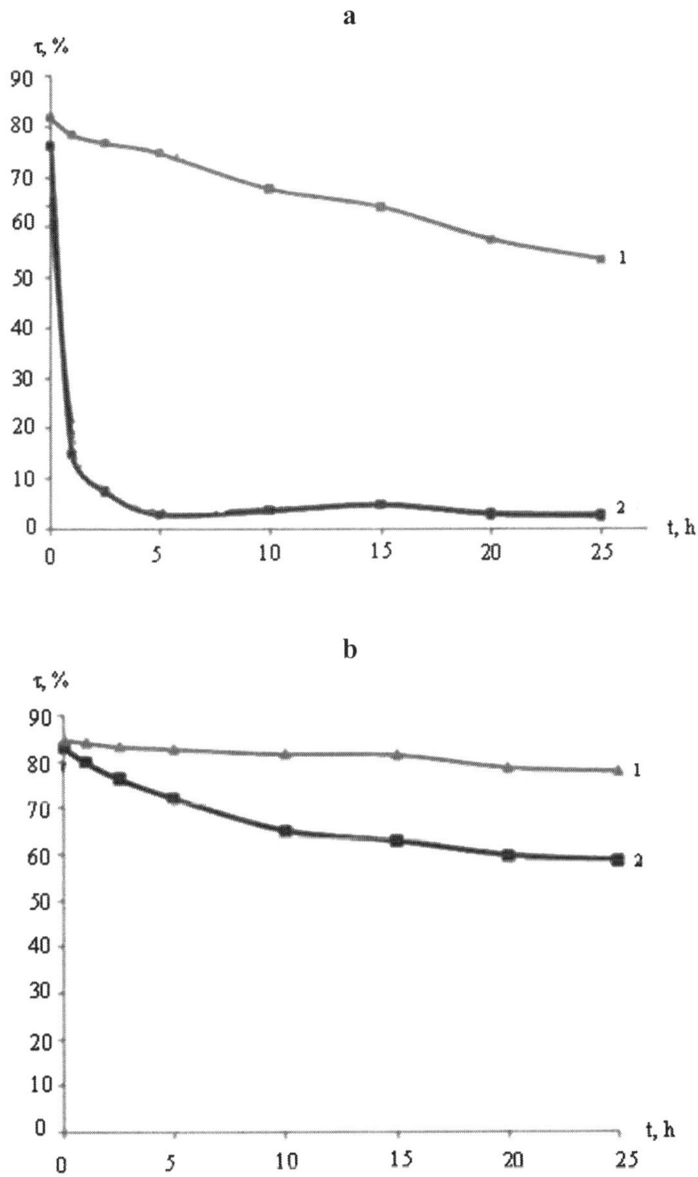

FIGURE 17-3 The dependence τ = f(t) obtained at λ = 320 nm (a) and λ = 400 nm (b). 1 – medium-barrier film, 2 – high-barrier film.

The photochemical changes of the films were recorded through the IR spectra. For instance, after a long (25-hour) UV irradiation time, the IR spectrum of the 50 mμ film has shown the significantly increased intensity of the absorption band at 1262 cm^{-1} and 1102 cm^{-1}, related to the C–OH stretching modes, as well as the emergence of a new band at 1721 cm^{-1} responsible for the C=O stretching modes.[10] The latter effect proves the process of photo-oxidative destruction of the polyethylene layer within the film.

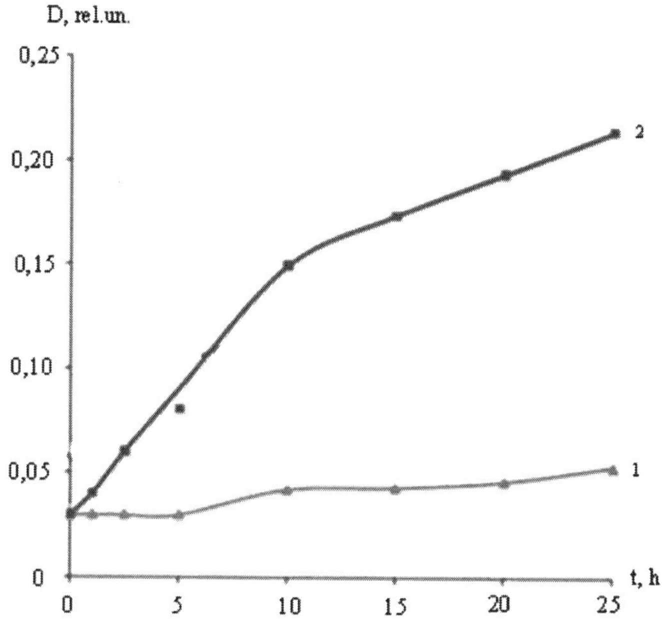

FIGURE 17-4 The dependence D = f(t) obtained at λ = 400 nm. 1 – medium-barrier film, 2 – high-barrier film.

Let us trace back how the strength properties of the films changed under the influence of UV irradiation. Their initial values are as follows: for 50 mμ film they are σ = 51.7 MPa, ε = 140%, for the 110 mμ film σ = 89.3 MPa, ε = 105%. Evidently, the high-barrier film is initially stronger and less elastic. It stayed same after the long light aging as well, which could be seen in the table of changing values of the strength properties.

In order to compare the resistance of the films to thermal oxidative destruction, we found out the onset temperature (T_o) and the peak temperature (T_p) based on the data from TGA. The TGA curves show a small decrease of the film mass (up to 1.5%) in the 50–100°C temperature interval, after which the mass does not change until T_p. The high-barrier film is characterized as the most resistant to the thermal oxidative destruction, which is proven by T_o (the most important attribute of the thermal properties of polymers) and T_p values of the film samples.

Thus, the research studied the structure and the performance properties of the 9-layer thermal shrink barrier film for cheese and meat packaging of 50 and 110 mμ, respectively. According to the changes in their optical characteristics after the UV irradiation procedures, the 50 mμ medium-barrier film is the most light resistant. However, having the added layers of EVOH, the 110 mμ high-barrier film showed smaller change in its strength properties after the UV aging. Moreover, this packaging film was also more resistant to thermal oxidative destruction.

ACKNOWLEDGMENT

The work has been supported by the Government of the Russian Federation (Russian Ministry of Education) as a part of the complex project of developing high-quality production under the contract № 02.G25.31.0037, according to the resolution № 218 of the Government of the Russian Federation from April, 9, 2010.

KEYWORDS

- **Absorbance**
- **Copolymer of Ethylene Vinyl Alcohol**
- **Kinetics**
- **Light Ageing**
- **Light Resistance**
- **Light Transmittance**
- **Multilayer Film**
- **Polyamide**
- **Polyethylene**
- **Strength Indicators**
- **Structure**
- **Thermal-Oxidative Destruction**

REFERENCES

1. Tretyakov, A. O. Packaging 6, P. 20–23 (2006).
2. Namur, T. The Production of packaging. New profit centers. Trans. from English. Moscow: Print Media center (2006).
3. Selke, S.; Cutler, D.; Hernandes, R. Plastics Packaging. Trans. from English. St. Petersburg: Professiya (2011).
4. Michael, L. Current Technologies in Food Packaging. Book Am. Soc. for Testing and Materials (1986).
5. Pastoriza, L.; Bernárdez, M. Journal of Food, 9, P. 126–130 (2011).
6. Serova, V. N.; Sugonyako, D. V.; Veriznikov, M. L.; Tuftin, A. A. Plastic Masses, 5–6, P. 54–56 (2014).
7. Serova, V. N.; Sugonyako, D. V.; Veriznikov, M. L.; Tuftin, A. A. Herald of Kazan Technological University, 17, 3, P. 104–107 (2014).
8. Serova, V. N. Polymer Optic Materials. St. Petersburg: Fundamentals and Technologies (2011).
9. Renby, B.; Rabek, Ya. Photodegradation, Photooxidation and Photostabilization of Polymers. New York: John Wiley (1975).
10. Kupcova, A. Kh.; Zhizhina, G. N. Fourier-Raman and Fourier-IR Spectra of Polymers: Encyclopedia. – Moscow: Phismatlit. (2001).

R. M. Garipov – D. Chem., Prof., Chairman of the Department of Polygraphic processes and Film-photographic Materials Technology of KNRTU, rugaripov@mail.ru; V.N. Serova – D. Chem., Prof., Chairman of the Department of Polygraphic processes and Film-photographic Materials Technology of KNRTU; A.I. Zagidullin – Cand. of Tech. Sc., A.P. of the Department of Polygraphic processes and Film-photographic Materials Technology of KNRTU; A.I. Khasanov – Cand. of Tech. Sc., A.P. of the Department of Polygraphic processes and Film-photographic Materials Technology of KNRTU; A.A. Efremova – Cand. of Tech. Sc., A.P. of the Department of Polygraphic processes and Film-photographic Materials Technology of KNRTU.

CHAPTER 18

MOLECULAR NITROGEN FIXATION WITH HYDROPEROXYL RADICALS: A THEORETICAL AND QUANTUM CHEMICAL STUDY

A.A. IJAGBUJI[1*], E.V. POSHTARËVA[1], A.N. REISSER[2], V.V. SCHWARZKOPF[2], T.C. PHILIPS[1], M.B. JEFFEREY[3], W.W. MCCARTHY[3], and I. I. ZAKHAROV[1]

[1]*Institute of Technology, East Ukrainian National University, Severodonetsk, 93400, Ukraine.*

[2]*Moscow State University, Moscow, Russia*

[3]*University of Melbourne, Parkville, Victoria, Australia*

**E-mail address: dejiijagbuji@yahoo.com*

CONTENTS

ABSTRACT

The relatively low-lying first electronic excited state of hydroperoxyl radical ($^2A'$) has been suggested to play a significant role in molecular nitrogen (N_2) fixation to nitrogen protoxide (N_2O). Quantum chemical calculations for the electronic and geometric structure of HOO-*N=N*-OOH intermediate are computed using the density functional theory (DFT) method at the B3LYP/6-311++G(3*df*,3pd) level. From the results obtained in this work, the binding of molecular N_2 by OOH* (*an example of a self-organizing chemical system*) apparently finds quantum chemical justification.

18.1 INTRODUCTION

Molecular nitrogen is intrinsically unreactive, so much so that it has confounded chemists for decades in attempts to functionalize this abundant diatomic molecule. The exceptional chemical inertness of N_2 molecule is partly due to its triple bond, as evidenced by a dissociation energy of 944 kJ/mol, low affinity (– 84 kcal/mol), and high ionization potential (367 kcal/mol). However, these factors are not solely responsible for N_2 inertness because other triply-bonded molecules, notably CO, readily undergo a wide variety of chemical transformations. Rather, N_2 inertness arises from lack of a dipole moment, and the large HOMO–LUMO gap of about 22.9 *e*V, causing the molecule to be resistant to electron transfer and Lewis acid/base reactivity.[1] *While biological systems and industrial processes can fix nitrogen to form ammonia, the discovery of processes that can utilize N_2 as a feedstock to generate organo-nitrogen materials, are of crucial importance.* The oxidation of molecular N_2 by excited singlet oxygen molecules in aqueous solution has been previously investigated in works.[2,3] Furthermore, hydrogen peroxide (H_2O_2) is believed to be a molecule of considerable interest in many fields including oxidation reactions,[2,3] combustion chemistry,[4,5] biological processes,[6] and photo-dissociation dynamics[7–9], owing to its favorable initiation properties in chain oxidation, in polymerization reactions, its intermediate position in redox ternary, and perhaps because it is the only peroxide for which accurate enthalpy, entropy, and heat capacity data exist.[10] *Traube's* observation[11] have been of significant interest from two positions: (i) Oxidation by H_2O_2 is a slow process, requiring the presence of a catalyst; (ii) H_2O_2 may obtain the properties of a strong oxidant only if it preliminarily dissociated to its radicals such as hydroxyl (˙OH) and hydroperoxyl (•OOH) radicals. In addition to that, the unusual reactivity of peroxides is generally attributed to weakness of the O–O bond linkage, and hence, the ease with which it is homolytically cleaved. Nevertheless, an understanding of the basic chemical principles and energetics of the oxygen-oxygen bond is fundamental to oxygen atom transfer from an oxygen donor under industrial, laboratory, biochemical, combustion, or atmospheric conditions.

At broad temperature range of 25–700°C, a distinctive mechanism for the homogeneous dissociation of H_2O_2, as suggested by *Satterfield* and *Stein*[12] , is thought to be initiated by O–O bond cleavage to give hydroxyl (•OH) radicals (reaction 18-1), whereas the formation mechanism of hydroperoxyl radical (HO_2•) was believed to be that of reactions (18-2*a*,18-2*b*). In relatively unpolluted atmospheric region, hydroperoxyl radical is produced by the reaction of the hydroxyl radical with volatile organic compounds (VOCs) such as carbon monoxide, and subsequent combination with oxygen molecule (reaction 18-2a). Further, the reaction of hydroxyl radical with H_2O_2 generates hydroperoxyl radical and water molecule (reaction 18-2b):

$$H_2O_2 \rightarrow 2 \bullet OH \quad (18\text{-}1)$$

$$\bullet OH + CO + O_2 \rightarrow CO_2 + HO_2\bullet \quad (18\text{-}2a)$$

$$\bullet OH + H_2O_2 \rightarrow H_2O + HO_2\bullet \quad (18\text{-}2b)$$

The experimental evidence of HO_2• radical in the gas-phase reactions was first proposed by *Marshall*[13] in 1926, and has since been found to play a prominent role in atmospheric chemistry. There have been numerous kinetic studies on hydroperoxyl radical because of the significant role it plays in atmosphere and combustion chemistry[14,15] and particularly its ability to recombine to form H_2O_2 (reaction 18-3), which later disappears by a 2nd order process.

$$HO_2 + HO_2 \rightarrow H_2O_2 + O_2 \quad (18\text{-}3)$$

In this review, the activation of molecular N_2 by H_2O_2 *radical is reviewed with the view to present new kinds of transformations for coordinated di-nitrogen. Moreover, some reaction types that are as yet unknown are outlined to try and stimulate further research in this area.* Therefore, the main objective was to investigate the activation of molecular N_2 by high-energy hydroperoxyl radical, and to perform quantum-chemical modeling of the reaction.

18.2 STRUCTURAL AND ELECTRONIC PARAMETERS OF HO_2 RADICAL IN DIFFERENT ELECTRONIC STATES

Since peroxides, especially in alkaline solution, can behave quite similarly to HO_2/O_2^-, studies on the properties of this radical are better done in the absence of H_2O_2.

The ground states of hydroperoxyl radical ($HO_2^\bullet$) have the free electron mostly localized in a p-orbital on the terminal oxygen atom that is perpendicular to the plane of the radical, that is, (^{2}A" for radicals with C_s symmetry), therefore there would be no net overlap and little reaction. However, the first excited state has the free electron in a p-orbital on the terminal oxygen that is in the plane of the radical, that is, (^{2}A' for radicals with C_s symmetry). At linear geometries, the electronic ground state is $^2\Pi$, and this state splits into the nondegenerated $X\,^2A''$ and $\tilde{A}\,^2A'$ states when the molecule bends. Both of these states have strongly bent equilibrium geometries. Since they correlate with a $^2\Pi$ state at linearity, the X $^2A''$ and Ã $^2A'$ states exhibit the Renner effect.[16] The $\tilde{A}$—X electronic band system of HO_2 in the gas phase was studied experimentally in 1974, both in emission and absorption[17,18] and it was realized that forbidden $\Delta K_a = 0$ transitions are present. Tuckett *et* al.[19] suggested that these transitions result from Renner interaction.

Several spectroscopic studies followed refs. [19 – 25]. Recently, *Fink* and *Ramsay* carried out a high-resolution reinvestigation of the $X/\tilde{A}$ (υ_1, υ_2, υ_3) = $\tilde{A}$ (0, 0, 0) → X (0, 0, 0) band. They also observed forbidden $\Delta K_a = 0$ transitions, and concluded that these are magnetic dipole transitions. The electric dipole transition moment for the electronic transition was found to be very small. To confirm the analysis by *Fink* and *Ramsay*[25], Osmann *et* al.[26] calculated the *ab-initio* potential energy, transition electric dipole, and transition magnetic dipole surfaces of the $X\,^2A''$ — $\tilde{A}\,^2A'$ system and simulated the $\tilde{A}\,^2A' \rightarrow X\,^2A''$ emission spectrum with the help of program *RENNER*[27 -29]. Later, Jensen *et* al.[30] calculated more points on the potential energy surfaces using the *ab initio* method of Ref. [28] to cover a wider range of bending geometries. They adjusted shapes of the surfaces in a least squares refinement to the energies of rovibronic states involving both electronic states. Their results provide an accurate representation of the surfaces in this energy region and confirmed the conclusions by *Fink* and Ramsay,[25] in particular, the assignment of the forbidden $\Delta K_a = 0$ transitions as magnetic dipole transitions.

Also, Jensen *et* al.[30] explained a perturbation observed by *Fink* and *Ramsay*[25] in the $\tilde{A}$ (0, 0, 0) → X (0, 0, 0) band for states with J ≈ $^{51}/_2$. The theoretical identification of the perturbing state has been confirmed in a very recent reanalysis of the experimental data.[31] . In another recent experimental study,[32] of $X\,^2A''$ ($HO_2^\bullet$), the $2\nu_1$ band has been investigated by diode laser spectroscopy. Compared to that of an `ordinary' triatomic molecule with an isolated electronic ground state, (e.g., H_2O), the theoretical description of the rotation and vibration in the electronic ground state $X\,^2A''$ of $HO_2^\bullet$ is subject to two complications:

1. The presence of the Renner effect which causes the $X\,^2A''$ state to be degenerated with the first excited state $\tilde{A}\,^2A'$ at linear geometries. In consequence, the Born-Oppenheimer approximation[33] breaks down, thereby, the $X\,^2A''$ and $\tilde{A}\,^2A'$ states must be treated together.
2. In both the $X\,^2A''$ and $\tilde{A}\,^2A'$ electronic states, $HO_2^\bullet$ has two versions, that is, nonsuperposable equilibrium structures that differ only in the numbering of identical nuclei.[34] The two versions have the H nucleus bound to one, or the other, of the two O nuclei, at the strongly bent equilibrium geometries of the X^2A'' and $\tilde{A}\,^2A'$ states. In highly excited bending states, there is tunneling between two equivalent minima corresponding to the two versions, and the two linear geometries HO-O and O-O-H, each one associated with a doubly degenerated, $^2\Pi$ electronic state, become accessible to the molecule. Thus, $HO_2^\bullet$ affords an example of the so-called double Renner effect.[35,36]

18.3 REACTIVITY OF THE HIGH-ENERGY HOO$^\bullet$ (^{2}A′) INTERMEDIATE IN N_2 ACTIVATION PROCESS

Over three and a half decade ago, Azerbaijan chemists succeeded in the oxidation of N_2 with H_2O_2 vapors at temperatures ≈ 800 K.[37,38] Based on the DFT method at the B3LYP/6-311++G(3*df*,3*pd*) level, the reactivity of high-energy state of hydroperoxyl radical HOO*(^{2}A') for N_2 activation process and binding to N_2O have been analyzed (*Nagiev's effect*). This was achieved by: the thermal decomposition of H_2O_2 to produce hydroxyl radicals (*reaction* 18-4); reaction of hydroxyl radical with H_2O_2 (*reaction* 18-5); self-reaction of HO_2 radicals to reform H_2O_2 and $O_2(^1\Delta_g)$ (*reaction* 18-6); reaction of the $O_2(^1\Delta_g)$ with hydroperoxyl radical (*reaction* 18-7); and the interaction of N_2 with excited state of hydroperoxyl radicals (*reaction* 18-8):

$$\mathrm{HOOH} \rightarrow {}^\bullet\mathrm{OH} + {}^\bullet\mathrm{OH}, \tag{18-4}$$

$$\mathrm{OH} + \mathrm{HOOH} \rightarrow \mathrm{H_2O} + \mathrm{OOH}(^2\mathrm{A''}), \tag{18-5}$$

$$\mathrm{HOO}\ (^2A'') + \mathrm{HOO}\ (^2A'') \rightarrow \mathrm{HOOH}\ (^1A_1) + O_2^*\ (^1\Delta_g), \qquad (18\text{-}6)$$

$$O_2^*\ (^1\Delta_g) + \mathrm{HOO}\ (^2A'') \rightarrow O_2(^3\Sigma_g) + \mathrm{HOO}^*\ (^2A'), \qquad (18\text{-}7)$$

$$\mathrm{HOO}^* + N_2 + \mathrm{HOO}^* \rightarrow H_2O + O_2 + N_2O. \qquad (18\text{-}8)$$

Based on the kinetic assessment of H_2O_2 dissociation to $^{\cdot}OH$ and $HO_2^{\cdot}$ radicals (*reaction* 18-9)

$$\frac{[\underline{HO_2^{\cdot}}] = [\underline{K_2}]\ [H_2O_2]}{[^{\cdot}OH]\ [K_1]} \qquad (18\text{-}9)$$

Syrkin and *Moiseev*[39] have shown that: (i) In some cases, $HO_2^{\cdot}$ radicals may act as catalysts for molecular transformation of the reagents; (ii) Due to low energy of the O-O bond, molecular reactions by the interaction of peroxide compounds can display the activation energy to be much lower than the effective energy of radical reaction. The quantitative assessment of active sites in the system displayed a ratio $[HO_2^{\cdot}]/[^{\cdot}OH]$ between 10^5 and 10^6 (*overtly dependent on the contact time*), that is, $HO_2^{\cdot}$ radical concentration is higher by many orders of magnitude than the $^{\cdot}OH$ radical. Under experimental conditions, the [•OOH]/[•OH] ratio was ~10^3–10^7[7,39] thereby suggesting $HO_2\cdot$ radical to be the predominantly active site in the H_2O_2–H_2O–N_2 system. Nevertheless, we carried out a quantum chemical study to verify the possibility of N_2 activation by •OOH radicals.

18.4 COMPUTATIONAL METHODS

The structure of the studied compounds (Figures 18-1–18-4) were optimized based on the DFT method using a combination of *Becke's three-parameter hybrid functional*[40] with *Lee-Yang-Par exchange-correlation functional*[41] with possible symmetry constraints control in the 6-31G++(*d*) valence-split basis set[42] by using the GAUSSIAN 03 software.[43] The quantum chemical calculations were carried out with complete geometry optimization. The calculation of the vibration frequencies and the zero-point energies for all species were carried out for each stationary point on the potential energy surface. The absence of imaginary frequencies characterizes the molecular structure as stable intermediate; the presence of one imaginary frequency in IR spectrum corresponds to the structure of a transition state (TS). The atomic charges, spin densities, and dipole moments were calculated based on the Mulliken analysis of electron density.

TABLE 18-1 Total energy (E_{total}, *a. u.*), structural parameters (R, φ), spin density on atoms q_s, dipole moments μ (D), vibration frequency iω (cm^{-1}), and 0-0 transition energy (T_0, eV) for HO_2 molecule in the ground ($^2A''$) and excited state ($^2A'$) calculated by the B3LYP/6-311++G (*3df,3pd*) approach.

HO_2 Molecule	Symmetry - C_S Ground - $^2A''$	Symmetry - C_S Excited - $^2A'$
E_{tot}	−150.968332	−150.935127
R_{OO} (Å)	1.324	1.384
R_{HO} (Å)	0.976	0.972
φ	105.6°	103.8°
q_s(H)	−0.01	+0.03
$q_s(O_1)$	+0.27	+0.05
$q_s(O_2)$	+0.74	+0.92
μ	2.23	1.91
ω_1	1114.1 (A') 27.2 [a]	949.8 (A') 56.3 [a]
ω_2	1336.5 (A') 39.5	1150.7 (A') 53.4
ω_3	3420.6 (A') 23.9	3523.0 (A') 91.6
T_0 [b]	0.0	0.90

[a] Calculated IR absorption intensity (*km/mol*) for each vibration mode. Vibration frequencies are shifted by the scaling factor =0.9498 which is in agreement with the experimental data for N_2O_3 molecule.

[b] Experimental T_0 value for excited state $^2A'$ equals to 0.87eV.[44]

One of the termination reactions of auto-oxidation is the reaction wherein two peroxyl radicals (*via self-reaction*) are converted into nonradical products (*see reaction* 18-6).

The path of the termination reaction of peroxyl radicals has been previously studied by means of perturbation method.[45]. More importantly, due to the rotation of ·OH and HO_2· groups in the HO_4H chain molecules, there are several isomers with similar stabilities, and six chemically bonded chain-structures have been identified. They are LM1*a*, LM1*b*, LM2*a*, LM2*b*, LM3*a*, and LM3*b* as shown in Figure 18-1. These isomers, LM1*a* and LM1*b*, LM2*a* and LM2*b*, and LM3*a* and LM3*b*, are mirror isomers of each other. LM1*a*, LM2*a*, and their mirror-isomers have C_2 symmetry, the apparent structure differences are that the bridging O–O bonds in LM2*a* and LM2*b* are 0.024 Å shorter than those in LM1*a* and LM1*b*; however, the O–O bonds in the HO_2 groups of LM2*a* and LM2*b* are 0.016 Å longer than those in LM1*a* and LM1*b*.[45]

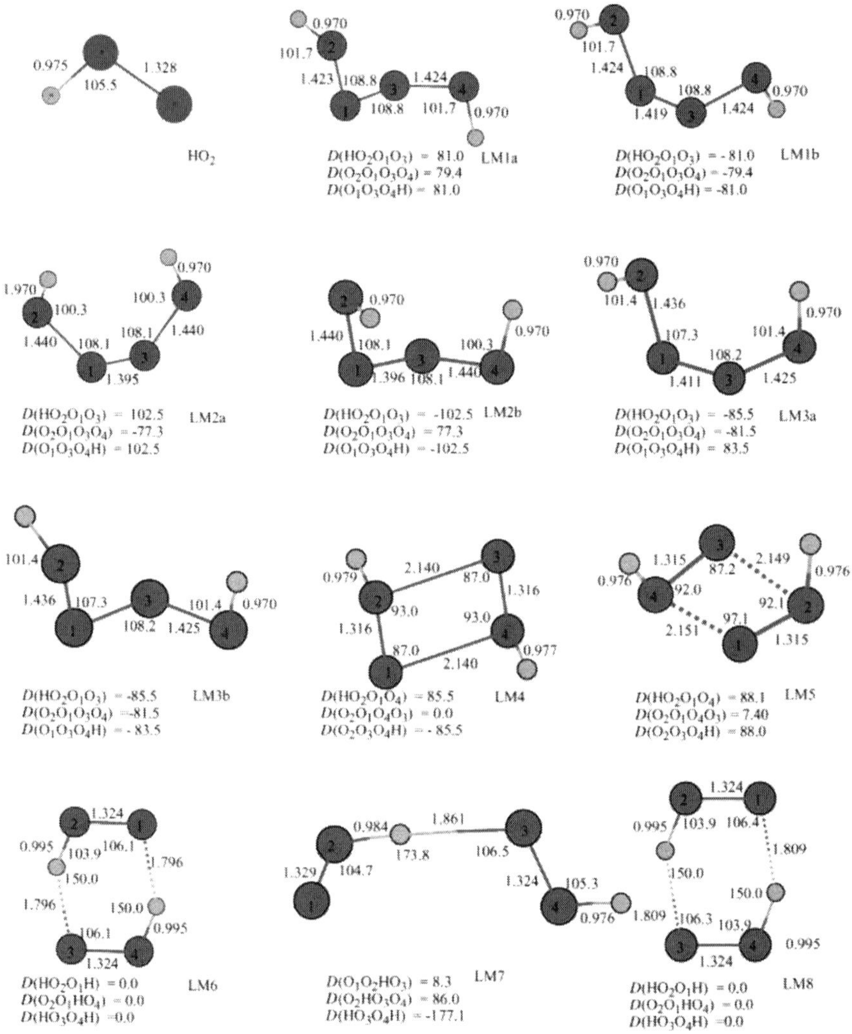

FIGURE 18-1 The optimized geometries of the reactants, singlet, and triplet intermediates in the $HO_2 + HO_2$ reaction at the B3LYP/6-311G(d, p) level.

In addition, in LM2a and LM2b, there are intramolecular hydrogen bonds (2.824 Å), which LM1a and LM1b lack. LM3a and LM3b have C_1 symmetry with slightly stronger intramolecular hydrogen bonds (2.661 Å) compared with those in LM2a and LM2b. From Figure 18-1, it is shown that the most stable chain structure intermediates are LM3a and LM3b.

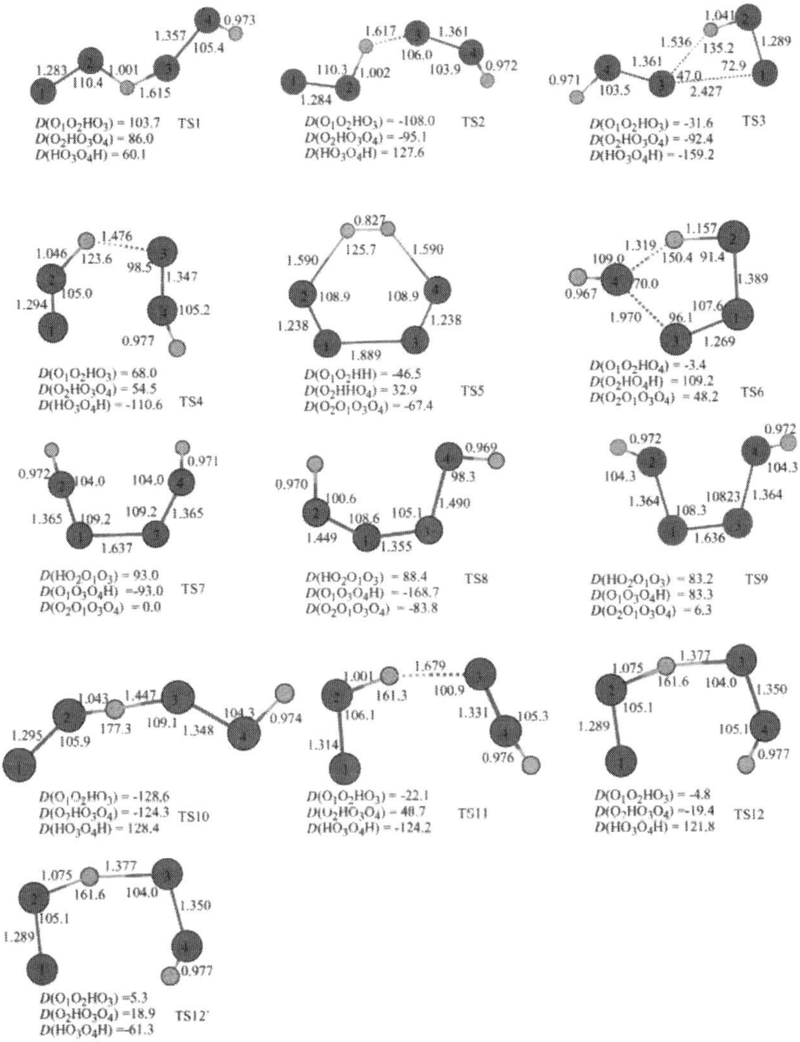

FIGURE 18-2 Optimized geometries of the singlet and triplet transition states for HO_2 + HO_2 reaction at the B3LYP/6-311G(*d*, *p*) level.

The calculated ground ($^2A''$) and excited-state (^2A) properties of hydroperoxyl radical (HO_2,C_s symmetry) is shown in Table 18-1. The calculated adiabatic excitation energy of 0.9 eV is broadly consistent with the experimental value of 0.87 eV,[44] as well as many other spectroscopic parameters. Thus, the chosen computational method seems to be quite reliable for the calculation of doublet-excited state in radicals.

The possibility for the formation of singlet oxygen O_2 ($^1\Delta_g$) in the dimerization of HOO peroxide radicals was first demonstrated by *Goddard and co-workers*.[46] In Figure 18-3, we present the B3LYP/6-311++G(3*df*,3*pd*) spin-polarized calculations for the dimerization of hydroperoxyl radical ($HOO^{\bullet}$) with the formation of cyclic dimer (CD) and generation of singlet oxygen O_2^* ($^1\Delta_g$):

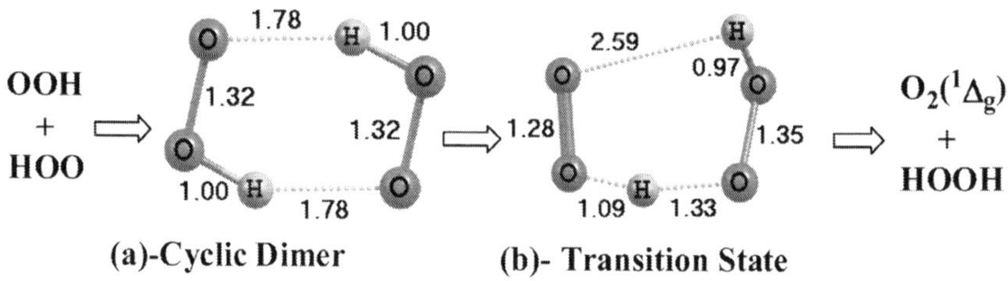

FIGURE 18-3 B3LYP/6-311++G(3*df*,3*pd*) spin-polarized calculation for the dimerization reaction of HOO (hydroperoxyl radical) with the generation of singlet oxygen O_2 ($^1\Delta_g$): (*a*) calculated structure of CD; (*b*) structure of TS.

From the calculated and experimental results for the heat of formation for reagents and products in *reaction* (18-6), it should be noted that this process is spontaneous ($\Delta_r G^0_{298} \approx -40$ kJ) and does not require any activation energy. The computed binding energy at the B3LYP/6311++G(*3df,3pd*) level for O_2^* ($^1\Delta_g$) + HOO ($^2A''$) → O_2 ($^3\Sigma_g$) + HOO* ($^2A'$) to form triplet oxygen, and high-energy hydroperoxyl radical, yielded medium hydrogen-bond interactions in the E_a = 80–90 kJ/mol range.

In this work, we present the quantum chemical substantiation for the process *(Nagiev's effect)*. Our quantum chemical studies on the possibility of bonding molecular N_2 to OOH hydroperoxyl radicals resulted in a stable molecular structure: HOO—N=N—OOH (Figure 18-4). The absence of imaginary frequencies in the calculated IR spectrum of the intermediate characterizes this molecular structure as stable (real), and gives us grounds to suggest that under the experimental conditions,[37,38] the following reaction occurs: HOO* + N_2 + OOH* → TS → HOO—N=N—OOH → H_2O + N_2O + O_2, as depicted in Figure 18-4.

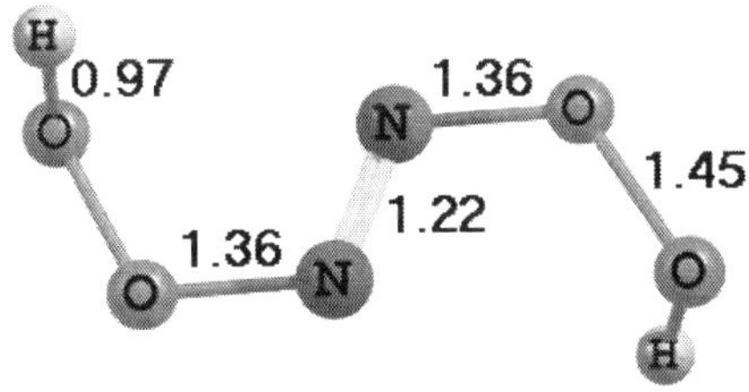

FIGURE 18-4 Structure of the HOO—N=N—OOH intermediate when molecular N_2 is fixed by •OOH hydroperoxyl radicals, C_i symmetry, B3LYP/6311++G(*3df,3pd*) calculation.

Figure 18-5 illustrates the B3LYP/6-311++G(*3df,3pd*) calculation result for the activation reaction energy profile of molecular N_2 in the presence of •OOH (hydroperoxyl radicals), which is characterized by the formation of the high energy HOO—N=N—OOH intermediate:

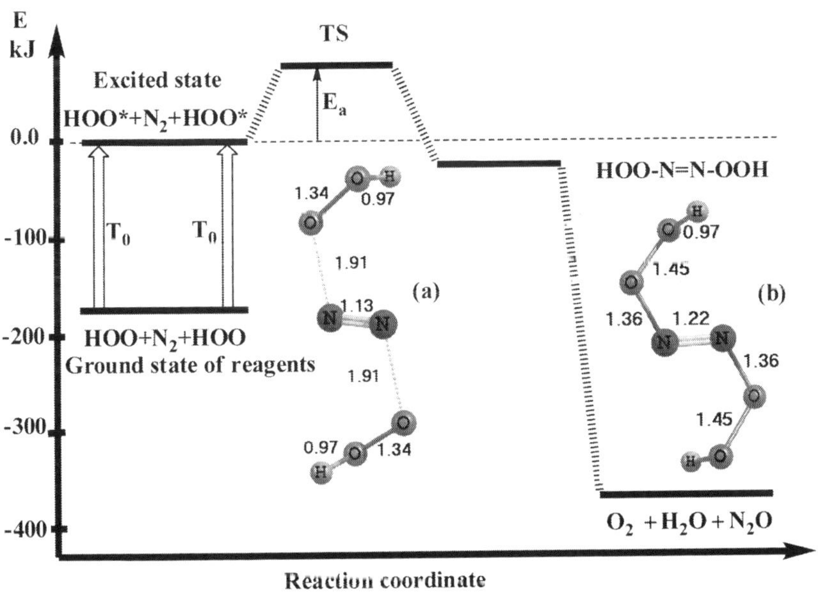

FIGURE 18-5 B3LYP/6-311++G(*3df,3pd*) calculated energy profile for the activation reaction of molecular N_2 by the HOO^* ($^2A'$). The calculated structure (*a*) characterizes activation TS; the structure (*b*) characterizes the reaction intermediate HOO—N=N—OOH.

From the analysis of quantum chemical calculation of the TS, activation of molecular N_2 by HOO hydroperoxyl radicals in Figure 18-4, we depict the scheme of the molecular orbital interaction HOO + N_2 + OOH via the *singlet coupling of radicals in the presence of* N_2. Here, the lone pairs of HOO radicals are involved in π conjugation with the anti-bonding π_y^* orbital of N_2, while the unpaired electrons of HOO radicals are involved in π conjugation with the bonding π_x orbital of N_2.

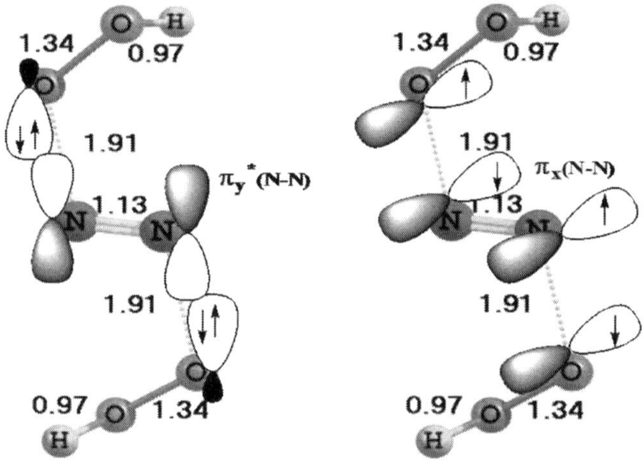

FIGURE 18-6 Scheme of molecular orbital activation of N≡N by peroxide radicals in the transition activated complex: lone pairs of HOO$^\bullet$ radicals are involved in π conjugation with the anti-bonding π_y^* orbital of N_2 (*a*); unpaired electrons of HOO radicals are involved in π conjugation with the bonding π_x orbital of N_2 (*b*). Calculated geometric parameters (bond lengths) of the TS are in Å.

From the molecular orbital scheme shown in Figure 18-6), it follows that the electronic state of the peroxide radical corresponds to the excited HOO* ($^2A'$) state rather than that of the ground state ($^2A''$), when the unpaired electron of the radical is in the σ molecular orbital of HOO$^\bullet$ (rather than in π MO, as it is for the ground state of HOO). How can peroxide radicals appear in the excited state HOO* ($^2A'$) under experimental conditions in works?[37,38] Moreover, the energy difference of these states (T_0 = 0.87 eV [47]) can be overcome only by means of a sensitizer due to a very small moment of the transition $X\,^2A'' - \tilde{A}\,^2A'$. In this case, the most appropriate sensitizer can be singlet oxygen O_2 ($^1{}_g$) because its energy T_0 = 0.98 eV relative to the ground O_2 ($^3\Sigma_g$) state, which is just higher than 0.87 eV. In addition, singlet oxygen O_2^* ($^1\Delta_g$) is characterized by a long radiation decay time in the gas phase ($\tau_d \approx$ 45 min). That does not reveal the main step of nitrogen bonding process, namely its activation, because H_2O_2 in the molecular form does not interact with molecular N_2. Moreover, in their perception of the process, the authors[37–38] ignored yet another final reaction product (18-12): molecular oxygen that they detected in the observed mass spectra.[37] All these neglections, in our opinion, did not allow the researchers to focus on the obtained results (unnoticed) in scientific literature for over 40 years.

TABLE 18-2 B3LYP/6-311++G(3*df*,3*pd*) calculation results of harmonic vibrations (cm^{-1}) for different isotopomers of H—O—O—N=N—O—O—H intermediate (C_i symmetry, Figure 18-1)

Type of vibrations	ISOTOPOMER $^{1}H—^{16}O—^{16}O—^{14}N=^{14}N—^{16}O—^{16}O—^{1}H$	ISOTOPOMER $^{1}H—^{16}O—^{16}O—^{15}N=^{15}N—^{16}O—^{16}O—^{1}H$
	A_g symmetry	
ν_s(O—H)	3717 (0,0)*	3717 (0,0)*
ν_s(N=N)	1588 (0,0)	1534 (0,0)
δ(O—O—H)	1378 (0,0)	1378 (0,0)
ν_s(N—O)	1002 (0,0)	971 (0,0)
ν_s(O—O)	870 (0,0)	869 (0,0)
δ(O—O—N)	720 (0,0)	720 (0,0)
δ(N—O—O)	364 (0,0)	357 (0,0)
δ(O—N=N)	309 (0,0)	308 (0,0)
τ(HOON)	194 (0,0)	193 (0,0)
	A_u symmetry	
ν_{as}(O—H)	3715 (99,0)	3715 (99,0)
δ(O—O—H)	1378 (90,0)	1378 (90,1)
ν_{as}(N—O)	*1012 (98,6)*	*998 (94,9)*
ν_{as}(O—O)	803 (33,0)	800 (32,8)
ν_{as}(N—O;O—O)	598 (50,4)	587 (49,6)
τ(ONNO)	416 (16,5)	410 (17,4)
τ(HOON)	233 (53,2)	231 (56,1)
τ(NOOH)	180 (143,8)	179 (140,3)
τ(OONN)	106 (24,1)	106 (23,5)

**The calculated intensities of IR spectral bands are given in parenthesis (km/mol).*

TABLE 18-3 Results of quantum chemical calculation for the total energy (E_{total}), zero vibrational energy (E_0), and absolute entropy S^0_{298} for reagents, products, TS, and their thermodynamic parameters ($\Delta_r G^0_{298}$, $\Delta_r H^0_{298}$, $\Delta_r S^0_{298}$) for reaction: HOO($^2A''$) + HOO($^2A''$) → CD → TS → HOOH($^1A_1$) + O_2($^1\Delta_g$)

Molecular System (electronic state)	B3LYP/6-311++G(3*df*,3*pd*) Calculations			Experimental Values[c]
	Total Energy E_{total}, *a. u.*	**Energy, E_0*, kJ/mol [a]**	**Enthalpy $\Delta_r H0_{298}$, [b] kJ/mol**	**Entropy $S0_{298}$,[c] J/(mol.K)**
HOO (C_s–$^2A''$)	– 150.96833	37.1 (0)	+7.2 [b] +2.1[c]	228.7 [229.1]c
+				
HOO (C_s–$^2A''$)	– 150.96833	37.1 (0)	+7.2 [b] +2.1[c]	228.7 [229.1]c
⇓				
CD (C_{2h}–^1A) Fig. 3*a*	– 301.95591 E_{dim} = –50.5 kJ/mol	85.6 (0)	–	357.0
⇓				
TS 1 (C_1–^1A) Fig. 3*b*	– 301.94126 E_a = 16.0 kJ/mol	70.3 (1) $i\omega$ = 1110 cm^{-1}	–	303.7
⇓				
HOOH (C_2–$^1A_1$)	– 151.61319	69.8 (0)	– 107.7 [–136.1]c	227.0 [233.0]
+				
O_2($^1\Delta_g$)	– 150.36347	9.8 (0)	+ 42.1 [+ 94.6]c	195.8 [205.0]
$\Delta_r G^0_{298}$ = – 69.7 kJ [– 39.7 kJ][c]	E_{total} = – 112.9 kJ	—	$\Delta_r H^0_{298}$ = –80.0 kJ [– 45.7 kJ][c]	$\Delta_r S^0_{298}$ = –34.6 J/K [–20.2 J/K][c]

Footnotes:

[a] The number of imaginary frequencies in the vibrational spectrum of molecules are given in parenthesis. The absence of imaginary frequencies characterizes the molecular structure as stable (real). The presence of one imaginary frequency (iω) characterizes the structure as TS.

[b] The heats of formation are calculated relative to the energy level of simple substances N_2 ($1\Sigma_g$), O_2 ($3\Sigma_g$), and H_2 ($1\Sigma_g$) with regard to zero vibrational energy E_0: $\Delta H = \Delta E_{total} + \Delta E_0$. The used ratios of energy units: 1 a.u.= 627.544 kcal; 1 kcal = 4.184 kJ.

[c] Experimental values of the thermodynamic parameters are taken from the National Institute of Standards and Technology (NIST) database: http://webbook.nist.gov.chemistry.

TABLE 18-4 Result of the quantum chemical DFT calculation for total energy E_{total}, zero vibrational energy E_0 for reagents, products, transition state, and thermodynamic values $\Delta_r G^0_{298}$, $\Delta_r H^0_{298}$, $\Delta_r S^0_{298}$ for molecular N_2 activation HOO* + N_2 + HOO* → TS → HOO-N_2-OOH → H_2O + N_2O + O_2

Molecular System (electronic state)	B3LYP/6-311++G(3df,3pd) Calculations			Experimental Values [c]
	Total Energy E_{total}, a. u.	ZPE E_0^*, kJ/mol [a]	Enthalpy $\Delta_r H^0_{298}$,[b] kJ/mol	Entropy S^0_{298},[c] J/(mol.K)
HOO* (C_s–$^2A'$)	– 150.93513	35.4 (0)	+102.7[b] [+86.2] [c]	229.7 [229.1] [c]
+				
N_2 ($^1\Sigma_g$)	– 109.56737	14.6 (0)	0.0 [0.0]	191.4 [191.5]
+				
HOO* (C_s–$^2A'$)	– 150.93513	35.4 (0)	+102.7[b] [+86.2] [c]	229.7 [229.1] [c]
⇓				
TS 1 (C_i–$^1A_{1g}$)	– 411.41492	105.0 (1)	–	357.0
	E_a = 79.2 kJ/mol	$i\omega$[a] = 182 cm^{-1}		
⇓				
HOO—N_2—OOH (C_i–$^1A_{1g}$) Fig. 4	– 411.44790	111.1 (0)	+ 204.0	336.0
⇓				
H_2O (C_{2v}—$^1A_1$)	– 76.46451	55.9 (0)	– 224.10 [– 241.8]	188.6 [188.8]
+				
N_2O ($^1\Sigma_g$)	– 184.73335	29.4 (0)	+ 72.3 [+ 81.6]	219.3 [220.0]
+				
O_2 ($^3\Sigma_g$)	– 150.37949	9.8 (0)	0.0 [0.0]	204.9 [205.0]
$\Delta_r G^0_{298}$ = – 345.9 kJ [– 321.9 kJ][c]	—	—	$\Delta_r H^0_{298}$ = – 357.2 kJ [–332.6 kJ][c]	$\Delta_r S^0_{298}$ = – 38.0 J/K [– 35.9 J/K][c]

Footnotes:

[a] The number of imaginary frequencies in the vibrational spectrum of molecules are given in parenthesis. The absence of imaginary frequencies characterizes the molecular structure as stable (real). The presence of one imaginary frequency (iω) characterizes structures as TS.

[b] Heats of formation are calculated relative to the energy level of simple substances N_2 ($^1\Sigma_g$), O_2 ($^3\Sigma_g$), and H_2 ($^1\Sigma_g$) with regard to zero vibrational energy E_0: $\Delta H = \Delta E_{total} + \Delta E_0$. The used ratios of energy units: 1 *a. u.*= 627.544 kcal; 1 *kcal* = 4.184 kJ.

[c] Experimental values of the thermodynamic parameters are taken from the National Institute of Standards and Technology (NIST) database: *http://webbook.nist.gov.chemistry*.

Note that the authors37,38 interpreted the bonding process of molecular N_2 by another chemical equation that does not reveal the main step of the nitrogen bonding process, namely its activation, because H_2O_2 in the molecular form does not interact with molecular N_2. Moreover, in their perception of the process, the authors37,38 ignored yet another final product of reaction 12: molecular oxygen that was detected in the observed mass spectra.[37] Nitrous oxide (N_2O), widely utilized in *rocket motors*, *internal combustion engine*, *rocket propellant*, and *medicine as an anesthetic agent*, is generated as a by-product during nitric acid production via low-temperature ammonia combustion in the gas-phase (*reaction* 10), however, a commonly known method[48] for N_2O synthesis is by the pyrolysis of ammonia nitrate at a temperature of 523–533 K (*reaction* 11):

$$4NH_3 + 4O_2 \rightarrow 2N_2O + 6H_2O \tag{18-10}$$

$$NH_4NO_3 \leftrightarrow N_2O + 2H_2O \tag{18-11}$$

$$N_2 + H_2O_2 \rightarrow N_2O + H_2O \tag{18-12}$$

In addition to that, from the experimental investigation[5] of molecular N_2 activation reaction with H_2O_2 (reaction 18-12), nitrous oxide (N_2O) as the primary fixation product is formed, as well as, acids suggested to be hyponitrous acid ($H_2N_2O_2$), nitrous acid (HNO_2), and nitric acid (HNO_3) were also detected as secondary products. Thus, N_2O synthesis from N_2 requires the presence of sufficient amount of $HO_2^{\bullet}$ free radical in the system. However, this amount can be significantly decreased by quadratic chain termination ($HO_2^{\bullet}$ radical recombination), which leads to molecular oxygen formation owing to the common knowledge that molecular oxygen interacts with nitrogen only at high temperatures. Therefore, accumulation of oxygen in the system, due to increase in the primary reaction rate at high temperature, reduces the target product yield. That been said, the conjugated reaction of N_2 oxidative fixation by H_2O_2 mostly produces nitrous oxide in concentrations below 19%, and in the quenching zone, most of the nitrous oxide is converted to $H_2N_2O_2$ due to the action of active sites ($HO_2^{\bullet}$ and $^{\bullet}OH$) and H_2O_2. Hence, chemical induction is the major factor promoting N_2 fixation in a N_2–H_2O_2–H_2O system, and it manifests itself owing to the conjugation of H_2O_2 dissociation reactions, which generates the intermediate ($HO_2^{\bullet}$ radical) in the system. This intermediate transfers induction action of the primary reaction to the N_2 oxidation process. In this case, H_2O_2 is injected in an amount much greater than demanded by N_2 oxidation, because the main requirement for effective chemical conjugation is the presence of $HO_2^{\bullet}$ in high concentration.

Result of quantum chemical calculations for the total energy (E_{total}), zero-point vibrational energy (E*zpe*) for reagents, products, TS, and their thermodynamic parameters ($\Delta_r G^0_{298}$, $\Delta_r H^0_{298}$, $\Delta_r S^0_{298}$) for the *reaction*: HOO($^2A''$) + HOO($^2A''$) → CD → TS → HOOH($^1A_1$) + O_2($^1\Delta_g$) has been presented in Table 18-3. The calculated values for the dimerization energy (E_{dim} = – 50.5 kJ/mol), activation energy (E_a = 16.0 kJ/mol), and the total energy (E_{total} = – 112.9 kJ/mol) of $HO_2^{\bullet}$ at the B3LYP/6-311++G(3*df*,3*pd*) level are broadly consistent with those reported in work;[45] wherein, the values E_{dim} = – 47.8 kJ/mol, E_a = 16.3 kJ/mol, and E_{total} = – 94.1 kJ/mol were obtained at the MP2/6-311+G(3*df*,2*p*) level. The structure, energetics, and infrared (IR) spectrum of the H_2O_2–N_2 complex have been studied by ab initio calculations and the FTIR matrix-isolation technique. Computationally, only one stable conformation was found, showing an almost linear hydrogen bond. The interaction energy of the H_2O_2–N_2 complex were calculated to be within 4.92 kJ/mol[49] to 5.7 kJ/mol[50], and 0.34 kJ/mol [49] at the MP2/6-31++G(2*d*,2*p*), and B3LYP/6-311++G(3*df*,3*pd*) at DFT theoretical levels. Experimentally, the interaction was found to be strong enough to induce substantial perturbations on the fundamental vibrational modes of H_2O_2 upon complexation.[50]

To ascertain the experimental claims,[37,38] the calculated IR spectra of various isotopomers of the H–O—O—N=N—O—O—H (C_i–$^1A_{1g}$) that may be formed during the oxidation of both natural ($^{14}N_2$) and heavy molecular nitrogen ($^{15}N_2$) in the reaction medium with H_2O_2 is presented in Table 18-2. Our calculations and interpretation of the spectra of these isotopomers show that the most intense absorption is the asymmetric vibration ν_{as}(N–O) at ~ 1000 cm^{-1} range.

Results of the quantum chemical DFT calculation for total energy E_{total}, zero-point vibrational energy E_0 for reagents, products, transition state, and thermodynamic values $\Delta_r G^0_{298}$, $\Delta_r H^0_{298}$, $\Delta_r S^0_{298}$ for molecular N_2 activation HOO* (C_s–$^2A'$) + N_2 ($^1\Sigma_g$) + HOO* (C_s–$^2A'$) → TS (C_i–$^1A_{1g}$) → HOO–N_2–OOH (C_i $^1A_{1g}$) → H_2O (C_{2v}—$^1A_1$) + N_2O ($^1\Sigma_g$) + O_2 ($^3\Sigma_g$) are computed in Table 18-4. From the obtained results in Table 18-4, it can be seen that the value of the calculated activation energy (E_a = 79.2 kJ/mol) at the B3LYP/6-311++G(3*df*) level. Further, the calculated values for the Gibb's free energy ($\Delta_r G^0_{298}$ = – 345.9 kJ), the heat of formation ($\Delta_r H^0_{298}$ = – 357.2 kJ), and entropy ($\Delta_r S^0_{298}$ = – 38.0 J/K) of the reaction are in excellent agreement with the experimental data ($\Delta_r G^0_{298}$ = – 321.9 kJ), ($\Delta_r H^0_{298}$ = – 332.6 kJ), and ($\Delta_r S^0_{298}$ = – 35.9 J/K). That been said, from the calculated IR absorption spectra for the studied molecular reactions, the absence of negative eigen values in the corresponding IR suggests that all stationary points correspond to genuine minima. Thereby, in our view-point, based on isotope shift in the high-frequency N–O vibration for the IR spectrum of intermediate molecular structures (Table 18-2), the calculated value for the activation energy (E_a = 79.2 kJ/mol) at the B3LYP/6-311++G(3*df*) level, and the negative value of Gibb's free energy ($\Delta_r G^0_{298}$ = – 345.9 kJ), the N≡N oxidation reaction by H_2O_2 is, therefore, kinetically and/or thermodynamically feasible, and that the intermediate H—O—O—N=N—O—O—H (C_i–$^1A_{1g}$) plays a key role in molecular N_2 activation.

To conclude the discussion, it should be noted that the oxidative fixation of molecular N_2 *with* H_2O_2 *is fairly simple for process-engineering design, but usually, however, proceeds under homogeneous conditions without the low-temperature ammonia oxidation or any catalyst under atmospheric pressure to produce high yield of fixed nitrogen.* Hence, it may be concluded that the elementary stage *(reaction* 18-12) is responsible for N_2O synthesis because at this same stage, the role of $HO_2^\bullet$ radical is limited only by the catalytic effect.

18.5 CONCLUSIONS

We have presented the quantum chemical substantiation for the activation process and binding of molecular N_2 to N_2O via the interaction of N_2 with HOO^* radicals.

In our opinion, the considered process for molecular N_2 binding (*Nagiev's effect*), which is considered as one of the mechanisms for *synergistic effect* (vivid example of a self-organizing chemical system) for the activation of chemically-inert N_2 was achieved by:

i. Thermal decomposition of H_2O_2 to produce hydroxyl radicals
ii. Reaction of hydroxyl radical with H_2O_2
iii. Self-reaction of $HO_2^\bullet$ radicals to form singlet oxygen and also regenerate H_2O_2
iv. Interaction of singlet oxygen with hydroperoxyl radical
v. Activation of N_2 with excited state of hydroperoxyl radicals

In connection with the development of chemical engineering processes, conjugated reaction principles are employed for nonspontaneous oxidative nitrogen fixation by chemical induction.

The application of this principle for nitrogen fixation, mainly shaped as nitrogen protoxide (N_2O), under mild conditions, is characterized as an *environmentally-benign* and an *energy-saving method*, and has also been investigated as a possible commercial route for the manufacture of N_2O.

ACKNOWLEDGMENTS

The authors are grateful to the Ukrainian–American Laboratory of Computational Chemistry at the Scientific-Technical Complex, "Institute of Single Crystals, *National Academy of Sciences of Ukraine*," for the assistance rendered to perform the quantum chemical calculations.

KEYWORDS

- **Conversion to N_2O**
- **OOH* Excited States**
- **Reaction Mechanisms for N_2 Oxidation**
- **Structure of Intermediate HOO-N=N-OOH**
- **Quantum Chemical Calculations**

LITERATURES CITED

1. Bazhenova, T. A.; Shilov, A. E. *Coord. Chem. Rev.* 144 (1995), p. 69–145.
2. Khan, A. V.; Kasha, M. *J. Chem. Phys.*, 39, 2105 (1963), p. 413–418.
3. Browne, R. J.; Ogryzlo, E. A. *Proc. Chem. Soc.*, 117 (1964), p. 241–245.
4. Gray, D.; Lissi, E.; Heicklen, J. *J. Phys. Chem.*, (1972), 76 (14), p. 1919– 1924.
5. Nagiev, T. M. Coherent synchronized oxidation reactions by hydrogen peroxide. Elsevier (2007) Amsterdam.
6. Wazawa, T.; Matsuoka, A.; Tajima, G.; Sugawara, Y.; Shikama, K. *Biophys. J.*, (1992), 63(2), p. 544–550.
7. Schinke, R.; Staemmler, V. *Chem. Phys. Lett.*, (1988), 145(6), p. 486–492.
8. Bersohn, R.; Shapiro, M. *J. Chem. Phys.*, 85 (1986), p. 1396.
9. Melnikov, V. V.; Odaka, T. E.; Jensen, P.; Hirano, T.; *J. Phys. Chem.*, 128 (2008), p. 306–315.
10. Benson, S. W.; Shaw, R. In Organic Peroxides, D. Swern, (Ed.) Wiley, New York, (1970) Vol. 1.
11. Traube, I. Berichte., (1883), 16, 123.

12. Satterfield, N. C.; Stein, W. T.; *J. Phys. Chem.,* 61 (1957), p. 537–540.
13. Marshall, A. L. *J. Phys. Chem.*, 30, 44, (1926).
14. Zhu, R.; Lin, M. C. *Phys. Chem. Comm.*, 23, 1 (2001).
15. Viti, S.; Roue, E.; Hartquist, T. W.; Pineau des Foriets, G.; Williams, D. A. *Astron. Astrophys.*, 370, 557 (2001).
16. Renner, R. *Z. Physik*, 92, 172 (1934).
17. Hunziker, H. E.; Wendt, H. R. *J. Chem. Phys.*, 60, 4622 (1974).
18. Becker, K. H.; Fink, E. H.; Leiss, A.; Schurath, U. *J. Chem. Phys.*, 60, 4623 (1974).
19. Tuckett, R. P.; Freedman, P. A.; Jones, W. J. *Mol. Phys.*, 37, 379 (1979).
20. Saito, S.; Matsumura, C. *J. Mol. Spectrosc.*, 80, 34 (1980).
21. Yamada, C.; Endo, Y.; Hirota, E. *J. Chem. Phys.*, 78, 4379 (1983).
22. Burkholder, J. B.; Hammer, P. D.; Howard, C. J.; Towle, J. P.; Brown, J. M. *J. Mol. Spectrosc.*, 151, 493 (1992).
23. Chance, K. V.; Park, K.; Evenson, K. M.; Zink, L. R.; Stroh, F. *J. Mol. Spectrosc.*, 172, 407 (1995).
24. Chance, K. V.; Park, K.; Evenson, K. M.; Zink, L. R.; Stroh, F.; Fink, E. H.; Ramsay, D. A. *J. Mol. Spectrosc.*, 183, 418 (1997).
25. Fink, E. H.; Ramsay, D. A. *J. Mol. Spectrosc.*, 185, 304 (1997).
26. Osmann, G.; Bunker, P. R.; Jensen, P.; Buenker, R. J.; Gu, J. P.; Hirsch, G. *J. Mol. Spectrosc.*, 197, 262 (1999).
27. P. Jensen, M. Brumm, W.P. Kraemer and P.R. Bunker, *J. Mol. Spectrosc.*, 171, 31 (1995).
28. Kolbuszewski, M.; Bunker, P. R.; Jensen, P.; Kraemer, W. P. *Mol. Phys.*, 88, 105 (1996).
29. Osmann, G.; Bunker, P. R.; Jensen, P.; Kraemer, W. P. Chem. Phys., 225, 33 (1997).
30. Jensen, P.; Buenker, R. J.; Gu, J. P.; Osmann, G.; Bunker, P. R. *Can. J. Phys.*, 79, 641 (2001).
31. Fink, E. H.; Ramsay, D. A. *J. Mol. Struct.*, 795, 155 (2006).
32. DeSain, J. D.; Ho, A. D.; Taatjes, C. A. *J. Mol. Spectrosc.*, 219, 163 (2003).
33. Bunker, P. R.; Jensen, P. in: *Computational Molecular Spectroscopy*, Wiley, Chichester, (2000).
34. Bone, R. G. A.; Rowlands, T. W.; Handy, N. C.; Stone, A. J. *Mol. Phys.*, 72, 33 (1991).
35. Odaka, T. E.; Hirano, T.; Jensen, P. *J. Mol. Struct.*, 795, 14 (2006).
36. Odaka, T. E.; Melnikov, V. V.; Jensen, P.; Hirano, T.; Lang, B.; Langer, P. *J. Chem. Phys.*, 126 , 094301 (2007).
37. Nagiev, M. F.; Nagiev, T. M.; Aslanov, F. A.; Bayramov, V. M.; Iskenderov, R. A. *Dokl. AN. SSSR.*, 213 (1973), 1096.
38. T.M. Nagiev, *Russ. Chem. Rev.*, 54 (1985) 974 – 985.
39. Syrkin, Ya. K.; Moiseev, I. I. *Russ. Chem. Rev.* (1960), 29 (4), p.193 – 214.
40. Becke, A. D. *J. Chem. Phys.*, 98 (1993), P. 5648–5652.
41. Lee, C.; Yang, W.; Parr, R. G. *Phys. Rev. B*, 37 (1988), p. 785–797.
42. Krishnan, R.; Binkley, J. S.; Seeger, R.; Pople, J. A. *J. Chem. Phys.*, 72 (1980), p. 650–5652.
43. Frisch, M. J.; Trucks, G. W.; Schlegel, H. B.; Scuseria, G. E.; Robb, M. A.; Cheeseman, J. R.; Montgomery Jr., J. A.; Vreven, T.; Kudin, K. N.; Burant, J. C.; Millam, J. M.; Iyengar, S. S.; Tomasi, J.; Barone, V.; Mennucci, B.; Cossi, M.; Scalmani, G.; Rega, N.; Petersson, G. A.; Nakatsuji, H.; Hada, M.; Ehara, M.; Toyota, K.; Fukuda, R.; Hasegawa, J.; Ishida, M.; Nakajima, T.; Honda, Y.; Kitao, O.; Nakai, H.; Klene, M.; Li,

X.; Knox, J. E.; Hratchian, H. P.; Cross, J. B.; Adamo, C.; Jaramillo, J.; Gomperts, R.; Stratmann, R. E.; Yazyev, O.; Austin, A. J.; Cammi, R.; Pomelli, C.; Ochterski, J. W.; Ayala, P. Y.; Morokuma, K.; Voth, G. A.; Salvador, P.; Dannenberg, J. J.; Zakrzewski, V. G.; Dapprich, S.; Daniels, A. D.; Strain, M. C.; Farkas, O.; Malick, D. K.; Rabuck, A. D.; Raghavachari, K.; Foresman, J. B.; Ortiz, J. V.; Cui, Q.; Baboul, A. G.; Clifford, S.; Cioslowski, J.; Stefanov, B. B.; Liu, G.; Liashenko, A.; Piskorz, P.; Komaromi, I.; Martin, R. L.; Fox, D. J.; Keith, T.; Al-Laham, M. A.; Peng, C.Y.; Nanayakkara, A.; Challacombe, M.; Gill, P. M. W.; Johnson, B.; Chen, W.; Wong, M. W.; Gonzalez, C.; Pople, J. A. Gaussian 03, Revision A.1, Gaussian, Inc., Pittsburgh PA, 2003.

44. Jacox, M. E. *J. Chem. Phys. Ref. Data*, Monograph № 3, 1994.
45. Zhu, R.; Lin, M. C. *Phys. Chem. Comm.*, 2001, 23, p. 1–6.
46. Xu, X.; Muller, R. P.; Goddard III, W. A. *Proc. Nat. Acad. Sci.*, 2002, USA, 99, 376.
47. Minaev, B. F.; Zakharov, I. I.; Zakharova, O. I.; Tselishtev, A. B.; Filonchook, A. V.; Shevchenko, A. V. *Chem. Phys. Chem.*, 11 (2010), p. 4028.
48. Holleman, A. F.; Wiberg, E. (2001, *Inorganic Chemistry* San Diego Academic Press ISBN 0-12-352651-5).
49. Molina, J. M.; Dobado, J. A.; Daza, M. C.; Villaveces, J. L. *J. Molec. Struct.*, 580 (2002), p. 117–126.
50. Lundell, J.; Pehkonen, S.; Pettersson, M.; Räsänen, M. *Chem. Phys. Lett.* 286, Issues 5-6, 17 (1998), p. 382 – 388.

CHAPTER 19

SYNTHESIS, STRUCTURE OF NEW PHOSPHORYL METHYL DERIVATIVE AMINOACIDS AND THEIR MEMBRANE TRANSPORT PROPERTIES RELATED TO ALKALI METALS

SERGEY ALEKSEEVICH KOSHKIN[1],
AIRAT RIZVANOVICH GARIFZYANOV[1],
NATALIA VIKTOROVNA DAVLETSHINA[1],
RUSTAM RIFKHATOVICH DAVLETSHIN[1],
OLEG VLADISLAVOVICH STOYANOV[2*], and
RAFAEL ASKHATOVICH CHERKASOV[1**]

[1]*Kazan Federal University, The Russian Federation, 420008, Kazan.*

[2]*Kazan National Research Technological University, The Russian Federation, 420015, Kazan.*

**E-mail: ov_stoyanov@mail.ru*

***E-mail: rafael.cherkasov@kpfu.ru*

CONTENTS

ABSTRACT

The new phosphoryl methyl derivative aminoacids have been synthesized and their structure has been investigated by the NMR-method. The membrane-transport properties of synthesized compositions have been studied.

19.1 INTRODUCTION

The importance of the development of extraction, concentration, and separation methods of alkali metals is stipulated by the necessity to solve problems resulting from depletion of natural sources of micro and macro elements. The particular importance belongs to a range of issues associated with mining and use of lithium. The urgency of indicated problem is associated also with ever-increasing necessity to develop the environmentally acceptable methods of disposal of used lithium power sources. The modern methods of processing of natural sources of alkali metals and especially lithium as well as its extraction from used materials apply extensively membrane technologies including transport of alkali metals ions through impregnated liquid membranes.[1–3]

Recently, it has been demonstrated that lipophilic α-amino phosphine oxides approve themselves as effective carriers of monobasic organic acids, rare and trace elements ions, and a number of other substrates of natural and industrial origin.[4–6] At the same time transmembrane mass transfer is inefficient due to lack of basic centers of coordination in the carrier[7] during transition to polyfunctional substrates like polyfunctional hydroxy acids. Another significant disadvantage of neutral amino phosphorylic membranous extraction agents is that they are unable to perform transmembrane transfer of alkali metals which in conditions of membrane extraction through liquid supported membrane virtually completely remain in donating aqueous phase through the whole pH interval.[8] It is suggested that receptor abilities of amino phosphoryl extracting agents can increase as a result of introduction of "additional" basicity centers, for example, carboxyl group capable of participation in the formation of hydrogen bonds together with phosphoryl and amino groups. In our opinion it is convenient to produce compounds of such structure from structurally various natural amino acids. However, the use of them in Kabachnik-Fields reactions or in addition reaction of hydrophosphoryl compounds to imines (Pudovic's reaction) having universal value for amino phosphoryl compounds synthesis results in base products with low yield.[9,10]

19.2 EXPERIMENTAL PROCEDURE

Nuclear magnetic resonance (NMR) ^{1}H and ^{31}P spectra were registered on Bruker AVANCE III HD NanoBay spectrometer with operating frequency 400 and 121.4 MHz, correspondingly. Mass-spectrometer measurements were carried-out on AB Sciex TripleTOF 5600 high-resolution spectrometer with ionization source turbo ion spray (TIS) in positive ionization mode (ionization source voltage is 5500V, nebuliser gas is nitrogen). Infrared spectra were measured for samples in petrolatum oil on Bruker Tensor 27 Fourier transform spectrometer. Diffractional measurements of monocrystal (S)-N-dicyclo hexyl phosphoryl methyl-1-aminopropionic acid were carried out on Bruker Kappa APEX DUO diffractometer ($\lambda_{MoK\alpha}$= 0.71073 Å, graphite monochromator, ω and ϕ- scanning) with use of APEX2 software at 100°K.

The study of membrane transport through impregnated membranes was carried out in a cell looking like two glasses inserted one inside the other. Donating phase was inserted into the external thermostatically controlled glass (Dewar vessel) and acceptor phase was put into the internal Teflon glass with impregnated liquid membrane acting as a bottom. Change of substrate concentration in acceptor phase was measured with conductivity sensor.

Impregnated membranes were prepared by means of impregnation (soaking) of porous Teflon filters produced by VLADiSART FMPTFE-0.2 with the size of pores 0.2 microns and thickness 50 microns 0.1 M by means of investigated carrier solution in phenyl cyclohexane (ACROS ORGANIC, Belgium) under reduced pressure of water-jet pump.

19.2.1 SYNTHESIS OF N-METHYL PHOSPHORYLATED AMINO ACIDS WITH USE OF AMINO ACIDS HYDROCHLORIDES AS CATALYSTS (GENERAL PROCEDURE)

0.009 mol of amino acid, 0.0189 mol (0.567 g) of paraform, 0.018 mol (4.94 g) of dioctyl phosphinite, 0.0045 mol of hydrochloride of corresponding amino acid, and 100 ml of acetonitrile were placed into two-neck flask fitted with mixer and backflow condenser and were heated when mixing on water-bath until acetonitrile boiling during three hours. Complete dissolution of amino acid was not observed. Cool reaction mass was filtered out of unreacted amino acid in glass filter, sediment was washed with chloroform, acetonitrile and chloroform were removed in vacuum of water-jet pump.

19.2.2 N,N-BIS(DIHEXYLPHOSPHORYLMETHYL)-β-ALANINE (I)

NMR $^{31}P\{H\}$ ($CDCl_3$, 121.4 MHz, δ, ppm): 49,8 (c).

NMR ^1H ($CDCl_3$, 400 MHz, δ, ppm, *J*/Hz): 0,89 (triplet, 12,H, $C\underline{H}_3$, $^3J_{HH}$=8 Hz), 1,2-1,94 (five multiplet, 40H, $(C\underline{H}_2)_5P$), 2,52 (triplet, 2H, $C\underline{H}_2COOH$, $^3J_{HH}$=8 Hz), 3,01 (doublet, 2H, PCH_2N, $^2J_{HP}$=4 Hz), 3,13 (triplet, 2H, NCH_2, $^3J_{HH}$=12 Hz), 7,26 (singlet, $CHCl_3$).

Calculated $[M+Na]^+$: 572, 3974 Yes. Found TIS-MS: $[M+Na]^+$ (м/z, 572, 3973 Yes).

19.2.3 (S)-N-DICYCLO HEXYL PHOSPHORYL METHYL-a-ALANINE (II)

NMR $^{31}P\{H\}$ (CH_3CN, 122,4 MHz, δ, ppm): 52,4 c.

NMR ^1H ($CDCl_3$, 400 MHz, δ, ppm, *J*/Hz):1,23-1,9 (multiplet, 25H, $2C_6H_{11}$, CH_3), 2,85 and 3.15 (doublet of doublets, 2H, $PC\underline{H}_A\underline{H}_B$, $^2J_{HH}$ = 15 Hz, $^2J_{PH}$ = 64,5 Hz), 3,27 (quadruplet, H, CH, $^3J_{HP}$=12 Hz), 7,0-7,25 (multiplet, admixture of aromatic solvent).

Infrared spectrum (petrolatum oil, ν/cm^{-1}): 3309 (NH); 1122 (P=O); 1447, 1461 (CH aromatic).

19.2.4 N,N-BIS(DI OCTYL PHOSPHORYL METHYL)-β-ALANINE (III)

It, separated from reaction mass in the form of superfine white amorphous substance, was purified with double recrystallization from acetone. White amorphous substance was obtained with $t_{пл}$=98°C, Yield is 81%.

Elemental analysis $C_{37}H_{77}NO_4P_2$. Found, %: C 67,28, H 11,91, O 9,72, N 2,01, P 9,08; Calculated, %: C 67,13, H 11,72, O 9,67, N 2,14, P 9,36.

NMR $^{31}P\{H\}$ after purification ($CDCl_3$, 121,4 MHz, δ, ppm): 49,8 (singlet, product), 58,1 (hydrochloride of the product).

NMR ^1H ($CDCl_3$, 400 MHz, δ, ppm, *J*/Hz): 0,89 (triplet, 12,H, $C\underline{H}_3$, $^3J_{HH}$=12 Hz), 1,28-1,88 (five multiplets, 56H, $(C\underline{H}_2)_7P$), 2,53 (triplet, 2H, C$\underline{H}$2COOH, $^3J_{HH}$=8 Hz), 3,03 (doublet, 2H, CH_2P, PCH_2N, $^2J_{HP}$=4 Hz), 3,15 (triplet, 2H, NCH_2, $^3J_{HH}$=12 Hz), 7,28 (singlet, $CHCl_3$).

NMR ^{13}C ($CDCl_3$, 300 MHz, δ, ppm, *J*/Hz): 14.09 (singlet, CH_3); 21,59 (doublet, $\underline{C}H_2CH_2CH_2P$, $^2J_{CP}$=4 Hz), 22,63 (singlet, $CH_3\underline{C}H_2$), 26,87 (doublet,

$P\underline{C}H_2$, $^1J_{CP}$= 64 Г), 29,12 (c, $CH_3CH_2CH_2\underline{C}H_2$ + $CH_3CH_2CH_2CH_2\underline{C}H_2$), 31,24 (doublet, $\underline{C}H_2CH_2P$, $^2J_{CP}$=14 Hz), 31,64 (c, $\underline{C}H_2COOH$), 31,79 (singlet, $CH_3CH_2\underline{C}H_2$), 53,68 (doublet, PCH_2N, $^1J_{CP}$=72 Hz), 54,37 (triplet, NCH_2), 77,13 (triplet, $CDCl_3$), 173,7 (singlet, COOH).

Calculated $[M+Na]^+$ (m/z, Yes): 684,5226. Found TIS-MS $[M+Na]^+$ (m/z, Yes): 684,5229.

19.2.5 N,N-BIS(DIOCTYL PHOSPHORYL METHYL)GLYCINE (IV)

It, extracted in the form of yellow oil, was transformed into hard salt by means of heating in plenty of lithium hydroxide in isopropanol. Salt was recrystallized from acetone. Refined salt was solved in benzene and was extracted first with 5% hydrochloric acid solution and then with water to neutral reaction of rinsing waters. Benzene fraction was dried above magnesium sulfate and benzene was removed in vacuum of water-jet pump. The obtained product in the form of colorless thick oil was dried when heated to 60°C in vacuum under 30 Pa during 10 hours.

Yield 84%. Elemental analysis $C_{36}H_{75}NO_4P_2$. Found, %: C 66,79, H 11,78, O 9,76, N 2,05, P 9,62; Calculated, %: C 66,73, H 11,67, O 9,88, N 2,16, P 9,56.

NMR $^{31}P\{H\}$ after purification ($CDCl_3$, 122,4 MHz, δ, ppm): 50.3 (c.)

NMR ^1H ($CDCl_3$, 400 MHz, δ, ppm): 0,89 (т, $4CH_3$, 12H, $^3J_{HH}$=4 Hz); 1,29, 1,39, 1,58, 1,75, 1,83 (multiplet, $4CH_3(C\underline{H}_2)_7$, 48H); 3,32 (doublet, PCH_2N, 2H, $^2J_{HP}$=4 Hz), 3,68 (singlet, NCH_2COO, 2H), 7,28 (singlet, $CHCl_3$), 7,37 (singlet, admixture, aromatic protons).

NMR $^{13}C\{H\}$ ($CDCl_3$, 100 MHz, δ, ppm): 172.22 (singlet, COOH), 77,17 (triplet, $CDCl_3$), 58,04 (singlet, $N\underline{C}H_2COO$), 53,6 (doublet, PCH_2N, $^1J_{CP}$=79 Hz); 31.77 (singlet, $4CH_3CH_2\underline{C}H_2^{Oct}$), 31.23 (doublet, $4PCH_2\underline{C}H_2$, $^2J_{CP}$=13 Hz), 29,09 (singlet, $4CH_3CH_2CH_2\underline{C}H_2$ + $4CH_3CH_2CH_2CH_2\underline{C}H_2$), 26,61 (singlet, $4P\underline{C}H_2CH_2$, $^1J_{CP}$=64 Hz); 22,6 (singlet, $4CH_3\underline{C}H_2$); 21.6 (singlet, $4\underline{C}H_2CH_2CH_2P$); 14.04 (singlet, $4CH_3$).

Calculated $[M+Na]^+$ (m/z, Yes): 670,5069. Found TIS-MS $[M+Na]^+$ (m/z, Yes): 670,5066.

This work was funded by the subsidy allocated to Kazan Federal University for the project part of the state assignment in the sphere of scientific activities (grant 13-03-00536).

19.3 RESULTS AND DISCUSSION

Previously, it was reported about the successful synthesis of N,N-bis methyl phosphorylated derivatives of glycine and β-alanine,[11] as well as mono-methyl phosphorylated derivative of (S)-α- alanine,[12] from hydrochlorides of corresponding amino acids according to the reaction of Kabachnik-Fields. In this article three-component variant of this reaction is provided with involvement of equimolar quantities of hydrochlorides of β-alanine, (S)-α-alanine, formaldehyde, dihexyl- and dicyclohexyle phosphinous acids correspondingly.

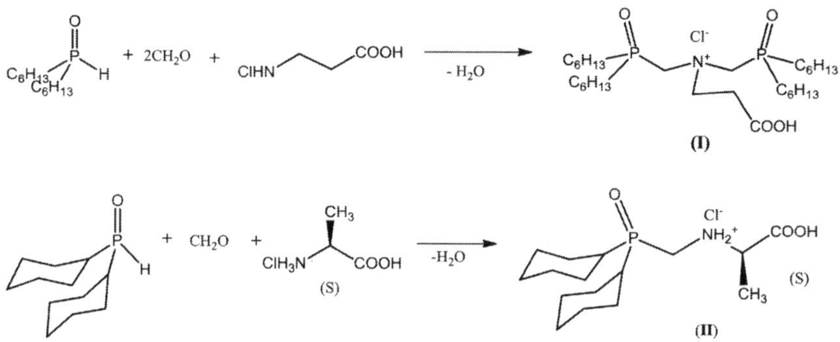

Interaction in boiling acetonitrile during three hours according to the data from NMR ^{31}P results in formation of the only products in every case—hydrochlorides of N,N-bis(dihexyl phosphoryl methyl)-2-aminopropionic acid (I) and (S)-N-dicyclo hexyl phosphoryl methyl-1-aminopropionic acid (II) with conservation of chiral centre configuration—carbon atom of original amino acid which is not affected in the course of the process. The observed in reaction (S)-α-alanine formation of exclusively mono- phosphine oxide is apparently specified by steric restrictions from cyclo-hexyl substituents of phosphorus atoms that obstruct the introduction of second methyl phosphine oxidic fragment.

Structure of products is specified in accordance with NMR spectrums ^{1}H, $^{13}C\{H\}$, IR, and mass spectroscopy results. The structure for compound (II) is specified also by means of X-ray diffraction analysis method. It is not possible to obtain description of structure by means of X-ray diffraction analysis method for compound (II) due to disorder of its structure.[12]

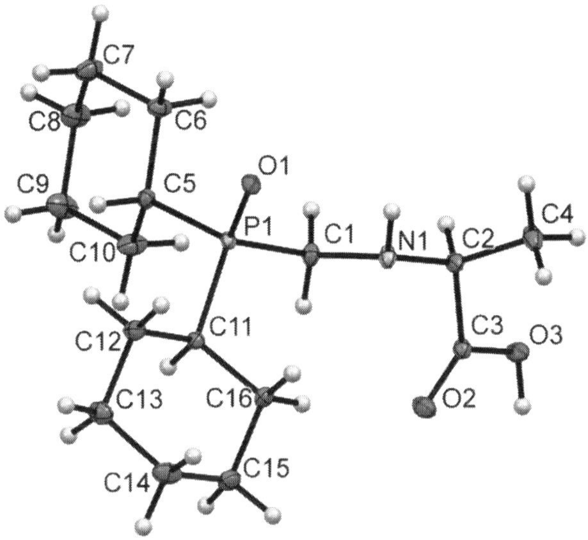

FIGURE 19-1 Molecular structure of (S)-N-(dicyclo hexyl phosphoryl methyl) methyl-1-aminopropionic acid (II) according to NMR (Nuclear Magnetic Resonance) data.

The processes of membrane extraction of alkali metals ions—Li(I), Na(I) and K(I) —were studied with the use of lipophilic methyl phosphorylated amino acids—N,N-bis(dioctyl phosphoryl methyl)-β-alanyne (III) and N,N-bis(dioctyl phosphoryl methyl)-glycine (IV) as transmembrane carriers. The obtained values of flows for these compounds are listed in Table 19 1.

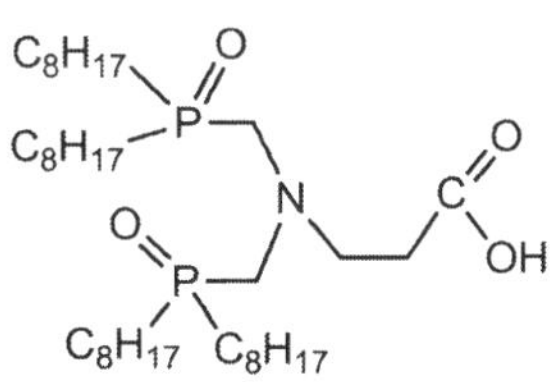

(III)

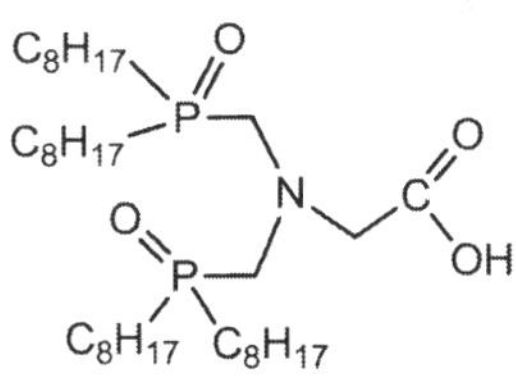

(IV)

Table 19-1 shows that there is a definite efficiency of transmembrane transfer of all alkali metals ions from neutral and alkalescent media with use of selected amino-acid carriers. This efficiency as mentioned earlier is not appropriate of neutral amino phosphoryl extracting agents. At the same time there is no evident selectivity in relation to lithium ion that has been demonstrated recently when using ions of alkali metals of synergetic carrier consisting of equimolar mixture of N,N-bis(dihexyl phosphoryl methyl) octyl amine and dioctyl mono or dithionic acids of phosphorus in the processes of membrane extraction.[13] It should be mentioned that the carrier (III) was found to be more effective in alkaline medium; values of the stream increase four times against the carrier (IV). It can be noted that when decreasing pH of donating solution of chemical agent (III), there is a significant almost 3-fold increase of stream values of all metals, although a selective transfer of any metal is not observed as well.

TABLE 19-1 Transfer flows (P, mol/m^2·min) of Li(I), Na(I), K(I) ions under different pH values of donating solution; $C_{VI} = C_{VII} = 0.1$ M

Carrier	pH	ΔC*			P·10^{-3}		
		Li(I)	Na(I)	K(I)	Li(I)	Na(I)	K(I)
(III)	7.47	1.44	1,81	2.09	1.63	2.05	2.37
	9.43	6.14	5.80	5.98	6.96	6.57	6.78
(IV)	7.47	2.38	2.16	2.19	2.70	2.45	2.49
	9.35	0.90	0.95	0.92	1.02	1.08	1.05

*Change in metal concentration in donating solution $dC_{M+}/dt \cdot 10^4$, mol/min

Previously,[11] the values of ionization constants of bis-phosphorylated acids, which make $pK_{BH^+_1} < 2$ and $pK_{BH^+_2}$ 5.42 and 5.02 for carriers (III) and (IV) correspondingly, were determined. It was found that the first dissociation constant corresponds deprotonation from protonated amino acid NH-group. The value of this constant due to strong electron-accepting influence of two methylene phosphoryl groups is beneath determination limit which is determined by means of potentiometric method and according to our estimate does not exceed two units of pK. At the same time, the second dissociation constant corresponds to proton abstraction from carboxyl group and indicates that it has a larger acidity compared to amino acids precursors.

According to obtained values and data from Table 19-1, it may be concluded that in the process of transfer, the protonated form of acid participates in a greater degree because this form is present in a solution in prevailing number when pH decreases.

By the example of the carrier (IV) change dependencies of stream values of selected ions on the carrier concentration in membrane phase (Figure 19-2) were obtained.

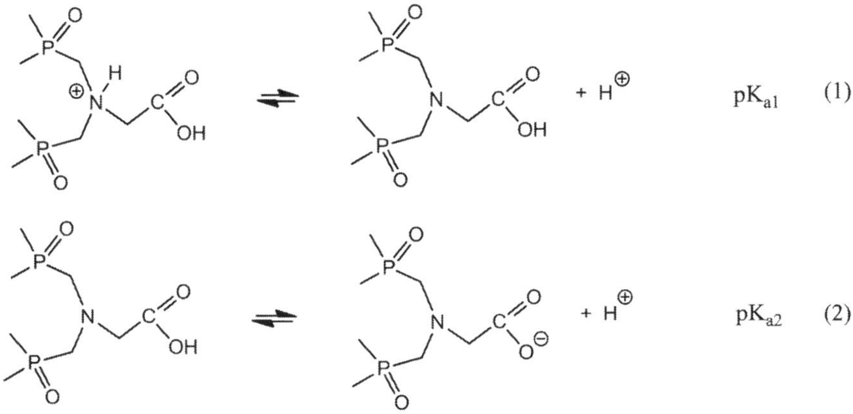

19-1, 19-2

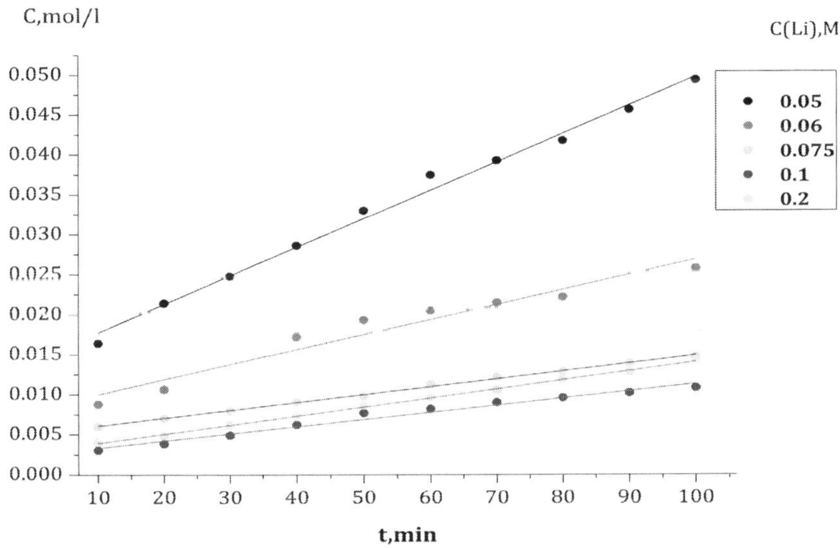

FIGURE 19-2 Kinetic dependencies for transfer processes of Li(I) when carrier (IV) concentration varies (IV), pH = 9.35–9.46;

When carrier concentration grows from 0.05 mol/L to 0.06 mol/L, the sharp decrease of all cations streams is observed especially of lithium ion (Figure 19-3).

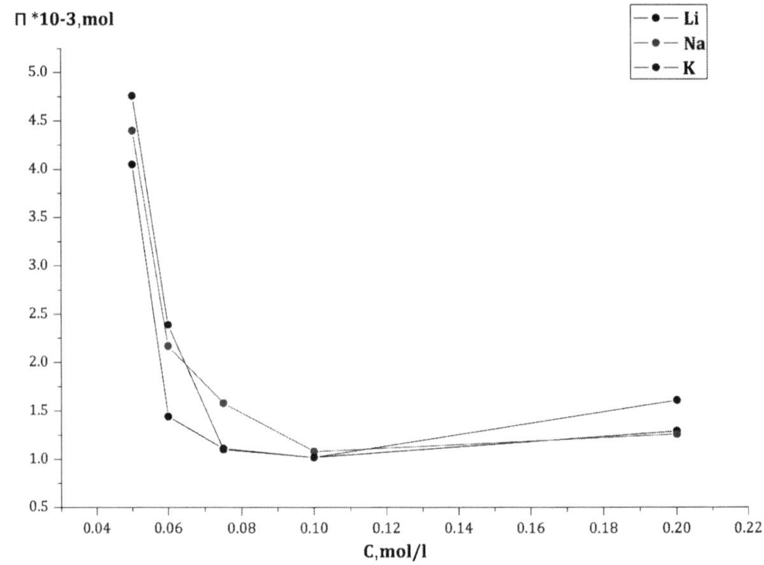

FIGURE 19-3 Dependence of flows of Li, Na, K ions on carrier concentration (IV) under pH = 9.35–9.46.

Further decrease of carrier concentration to 0.1 mol/L results in monotonous lowering of metals transfer streams and their insignificant increase at C=0.2 mol/L. Such type of the dependence can be explained by membranous phase viscosity growth under carrier concentration increase.

To increase stream values an attempt was taken to introduce demulsifying agent (DA) BUT-53 in the amount of 0.05 mol/L to membranous phase because it is possible that equal values of metals streams when pH equals 9.4 and pH 7.5 are caused by micellar transfer. Lipophilic decyl alcohol was also added to membranous phase because as is known alcohols can break formed emulsions, and in addition it was possible to make an assumption about the demonstration of synergistic effect of alcohol/amino phosphine oxides mixture.

Table 19-2 contains values of obtained streams for all metals ions that clearly testify that supply of demulsifying agent doesnot promote the increase of ions transfer flows of alkali metals nor yet decreases their values. It can be believed that in this case emulsions are not formed in the process of transmembrane transfer and degree of micellar transfer is very small. It is impossible to exclude the fact that used demulsifying agent is not suitable for organic system that has been studied before. Introduction of decyl alcohol adversely affects the efficiency of transmembrane transfer—flows values decrease almost twofold for all metals. Use of both decyl alcohol and demulsifying agent doesnot result in selective extraction of any metals. This is illustrated in Figure 19-4 wherein kinetic dependencies of lithium ion transfer are reported including transfer by means of pure amino phosphine oxide and its mixtures with demulsifying agent and decyl alcohol, respectively.

TABLE 19-2 Streams values (Π) of ions Li, Na, K for the carrier (IV), $C_{IV} = 0.1$ M

Carrier	pH_{don}	ΔC*	Stream M^+ $\Pi \cdot 10^{-3}$, mol/m²·min		
			Li(I)	Na(I)	K(I)
(VI)	9.43	6.14	6.96	6.57	6.78
(VI) + DA	9.52	0.94	1.07	1.17	1.02
(VI)+$C_{10}H_{21}OH$	9.43	2.56	2.90	5.45	2.90

*metal concentration change in donating solution: $dC_{M+}/dt \cdot 10^4$, mol/min.

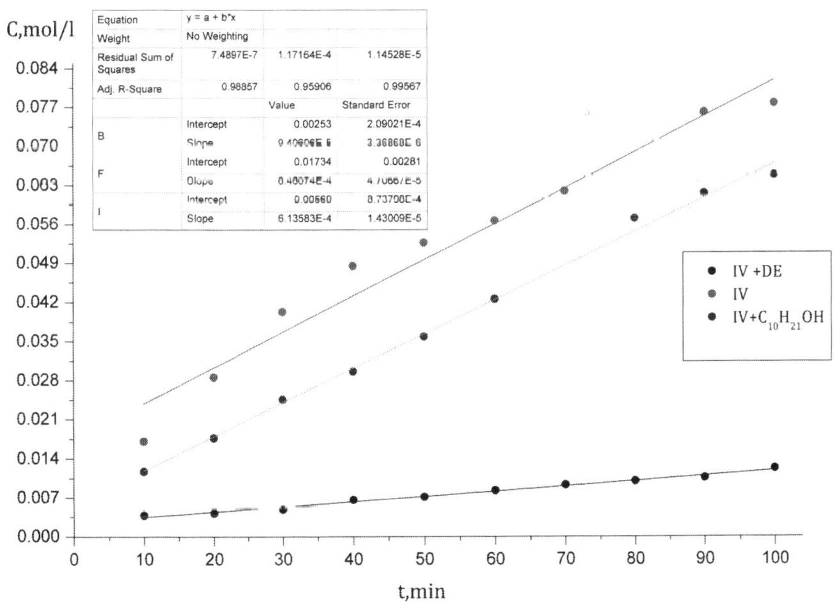

FIGURE 19-4 Kinetic dependencies for Li(I) transfer processes when varying mixture composition with carrier (IV) at pH = 9.43–9.52.

Thereby, despite the fact that bis-phosphorylated carrying agents (III) and (IV) based on amino acids demonstrate expressed efficiency of membrane extraction of alkali metals ions, however, they do not show selectivity related to ions of any of these metals. At the same time one important advantage of new amino phosphoryl carriers that have been studied before should be emphasized—their ability to perform membrane extraction of alkali metals from neutral media whereas described in literature membrane system based on β-diketones and trioctylphosphine oxide allows to extract lithium from alkali solutions only.[3] Whereas the main part of underground ore waters as well as sea water are characterized by pH values close to neutral, the studied membrane-transport systems appear as quite promising for practical use.

19.4 CONCLUSIONS

1. The new phosphoryl methyl derivative aminoacids have been synthesized and their structure has been investigated by NMR-method.
2. The membrane-transport properties of synthesized compositions have been studied.

KEYWORDS

- **Alkali Metals**
- **Aminoacids**
- **NMR-Method**
- **Membrane-Transport Properties**
- **Phosphoryl Methyl Derivatives**

REFERENCES

1. Bartsch, R. A.; Way, J. D. Chemical separations with liquid membranes: an overview. Chapter 1. Washington. ACS Symposium Series. Am. Chem. Soc. 1996. P. 1.
2. Sastre, A. M.; Kumar, A.; Shukla, O. P.; Singh, R. K. Sep. Purific. Rev. 1998. V. 27, № 2. P. 213.
3. Ma, P.; Chen, X. Sep. Sci. Technol. 2000. V.35, № 15. P. 2513.
4. Cherkasov, R. A.; Garifzyanov, A. R.; Galeev, R. R.; Kurnosova, N. V.; Davletshin, R. R.; Zakharov, S. V. Russ. J. Gen. Chem. 2011. V.81, № 7. P. 1114.

5. Garifzyanov, A. R.; Davletshin, R. R.; Davletshina, N. V.; Cherkasov, R. A. Russ. J. Gen. Chem. 2012. V. 82, № 10. P. 619.
6. Garifzyanov, A. R.; Davletshin, R. R.; Myatish, E. Y.; Cherkasov, R. A. Russ.J. Gen. Chem. 2013. V. 83, № 2. P. 213.
7. Cherkasov, R. A.; Garifzyanov, A. R.; Talan, A. S.; Davletshin, R. R.; Kurnosova, N. V. Russ. J. Gen. Chem. 2009. V. 79. № 9. P. 1480.
8. Cherkasov, R. A.; Garifzyanov, A. R.; Talan, A. S.; Davletshin, R. R.; Kurnosova, N. V. Russ. J. Gen. Chem. 2009. V. 79. № 10. P. 1835.
9. Dziełak, A.; Pawełczak, M.; Mucha, A. Tetrahedron Lett. 2011. V.52. P. 3141.
10. Wang, Q.; Zhu, M.; Zhu, R.; Lu, L.; Yuan, C.; Xing, S.; Fu, X.; Mei, Y.; Hang, Q. Eur. J. Med. Chem. 2012. V.49. P. 354.
11. Cherkasov, R. A.; Garifzyanov, A. R.; Koshkin, S. A.; Davletshina, N. V. Russ. J. Gen. Chem. 2012.V.82. № 8. P.1392.
12. Koshkin, S. A.; Garifzyanov, A. R.; Davletshina, N. V.; Katayeva, O. N.; Islamov, D. R.; Cherkasov, R. A.; Kolodyazhnaya, A. O.; Kolodyazhniy, O. I.; Valeeva, M. S. Russ. J. Gen. Chem. 2014. V.50. № 4. P. 607.
13. Garifzyanov, A. R.; Davletshina, N. V.; Garipova, A. R.; Cherkasov, R. A. Russ.J. Gen. Chem., 2014. V. 84. № 2. P. 293.

CHAPTER 20

THERMODYNAMIC ASPECTS OF THE CHANGES IN THE ELECTRICAL CONDUCTIVITY OF POLYETHYLENE FILLED CARBON BLACK

NINEL N. KOMOVA[1], DIMITRY I. ZIBIN[1], and GENNADY E. ZAIKOV[2]

[1]*Moscow University of Fine Chemical Technology, Vernadsky Prospekt, 86, Moscow, 119571 Russia*

[2]*Emanuel Institute of Biochemical Physics, Russian Academy of Sciences, ul. Kosygina 4, Moscow, 119991 Russia*

E-mail: komova_@mail.ru

CONTENTS

ABSTRACT

The effect of temperature change on the change in resistance of the high density polyethylene filled carbon black under different measurement conditions has been investigated in this paper. It is shown that the conductivity measurement conditions have a significant effect on the results. The thermodynamic model of the measurement conditions' influence on the resistance of the composite material is proposed in the process of analyzing the results.

20.1 INTRODUCTION

The mechanism of the electrical conductivity of composite materials, based on polymers filled with carbon black, is complex.This is due to the surface properties of carbon black, features of the interaction of the filler particles and the polymer structure formed in the polymer under the influence of electric fields. Because of the complex structure of the system of carbon particles in the polymer matrix, a theoretical calculation of the electrical conductivity of the composite with a predetermined concentration of filler it is a daunting task.

To explain the conductivity of polymer composites, there are two theories that can be used in varying degrees, to describe the observed regularities. According to the first theory, the conductivity determined by the process of electron emission is likely due to the tunnel effect between the particles, the distance between which is less than 5 nm.[1] The theory is valid for systems with low conductivity. According to the second theory, it is believed that contact between particles are resistive, and the probability of conducting chains is calculated. This theory is used for systems with high electrical conductivity and uses the geometrical factors connecting the electrical conductivity with a random set of conductive chains.

This principle makes it possible to determine the ratio of conductivities of composite σ and conductive component σs as multiplying the proportion of the conductive and nonconductive elements, as well as the probability of chaining (p) and the geometrical factor—the cross- sectional area conductive element (C): $\sigma/\sigma_c = f_\upsilon \cdot p \cdot C^2$ (f_υ–the volume fraction of filler).[2] Such an approach to the analysis of the conductivity of composite materials is similar to the description of systems with position fractal physics.[3]

From the perspective of the developed theories can be considered treason-resistance of composites with increasing temperature. In conductive polymer, composites may have significant changes in electrical resistance, characterized by both positive and negative temperature coefficient (PTC and NTC).[4,5] In the low temperature range (225–275K) with increasing temperature, the electrical resistance decreases exponentially, that is characteristic for semiconductor materials.[6]

At higher temperatures, increase in the electrical resistance is observed (PTC effect), due to the thermal expansion of the polymer and, as a consequence, increase in the distance between the conductive particles.[7–9]

In the analysis of the temperature dependence of the resistance of the composite material, we should take into account the changing nature of the distribution of carbon black and changing the crystalline phase in the polymer matrix as the temperature increases.[10–12] As a result, at the electrically conductive composite materials based on crystallizing polymers, a sharp increase has been detected in electrical resistance at temperatures near to the melting point. With further heating, sharp drop of the samples in electrical resistance occurs, which is explained by the formation of new conductive channels at the expense of de-agglomeration of conductive particles in the melt.[5,11,13] In general, because of the complex structural organization of these composites question about the reasons of sharp rise and fall of electrical resistance remains open.[14]

In refs. [15–17], the study of resistance patterns change over time under isothermal conditions are described. The resistivity of a conductive polymer composite changes with time during an isothermalannealing treatment. During the initial 2–5 minutes of the treatment, the resistivity of HDPE composition carbon black increases from the room-temperature value up to a maximum according to an inverse exponential law. The fact of increasing the initial resistance is explained as a result of the phenomenon of heat transfer in the sample, that is, achieve thermal equilibrium in the mass of the entire sample. Continued presence of the sample under isothermal conditions leads to a decrease in resistance in accordance with an exponential dependence and asymptotically approaches to the constant value. In ref. [17], this process is described by the equation following the type $\rho=\rho_0 + \Delta\rho\exp(-t/\tau)$, where $\rho_0 + \Delta\rho$ represents the initial value of maximum resistance, ρ_0 is a limiting resistivity constant value, and τ is a characteristic decay time.

In ref. [15], when assessing the results of the DSC, it was concluded that the process of reducing resistance is explained by the dynamic restructuring of the conductive filler network. In this case, the second conduction mechanism is realized based on the formation of chains of the conductive filler particles.

The analysis of the DSC spectra in ref. [17] showed that the rearrangements occurring as effects of the temperature and time are not governed by linear viscoelasticity. Refs. [15–17] state that, the thermal treatments, both at a constant heating rate and isothermal, induce filler rearrangements in the host polymer matrix.

In considering the problem of the influence of temperature on the resistivity or conductivity of composite materials should not neglect the fact that the process of measuring the resistance represents the "active measuring system," in which act on the sample (applied voltage) for the measured required value (resistance).[18]

Systems such as polymeric composite materials are inherently dissipative and composite systems. In the process of applying a voltage, they can undergo structural changes associated with a variety changes, both in the macromolecular and intermolecular structures. In addition, the filler particles in varying degrees undergo orientation in an electric field. Thus, there is a change in the forces and energy of interaction between the polymer molecules and filler particles and packing characteristics of the macromolecules on the surface of the filler particles.

As can be concluded from the behavior of solid-phase particles in an amorphous polymer matrix[10–13] during the electric field effect on the system comprising electrically conductive particles, a polarization of the particles emerges in a dielectric matrix, and forces arise directed at the orientation of polarized particles in the field. This gives rise to viscous (dissipative) stresses in the polymer matrix caused by internal friction forces and modifying the energy of the interaction of macromolecules with a surface of the filler particles. Viscous tension dependent on the intensity of the overlay field and disappear when the development of the deformation is completed. The logical development of changes in the system, as we have suggested, should lead to the following result. The work done by viscous stresses determines the change in the energy of the system during particles orientation. This energy is dissipated as heat. As a result, when an electric field acts on the polymer system containing conductive particle charge, local heating of the polymeric matrix is possible. The energy losses due to friction per unit and volume per unit of time determined by the so-called dissipative function[19] $R(M,t) = \mu/2(\partial q/\partial t)$ (μ-generalized coefficient of resistance, q- generalized coordinate), which defines the loss of energy in a small volume near the point M in a short time Δt. The amount of heat that is released in the entire volume V of the zone behind the changing time interval t_1–t_2 is equal[19] to

$$Q = \int_{t_1}^{t_2} \int_V 2R(M,t)\,dV\,dt \qquad (20\text{-}1)$$

As a result of local heating, the molecular mobility of the segments on the surface of the boundary layer increases, conformational entropy increases, and the heat capacity of the entire system undergoes a change. Understanding and accounting of such processes is necessary for the study of the temperature dependence of the electrical conductivity of the composites and their interpretation.

20.2 OBJECTS AND METHODS OF RESEARCH

As the objects of this study, compositions based on thermoplastic polymer, high-density polyethylene (HDPE) type Lupolen 5261 Z (production of BASELL) were used. This HDPE had a melt flow rate of 2 g/10 min (evaluated at 190°C and 21.6 kg) and a density of 0.954 g/cm^3 at 23°C. As the carbon black filler used CB-76 having a specific surface 180m^2/g, pH of an aqueous suspension 4.9, a bulk density of 330 g/m^3. Samples containing carbon black in an amount of 30% wf were prepared by melt blending in a closed rotary mixer "Brabender" with the volume of the working chamber 30 mL for 10 minutes at a rotor speed of 50 rev/min. The temperature of mixing was 160°C.

Amount injected filler is calculated in such a way that its value is allowed to exceed the percolation threshold–20 % wf.[20] The percolation threshold depends on factors such as the molecular structure of the polymer matrix, the temperature of a system in which conductivity is realized, the pressure and the magnitude of the applied voltage (in the case of alternating current the frequency of the voltage change).[21] Therefore, the concentration of filler in the studies was greater than the critical value.

The process of making the test specimens was performed in a hydraulic press at 200°C. When measuring the electrical resistance of the samples in the form of discs about 1mm thick and 50 mm diameter were placed between electrodes in the thermostat, temperature in the thermostat was maintained accurate to 1°C. The DC voltage applied to the electrodes was 100V and 10 V, respectively.

Resistance measurements were performed using tera-ohm meter E6-13A in two ways: First, in the absence of an electric field when the sample is heated to a predetermined temperature, and second, exposure of an electric field to the sample by heating.

20.3 RESULTS AND DISCUSSION

Figure 20-1 shows plots of the sample resistance changes in time at different temperatures and at a constant voltage of 100 V applied to the electrode plates. With an increase in temperature is observed an increase in resistance at the initial stages of the sample. This observation is consistent with the works[7–9,15] carried out in the study of similar systems. In these studies, the same results were obtained in the investigation of the kinetics of the change in the resistance of similar materials using a multimeter (Metex ME-32 Digital Multimeter) at 60 and 120°C and a voltage of 1. The influence of the voltage is not taken into account.

Our findings of resistance changes under a voltage of 100V at different temperatures show different character-obtained dependencies. As seen in Figure 20-1, the annealing treatments of composites system under the action of an electric field (100V) during more than 2 minutes changes the temperature dependence of the resistance. At 30°C the resistance increases and reaches its maximum value after 5 minutes. In future, the value remains unchanged. The change in resistance at higher temperatures (40–70°C) has an extreme character. Reduced resistance at prolonged exposure to the electric field at a voltage of 100 is proportional to the temperature. There are various shapes of the curves due to the fact that amorphous–crystalline state at lower temperatures (30°C) filler particles firmly fixed in the polymer matrix and the contribution of viscous component during deformation fragments macromolecules in the boundary layer is very small.

The explanation for this difference can be found in the model presented in ref. [10]. From the presented model for the particulate filler in an amorphous polymer matrix, it follows that when a dense interfacial layer (lower temperature) the diffusion of charge carriers to the filler particles having an induced dipole moment is difficult, so their transport between the opposing electrodes is much more intense than the charge carriers in a system with less dense interfacial structure. The model implies that the trajectory of the charge carriers by increasing the thickness of loosened nanolayers is much more complicated, which leads to an increase in resistance of the samples. The contribution of the viscous component of the deformation process fragments of macromolecules in the boundary layer increases, which leads to an increase of the dissipative processes.

For a description of the temperature dependence of the resistance upon heating, the sample can be used as a model based on the following considerations. The energy that is dissipated as heat in a unit volume in a nonpolar dielectric is expressed by[22]

$$w=\sigma E^2/8\pi \tag{20-2}$$

where σ is the conductivity dielectric and E is the field strength. The generalized model of the effective medium contracted to describe the conductivity in polymer composites[23] and composite properties are determined by the combination of the properties of its components. For composites comprising a mixture of particles with a broad size distribution, the conductive properties are described by the following equation (20-3):[20]

$$V_f \frac{\sigma_f - \sigma_m}{\sigma_f + A\sigma_m} + (1 - V_f)\frac{\sigma_1 - \sigma_m}{\sigma_1 + A\sigma_m} = 0, \tag{20-3}$$

where V_f is the volume fraction of filler, σ_f, σ_1, σ_m are electrically conductive filler and the matrix of the composite, A, is the parameter determining the concentration of a local field near the particles. It is obvious that the parameter A substantially depends not only on the properties and geometry of the filler but also depends on the structure of the polymer layer adjacent to the surface of the filler particles (the boundary layer).

Given the fact that the conductivity of the filler is significantly higher than the corresponding parameter HDPE (conductivity of carbon black marks UM-76 is $1{,}54\cdot10^3\cdot$Sm [$ohm^{-1}m^{-1}$], the conductivity of polyethylene about 10^{-17} Sm [23]), transforming equation (20-3), the value of the conductivity of the polymer relative to the conductivity of the filler can be neglected. As a result, the expression for the conductivity of the composition of the model takes the form

$$\sigma_m = \sigma_f (V_f - \frac{1 - V_f}{A}) \tag{20-4}$$

Since the parameter A depends on the structure of the interfacial layer in an explicit or implicit form should depend on the temperature of the system, and the conductivity should also depend on the temperature accordingly. From equation (20-4) it must be concluded that the increase option and the conductivity of the composition will increase.

The dependence of the same type can be obtained for the studied systems, using a thermodynamic approach, and taking into account the following presentation. The main role in the process of conduction plays a filler since its conductivity is several orders of magnitude higher than that of the polymer matrix. But the creation of conductive channel or tunnel conductivity depends on the structural characteristics of the matrix and its properties. Therefore, changes in the conformation and supramolecular structure of the polymer can significantly change the overall conductivity of the entire system. Such changes occur as a result of the orientation of the filler particles in the applied electric field, which causes the flow of dissipative processes, local heating, and the resulting change in the segmental mobility. These processes should be seen in the change of electrical capacitance and conductivity of the composite material.

The energy dissipated per unit volume of a substance is defined as w=C ρ ΔT, where ρ–density of the matrix material, which depends on the packing density and the presence of free volume in the formation of the supramolecular structure of polymers, ΔT–local change in temperature of the material. For further change dependencies assume that $C = C_v$ and take into account that the ratio of specific heats is determined by the formula Nersta-Lindemann:[26]

$$C_p\text{-}C_v=TV\alpha^2/K \text{ or } C_p\text{-}C_v=TA_o\alpha^2/T_m \tag{20-5}$$

where $V/K= A_o/T_m$, T_m is themelting point, A_O is universal constant in the semi-empirical equation Nernst-Lindemann, and is the coefficient of volume expansion.

$$\alpha = \frac{1}{V}\left(\frac{\partial V}{\partial T}\right)_{T,N} \tag{20-6}$$

where $\alpha=V/V\Delta T$ or $\alpha\,\Delta T=\Delta V/V$.

In the first approximation can be considered $C_p \approx \alpha$.[26] Since $\alpha = \gamma\dfrac{C_V\kappa T}{V}$, where κ–isothermal compressibility, γ–parameter equation Grüneisen ($\gamma \approx 1$),[23] after the appropriate conversion equation Nernst-Lindemann takes the form

$$C_v= \alpha - TA_o\alpha^2/T_m \tag{20-7}$$

As a result of the internal friction during the orientation of particles in an electric field, generated heat is absorbed by the polymer matrix. This leads to local heating, which allows to substitute equation (20-7) into (20-8) to obtain

$$\sigma E^2/8\pi = (\alpha - TA_o\alpha^2/T_m)\,\Delta T\,\rho \qquad (20\text{-}8)$$

After further transformation the obtained expression, the resistance of the model system will look like

$$\frac{1}{R} = \frac{8\pi\rho\Delta V}{E^2 l^2}(1 - T\frac{A_o\alpha}{T_m}) \qquad (20\text{-}9)$$

Thus, the conductivity (or resistance) is determined by the intensity of the electric field, the dimensions of the polymer layer between the filler particles (20-l), the sample temperature, and the property of the polymer matrix. The magnitude of change in the specific volume resistance from particle to particle (gradient of specific volume resistance) is proportional to R/l^2, where

$$grad\rho = \frac{E^2}{8\pi\rho\Delta V(1 - T\frac{A_0\alpha}{T_m})} \qquad (20\text{-}9`)$$

From (20-9) and (20-9`) followed at increase temperature leads to an increase in resistance and the gradient of the specific volume resistance. The proposed model of the temperature dependence of polymer filled resistance agrees with experimental data. From the derived relationships can be obtained the equation of depending on the density ρ of the matrix polymer composition, and the parameter A, which determines the concentration of a local field near the particles. From these relations one can derive an equation depending on the temperature of the density ρ matrix polymer composition, and the parameter A, which determines the concentration of the local field near the particles.

To get such a dependence we equate the expression for the conductivity of filled system obtained by the equation (20-4) and expression of the conductivity from the equation (20-8):

$$\sigma_f(V_f - \frac{1 - V_f}{A}) = \frac{8\pi}{E^2}\Delta T\rho(\alpha - \frac{TA_0\alpha^2}{T_m}) \qquad (20\text{-}10)$$

In drawing up the relationships considered that only two parameters (ρ and A) will directly depend on the temperature. As a result of sequences of algebraic manipulations, we obtain the following relation

$$\rho A = \frac{T_m E^2 (1-2V_f)}{8\pi\alpha\Delta T(1-V_f)} \frac{1}{TA_0\alpha - T_m} \tag{20-11}$$

Analysis of the expression shows that with an increase in temperature decreases the multiplication result of parameters ρ and A . This means that as the temperature rises in the filled polymer, the degree of localization field around the filler particles decreases and the density of the polymer matrix also reduces. As a result of these processes system resistance should increase with increasing temperature.

This relation is in good agreement with experiment, in which as the temperature rises in the initial stages of the process, there is an increase of resistance (Figure 20-1).

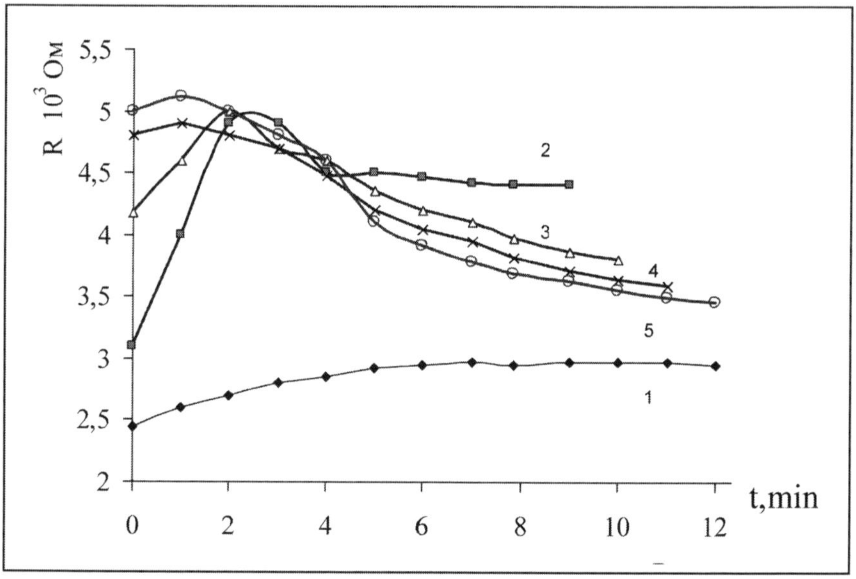

FIGURE 20-1 Kinetic curves of the change in resistance of HDPE filled with 30% by weight. Carbon black (N) obtained at a constant voltage of 100 V and a temperature of 1–30°C; 2–40°C; 3–50°C; 4–60°C; 5–70°C.

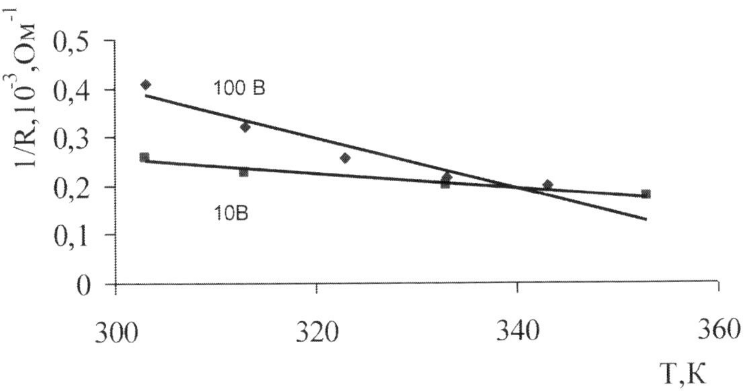

FIGURE 20-2 The dependence of the resistance of HDPE containing 30% by weight. Carbon black in the initial stages of measurement on the temperature at a constant applied electric field with a voltage across the electrodes 10 and 100 V.

Figure 20-2 shows the temperature dependence of the inverse of resistance measured without disabling voltage (100 and 10 V) during the heating of the sample. The experimental data are satisfactorily described in the coordinates of the equation (20-9). From this equation it follows that with increasing temperature, the resistance increases until $1 \approx T\frac{A_o\alpha}{T_m}$. This is possible when the temperature of the polymer system approaches the melting point of the polymer matrix, which agrees well with the experimental data in refs. [7–9,15]. Such dependence at different voltages are described by the following relationships:

1/R= 724- 1,6 T for 10 V and 1/R=1966,7 -5,2T for 100 V. (20-12)

From the found relations (20-12), we can determine the $A_o\alpha/T_m$. For the studied cases (10 and 100V), this value is nearly the same value 0.0023 and 0.0026, respectively.

The results of the change in resistance HDPE-CB with periodic power outage on the plates of the electrode between measurements show a similar pattern of resistance depending on the time at different temperatures (Figure 20-3). Kinetic curves of the resistance change are of exponential nature in the process of reduction in the case of periodic and constant voltage application. At a temperature of 30°C, relative resistance increases (Figure 20-1, curve 1, Figure 20-3, Curve 1). The coefficient of speed of this process is described by equations of the first order, positive and equal to 0.0008 s^{-1}. At higher temperatures, the fall in the relative resistance, a process which is well approximated by the equation of the first order. The coefficients of the rate of change of resistivity with temperature above 40°C decrease.

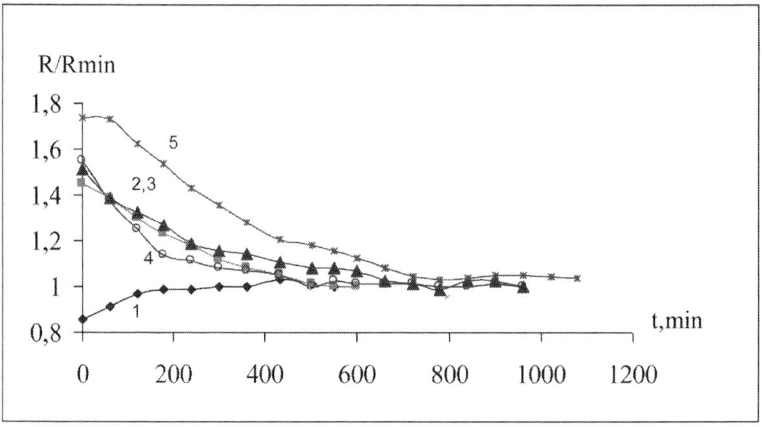

FIGURE 20-3 The kinetics of change relative to the resistance of the composite material with occasionally disconnected electrode voltage value of 100 V at temperatures: 1–30°C; 2–40°C; 3 - 50–°C; 4–60°C; 5–70°C.

The dependence of coefficients on the inverse temperature is exponential (Figure 20-4). The effective activation energy for this process is 13 kJ / mol, which corresponds to the order of magnitude of the activation energy of segmental mobility in amorphous-crystalline polymers,[27] but the numerical value is considerably less than the value defined for homopolymer (30 kJ / mol). This discrepancy can be explained by significant influence of the filler, boundary layer structure difference of the supramolecular structure of the unfilled polymer and temperature fluctuations contribute to the destruction of the conductive channel.

From these results, the process of change in resistance in a filled system PEVP-W can be represented as follows: with increasing temperature has increased segmental mobility of the filler particles surrounding the polymer layer, which is accompanied by a change in the degree of crystalline of the polymer matrix.[28] This increases the degree of freedom of the particles in the polymer medium that promotes arranging them in conductive chain.

One feature of this process is the fact that the change in resistance with time reaches a constant value faster at lower temperatures (Figure 20-1, curve 1 and 2). As the temperature increases, the process goes though at a slower rate but eventually the least resistance is set at high temperatures (Figure 20-1, curves 3–5).

Temperature dependence of the resistance changes under the influence of a constant electric field and periodically switches have a similar course (Figure 20-5) and the same activation energy. The difference in the character of the kinetic curves of the change in resistance in the initial stage of the process for the case of constant voltage (Figure 20-1), and in the case of the periodic inclusion of voltage (Figure 20-2) should be noted. This is due, in all probability, to the relaxation of the strain of the filler particles surrounding areas of macromolecules after termination of the electric field and decrease of directional orientation of the particles.

In the process of repeated periodic influence of the electric field, orientation of the particles reaches a state similar to permanent effects. In our experiments, by the method of periodic influence of the electric field on the system HDPE TU succeeded to reduce the resistance of a few units higher than under the action of a continuous field. This fact is a strong case for the proposed model, as periodic field increases the number of relaxation processes that lead to the accumulation of dissipative processes and increase local heating.

With decreasing the applied potential to the measurement system 10 times at 30°C over time, the resistance is growing at a faster rate (Figure 20-6). The speed factor of process for electric field 10 V to 2.2 greater than for 100 V.

The activation energy of the process reduces the resistance when applying a field 10 is 14.4 kJ / mol. The resulting value is close to that found in the same way the value of effective activation energy reducing resistance when the value of the applied voltage is 10 times greater (100 V).

As follows from the experimental data, the activation energy of the process is independent of the applied voltage. This is consistent with refs. [29,30], which shows that for nonpolar polymers that are HDPE, dependence of the conductivity on the field is very weak. For the composite material dependence of the conductivity on the field with increasing temperature will weaken due to the disordering effect of temperature.

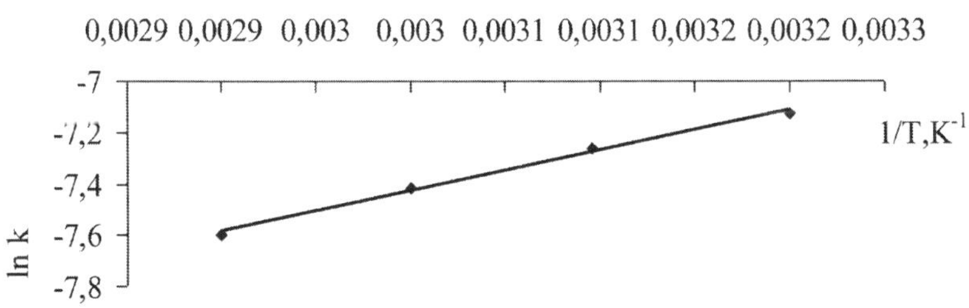

FIGURE 20-4 The dependence of the resistance drop rate (k) of HDPE filled with 30% by weight. Carbon black on the temperature for the time duration measurements over 2 min.

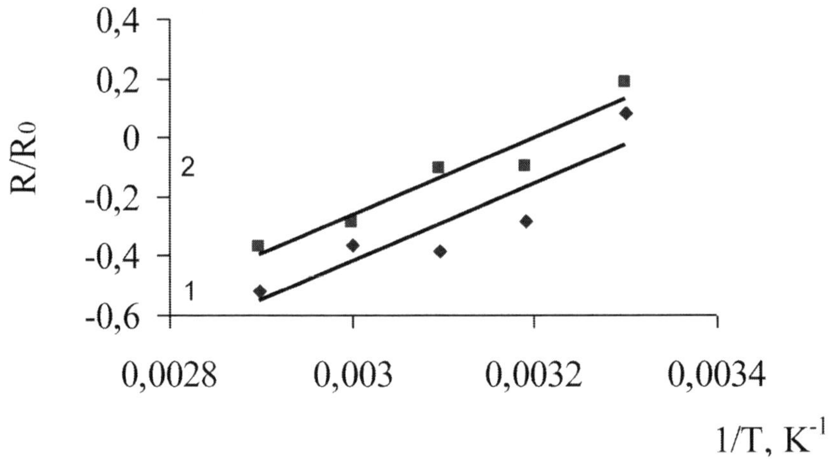

FIGURE 20-5 Relative of resistance carbon black filled HDPE on the temperature at a voltage of 100 V at the electrodes: 1 periodichesukoe application field, 2 dc field

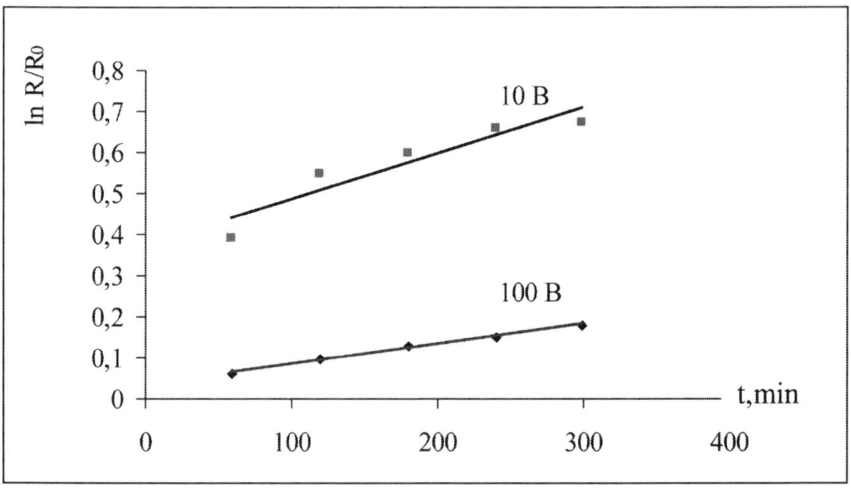

FIGURE 20-6 The kinetics of change in the relative of resistance of the composite material (HDPE filled with 30 wt. % carbon black) at 30°C and constant voltage to the electrodes 10 and 100 V.

20.4 CONCLUSIONS

Thus, these results show that the temperature dependence of the resistivity (conductivity) of HDPE filled with carbon black vary with time and depend on the history of the sample. The model time-temperature dependence of the resistance of the polymer composite gives reason to conclude that the conductivity of such systems is largely dependent on the conformational states of the interfacial boundary layer polymer-filler and the polymer matrix itself. These conditions are determined by external conditions, and have a relaxation character.

KEYWORDS

- **Activation Energy**
- **Carbon-Filled Polymers**
- **Mechanism of Conductivity**
- **Polyethylene Low Density**

REFERENCES

1. Komova, N. N.; Sirov, Yu.V.; Grigoriev, M. A. "Physical Character Conductivity of System, Including Ethylene-Propylene Co-Polymer and $SnCl_2$." Fine Chemical Technologies. Vol. 1. №5. P. 58–62.
2. Sommers, D. J. Polym. Plast. Technol. Eng. 1984. Vol. 23. № 1. P. 83.
3. Karpov, S. V.; Gerasimov, V. S.; Isaev, I. L.; Obushchenko, A. V. "Simulation of the Growth of Nanoparticle Aggregates Reproducing Their Natural Structure in Disperse Systems." Colloid Journal. Vol. 68. No. 4. 2006. P. 441-451.
4. Deng, H.; Lin, L.; Ji, M.; Zhang, S.; Yang, M.; Fu, Q. Progr. Polym. Sci. 2013. Vol. 7. № 10. P.1016.
5. Zhang, W.; Dehghani-Sanij, A. A.; Blackburn, R. S. J. Mater. Sci. 2007. Vol. 42. P. 3408.
6. Tawalbeh, T. M.; Saq'an, S.; Yasin, S. F.; Zihlif, A. M.; Ragosta, G. J. Mater. Sci.: Materials in Electronics. 2005. Vol. 16. P. 351.
7. Dafu, W.; Tiejun, Z.; Yi, X. S. J. Appl. Polym. Sci. 2000. Vol. 77. P. 53.
8. Tang, H.; Chen, X.; Tang, A.; Luo, Y. J. Appl. Polym. Sci. 1996. Vol. 59. P. 383.
9. Azulay, D.; Eylon, M.; Eshkenazi, O.; Toker, D.; Balberg, M.; Shimoni, N.; Millo, O.; Balberg, I. Phys. Rev. Lett. 2003. Vol. 90. P. 4.

10. Meng, H. Lean.; Wei-Ping, L. Chu. "Model for Charge Transport in Ferroelectric Nanocomposite Film." Journal of Polymers. Vol. 2015 (2015), Article ID 745056, 17 pages.
11. Mamunya, Ye. P. J. Macromol. Sci.-Phys. 1999. Vol. 38. № 5&6. P. 615.
12. Deng, H.; Skipa, T.; Zhang, R.; Lellinger, D.; Bilotti, E.; Alig, I.; Peijs, T. Polymer. 2009. Vol. 60. № 15. P. 3747.
13. Shu-Ang, Z. "Electrodynamics of Solids and Microwave Superconductivity" (Wiley Series in Microwave and Optical Engineering) Hardcover – July 23, 1999.
14. Dai, K.; Zhang, Y.C.; Tang, J. H.; Ji, X.; Li, Z. M. J. Appl. Polym. Sci. 2012. Vol. 125. P. 561.
15. Traina, M.; Pegoretti, A.; Penati, A. "Time–Temperature Dependence of the Electrical Resistivity of High-Density Polyethylene/Carbon Black CompositesJournal of Applied Polymer Science." DOI 10.1002/a. pp. 2065-2074.
16. Traina, M.; Pegoretti, A.; Penati, A. Presented at the 1st International Symposium on Nanostructured and Functional Polymer-Based Materials and Composites, Dresden, Germany, 2005.
17. Zhang, J. F.; Zheng, Q.; Yang, Y. Q.; Yi, X. S. J Appl. Polym. Sci. 2002, 83, 3112.
18. Cyranski, J. F. "Measurement Theory for Physics." Foundations of Physics. Vol. 9, October 1979. P. 641–671.
19. Valishin, A. A.; Mishchenko, D.V. "Features of Quasi-Brittle Fracture of Polymers and Composites Based on Them." Fine Chemical technologies. Vol. 18 № 6. P. 99–104.
20. Blythe, T.; Dloor, D. "Electrical Properties of Polymers." Cambridge University Press, 2005. 373 p.
21. Wing Mai, Y.; Zhen, Z. "Polymer Nanocomposites." Woodhead: Publishing Limited. 2006.
22. Tager, A. "Physical Chemistry of Polymers." Publisher: MIR Publishers, Moscow; 2nd Edition edition .1978. 578 p.
23. Sanditov, D. S.; Munkueva, S. B.; Batlaev, D. Z.; Sangadiev, S. S. "Parameter Grüneisen and Propagation Velocity of Acoustic Waves in Glassy Solids." Solid State Physics. Vol.54. № 8. P.1540–1545.
24. "Encyclopedia of Polymer Science and Technology." 1999—2014 by John Wiley and Sons.
25. Lipatov, Yu. S. "Thermal and Rheological Properties of Polymers: A Handbook." Kiev: Naukova Dumka. 1977. 244 p.
26. Cheban, Yu. V.; Lau, S. F.; Wunderlich, B. "Analysis of the Contribution of Skeletal Vibrations to the Heat Capacity of Linear Macromolecules in the Solid State." Colloid and Polymer Science. Vol. 260. 1982. P. 9–19.
27. Mark. J. "Physical Properties of Polymers." Cambridge University Press. 2004. 519 p.
28. 25. Yin, C. L.; Liu, Z. Y.; Gao, Y. J.; Yang, M. B. "Effect of Compounding Procedure on Morphology and Crystallization Behavior of Isotactic Polypropylene/High-Density Polyethylene/Carbon Black Ternary Composites." Polym. Adv. Technol. 2012. Vol. 23. P. 1112–1120.
29. Sotskov, V. A. "On the Influence of the Contact Resistance of the Particles in the Range of Percolation in Disordered Systems Close-Conductor-Insulator." Technical Physics. 2004. Vol. 4. No. 11. P.107–111.
30. Zhinua, L.; Chaobin, H.; Chung, T. "Conducting Blends of Polyaniline and Aromatic Main-Chain Liquid Crystalline Polymer, XYDAR SRT-900." Synthetic Metals. 2001. Vol. 123. P. 69–72.

CHAPTER 21

ENTROPIC AND SPATIAL-ENERGY INTERACTIONS

G. A. KORABLEV[1], V. I. KODOLOV[2], and G. E. ZAIKOV[3]

[1]*Izhevsk State Agricultural Academy*

[2]*Basic Research-Educational Center of Chemical Physics and Mesoscopy, UdSC, UrD, RAS*

[3]*Institute of Biochemical Physics, Russian Academy of Science*

CONTENTS

ABSTRACT

The concept of the entropy of spatial-energy interactions is used similarly to the ideas of thermodynamics on the static entropy.

The idea of entropy appeared based on the second law of thermodynamics and ideas of the adduced quantity of heat.

These rules are general assertions independent of microscopic models. Therefore, their application and consideration can result in a large number of consequences which are most fruitfully used in statistic thermodynamics.

In this research, we are trying to apply the concept of entropy to assess the degree of spatial-energy interactions using their graphic dependence, and in other fields. The nomogram to assess the entropy of different processes is obtained. The variability of entropy demonstrations is discussed in biochemical processes, economics, and engineering systems, as well.

21.1 INTRODUCTION

In statistic thermodynamics, the entropy of the closed and equilibrious system equals the logarithm of the probability of its definite macrostate:

$$S = k \ln W \tag{21-1}$$

where W is the number of available states of the system or degree of the degradation of microstates, and k is Boltzmann's constant,

or $$W = e^{s/k} \tag{21-2}$$

These correlations are general assertions of macroscopic character, they do not contain any references to the structure elements of the systems considered and they are completely independent of microscopic models.[1]

Therefore, the application and consideration of these laws can result in a large number of consequences.

At the same time, the main characteristic of the process is the thermodynamic probability W. In actual processes in the isolated system, the entropy growth is inevitable—disorder and chaos increase in the system, the quality of internal energy goes down.w

The thermodynamic probability equals the number of microstates corresponding to the given macrostate.

Since the system degradation degree is not connected with the physical features of the systems, the entropy statistic concept can also have other applications and demonstrations (apart from statistic thermodynamics).

It is clear that out of the two systems completely different by their physical content, the entropy can be the same if their number of possible microstates corresponding to one macroparameter (whatever parameter it is) coincide. Therefore, the idea of entropy can be used in various fields. The increasing self-organization of human society ... leads to the increase in entropy and disorder in the environment that is demonstrated, in particular, by a large number of disposal sites all over the earth.[2]

21.2 ON TWO PRINCIPLES OF ADDING ENERGY CHARACTERISTICS OF INTERACTIONS

The analysis of kinetics of various physical and chemical processes shows that in many cases the reciprocals of velocities, kinetic or energy characteristics of the corresponding interactions are added.

Some examples are as follows: ambipolar diffusion, resulting velocity of topochemical reaction, change in the light velocity during the transition from vacuum into the given medium, effective permeability of biomembranes.

In particular, such supposition is confirmed by the formula of electron transport possibility (W_∞)

due to the overlapping of wave functions 1 and 2 (in steady state) during electron-conformation interactions:

$$W_\infty = \frac{1}{2}\frac{W_1 W_2}{W_1 + W_2} \tag{21-3}$$

Equation (21-3) is used when evaluating the characteristics of diffusion processes followed by nonradiating transport of electrons in proteins.[3]

And also: "From classical mechanics it is known that the relative motion of two particles with the interaction energy U(r) takes place as the motion of material point with the reduced mass μ:

$$\frac{1}{\mu} = \frac{1}{m_1} + \frac{1}{m_2} \tag{21-4}$$

in the field of central force U(r), and general translational motion—as a free motion of material point with the mass:

$$m = m_1 + m_2 \tag{21-5}$$

Such things take place in quantum mechanics as well."[4]

The task of two-particle interactions taking place along the bond line was solved in the times of Newton and Lagrange:

$$\mathring{A} = \frac{m_1 v_1^2}{2} + \frac{m_2 v_2^2}{2} + U\left(\overline{r_2} - \overline{r_1}\right), \tag{21-6}$$

where E is the total energy of the system; first and second elements are kinetic energies of the particles; third element is the potential energy between particles 1 and 2, vectors $\overline{r_2}$ and $\overline{r_1}$, and characterizes the distance between the particles in final and initial states.

For moving thermodynamic systems, the first commencement of thermodynamics is as follows:[5]

$$\delta \mathring{A} = d\left(U + \frac{mv^2}{2}\right) \pm \delta A, \tag{21-7}$$

where

δE is the amount of energy transferred to the system

Element $d\left(U + \frac{mv^2}{2}\right)$ characterizes the changes in internal and kinetic energies of the system

$+\delta A$ denotes the work performed by the system

$-\delta A$ denotes the work performed with the system

As the work value numerically equals the change in the potential energy, then

$$+\delta A = -\Delta U \text{ and } -\delta A = +\Delta U \tag{21-8, 21-9}$$

It is probable that not only in thermodynamics but also in many other processes in the dynamics of moving particles interaction, not only the value of potential energy is critical but also its change as well. Therefore, similar to the equation (21-4), the following equation should be fulfilled for two-particle interactions:

$$\delta E = d\left(\frac{m_1 v_1^2}{2} + \frac{m_2 v_2^2}{2}\right) \pm \Delta U \tag{21-10}$$

Here, $$\Delta U = U_2 - U_1 \tag{21-11}$$

where U_2 and U_1 are potential energies of the system in final and initial states.

At the same time, the total energy (E) and kinetic energy $\left(\frac{mv^2}{2}\right)$ can be calculated from their zero value, then only the last element is modified in equation (21-6).

The character of the change in the potential energy value (ΔU) was analyzed by its sign for various potential fields and the results are given in Table 21-1.

TABLE 21-1 Directedness of the interaction processes

No	Systems	Type of potential field	Process	U	r_2/r_1 (x_2/x_1)	U_2/U_1	Sign ΔU	Sign δA	Process directedness in potential field
1	opposite electrical charges	electrostatic	attraction	$-k\frac{q_1q_2}{r}$	$r_2 < r_1$	$U_2 > U_1$	-	+	along the gradient
			repulsion	$-k\frac{q_1q_2}{r}$	$r_2 > r_1$	$U_2 < U_1$	+	-	against the gradient
2	similar electrical charges	Electrostatic	attraction	$k\frac{q_1q_2}{r}$	$r_2 < r_1$	$U_2 > U_1$	+	-	against the gradient
			repulsion	$k\frac{q_1q_2}{r}$	$r_2 > r_1$	$U_2 < U_1$	-	+	along the gradient
3	elementary masses m_1 and m_2	gravitational	attraction	$-\gamma\frac{m_1m_2}{r}$	$r_2 < r_1$	$U_2 > U_1$	-	+	along the gradient
			repulsion	$-\gamma\frac{m_1m_2}{r}$	$r_2 > r_1$	$U_2 < U_1$	+	-	against the gradient
4	spring deformation	field of elastic forces	compression	$k\frac{\Delta x^2}{2}$	$r_2 < r_1$	$U_2 > U_1$	+	-	against the gradient
			extension	$k\frac{\Delta x^2}{2}$	$x_2 > x_1$	$U_2 > U_1$	+	-	against the gradient
5	photoeffect	electrostatic	repulsion	$k\frac{q_1q_2}{r}$	$r_2 > r_1$	$U_2 < U_1$	-	+	along the gradient

From the table it is seen that the values—ΔU and accordingly + δA (positive work)—correspond to the interactions taking place along the potential gradient, and ΔU and – δA (negative work) occur during the interactions against the potential gradient.

The solution of two-particle task of the interaction of two material points with masses m_1 and m_2 obtained under the condition of the absence of external forces corresponds to the interactions flowing along the gradient, the positive work is performed by the system (similar to the attraction process in the gravitation field).

The solution of this equation via the reduced mass (μ) is[6] the Lagrange equation for the relative motion of the isolated system of two interacting material points with masses m_1 and m_2, which in coordinate x is as follows:

$$\mu \cdot x'' = -\frac{\partial U}{\partial x}; \qquad \frac{1}{\mu} = \frac{1}{m_1} + \frac{1}{m_2}.$$

Here, U is the mutual potential energy of material points; μ is the reduced mass. At the same time, $x'' = a$ (feature of the system acceleration). For elementary portions of the interactions, Δx can be taken as follows:

$$\frac{\partial U}{\partial x} \approx \frac{\Delta U}{\Delta x} \qquad \text{This is } \mu a \Delta x = -\Delta U. \qquad \text{Then:}$$

$$\frac{1}{1/(a\Delta x)} \frac{1}{(1/m_1 + 1/m_2)} \approx -\Delta U \quad ; \quad \frac{1}{1/(m_1 a \Delta x) + 1/(m_2 a \Delta x)} \approx -\Delta U$$

Or:

$$\frac{1}{\Delta U} \approx \frac{1}{\Delta U_1} + \frac{1}{\Delta U_2} \tag{21-12}$$

where ΔU_1 and ΔU_2 are the potential energies of material points on the elementary portion of interactions, ΔU is the resulting (mutual) potential energy of this interactions.

Thus

1. In the systems in which the interactions proceed along the potential gradient (positive performance), the resulting potential energy is found based on the principle of adding reciprocals of the corresponding energies of subsystems.[7] Similarly, the reduced mass for the relative motion of two-particle system is calculated.

2. In the systems in which the interactions proceed against the potential gradient (negative performance), the algebraic addition of their masses as well as the corresponding energies of subsystems is performed (by the analogy with Hamiltonian).

21.3 SPATIAL-ENERGY PARAMETER (P-PARAMETER)

From the equation (21-12) it is seen that the resulting energy characteristic of the system of two material points interaction is found based on the principle of adding reciprocals of initial energies of interacting subsystems.

"Electron with the mass m moving near the proton with the mass M is equivalent to the particle with the mass: $\mu = \frac{mM}{m+M}$."[8]

Therefore, when modifying the equation (21-12), we can assume that the energy of atom valence orbitals (responsible for interatomic interactions) can be calculated[7] by the principle of adding reciprocals of some initial energy components based on the following equations:

$$\frac{1}{q^2/r_i} + \frac{1}{W_i n_i} = \frac{1}{P_E} \text{ or } \frac{1}{P_0} = \frac{1}{q^2} + \frac{1}{(Wrn)_i}; \; P_E = P_0/r_i$$

(21-13), (21-14), (21-15)

Here: W_i–electron orbital energy;[9] r_i–orbital radius of i–orbital;[10] q–Z^*/n^*;[11,12] n_i– number of electrons of the given orbital; Z^* and n^*–nucleus effective charge and effective main quantum number; r–bond dimensional characteristics.

P_0 is called a spatial-energy parameter (SEP), and P_E – effective P–parameter (effective SEP). Effective SEP has a physical sense of some averaged energy of valence electrons in the atom and is measured in energy units, for example, electron-volts (eV).

The values of P_0-parameter are tabulated constants for the electrons of the given atom orbital.

For dimensionality SEP can be written down as follows:

$$[P_0] = [q^2] = [E]\cdot[r] = [h]\cdot[v] = \frac{kg \cdot m^3}{s^2} = J \cdot m$$

,

where [E], [h], and [υ] are dimensions of energy, Planck constant, and velocity, respectively. Thus P-parameter corresponds to the processes going along the potential gradient.

The introduction of P-parameter should be considered as further development of quasi- classical notions using quantum-mechanical data on atom structure to obtain the criteria of energy conditions of phase-formation. At the same time, for the similarly charged (e.g., orbitals in the given atom) homogeneous systems, the principle of algebraic addition of such parameters is preserved:

$$\sum P_E = \sum (P_0 / r_i); \sum P_E \frac{\sum P_0}{r} \qquad (21\text{-}16), (21\text{-}17)$$

or:

$$\sum P_0 = P_0' + P_0'' + P_0''' + \ldots; \quad r\sum P_E = \sum P_0 \qquad (21\text{-}18), (21\text{-}19)$$

Here P-parameters are summed on all atom valence orbitals.

To calculate the values of P_E-parameter at the given distance from the nucleus depending on the bond type either atom radius (R) or ion radius (r_I) can be used instead of r.

The calculations demonstrated that the values of P_E-parameters are numerically equal to (within 2%) the total energy of valence electrons *(U)* by the atom statistic model. Using the known correlation between the electron density (β) and interatomic potential by the atom statistic model,[13] we can obtain the direct dependence of P_E-parameter upon the electron density at the distance r_i from the nucleus.

The rationality of such approach is confirmed by the calculation of electron density using wave functions of Clementi[14] and its comparison with the value of electron density calculated via the value of P_E-parameter.

21.4 WAVE EQUATION OF P-PARAMETER

To characterize atom spatial-energy properties, two types of P-parameters are introduced. The bond between them is a simple one: $P_E = \frac{P_0}{R}$ where R– atom dimensional characteristic. Taking into account the additional quantum characteristics of sublevels in the atom, this equation can be written down in coordinate x as follows:

$$\Delta P_E \approx \frac{\Delta P_0}{\Delta x} \qquad \text{or} \qquad \partial P_E = \frac{\partial P_0}{\partial x}$$

where the value ΔP equals the difference between P_0-parameter of *i* orbital and P_{CD}–countdown parameter (parameter of main state at the given set of quantum numbers).

According to the established[7] rule of adding P-parameters of similarly charged or homogeneous systems for two orbitals in the given atom with different quantum characteristics and according to the energy conservation rule we have:

$$\Delta P''_E - \Delta P'_E = P_{E,\lambda}$$

where $P_{E,\lambda}$–spatial-energy parameter of quantum transition.

Taking for the dimensional characteristic of the interaction $\Delta\lambda = \Delta x$, we have:

$$\frac{\Delta P''_0}{\Delta\lambda} - \frac{\Delta P'_0}{\Delta\lambda} = \frac{P_0}{\Delta\lambda} \quad \text{or:} \quad \frac{\Delta P'_0}{\Delta\lambda} - \frac{\Delta P''_0}{\Delta\lambda} = -\frac{P_0\lambda}{\Delta\lambda}$$

Let us again divide by $\Delta\lambda$ term by term: $\left(\frac{\Delta P'_0}{\Delta\lambda} - \frac{\Delta P''_0}{\Delta\lambda}\right) \Big/ \Delta\lambda = -\frac{P_0}{\Delta\lambda^2}$ where

$$\left(\frac{\Delta P'_0}{\Delta\lambda} - \frac{\Delta P''_0}{\Delta\lambda}\right) \Big/ \Delta\lambda \sim \frac{d^2 P_0}{d\lambda^2} \qquad \text{i.e.:} \qquad \frac{d^2 P_0}{d\lambda^2} + \frac{P_0}{\Delta\lambda^2} \sim 0$$

Taking into account only those interactions when $2\pi\Delta x = \Delta\lambda$ (closed oscillator), we have the following equation:

$$\frac{d^2 P_0}{4\pi^2 dx^2} + \frac{P_0}{\Delta\lambda^2} = 0 \quad \text{or} \quad \frac{d^2 P_0}{dx^2} + 4\pi^2 \frac{P_0}{\Delta\lambda^2} \approx 0$$

Since $\Delta\lambda = \frac{h}{mv}$, then:

$$\frac{d^2 P_0}{dx^2} + 4\pi^2 \frac{P_0}{h^2} m^2 v^2 \approx 0$$

or

$$\frac{d^2 P_0}{dx^2} + \frac{8\pi^2 m}{h^2} P_0 E_k = 0 \tag{21-20}$$

where $E_k = \frac{mV^2}{2}$ –electron kinetic energy.

Schrodinger equation for the stationery state in coordinate x:

$$\frac{d^2\psi}{dx^2} + \frac{8\pi^2 m}{h^2}\psi E_k = 0$$

When comparing these two equations, we see that P_0-parameter numerically correlates with the value of Ψ-function: $P_0 \approx \Psi$, and is generally proportional to it: $P_0 \sim \Psi$. Taking into account the broad practical opportunities of applying the P-parameter methodology, we can consider this criterion as the materialized analog of Ψ-function.[15,16]

Since P_0-parameters like Ψ-function have wave properties, the superposition principles should be fulfilled for them, defining the linear character of the equations of adding and changing P-parameter.

21.5 STRUCTURAL EXCHANGE SPATIAL-ENERGY INTERACTIONS

In the process of solid solution formation and other structural equilibrium-exchange interactions, the single electron density should be set in the points of atom-component contact. This process is accompanied by the redistribution of electron density between the valence areas of both particles and transition of the part of electrons from some external spheres into the neighboring ones. Apparently, frame atom electrons do not take part in such exchange.

Obviously, when electron densities in free atom-components are similar, the transfer processes between boundary atoms of particles are minimal; this will be favorable for the formation of a new structure. Thus the evaluation of the degree of structural interactions in many cases means the comparative assessment of the electron density of valence electrons in free atoms (on averaged orbitals) participating in the process, which can be correlated with the help of P-parameter model.

The less the difference $(P'_0/r'_i - P''_0/r''_i)$, the more favorable is the formation of a new structure or solid solution from the energy point.

In this regard, the maximum total solubility, evaluated via the coefficient of structural interaction α, is determined by the condition of minimum value α, which represents the relative difference of effective energies of external orbitals of interacting subsystems:

$$\alpha = \frac{P'o/r_i' - P''_o/r_i''}{(P'o/r_i' + P''_o/r_i'')/2} 100\% \qquad \alpha = \frac{P'_C - P''_C}{P'_C + P''_C} 200\%$$

where P_S – structural parameter is found by the equation:

$$\frac{1}{P_C} = \frac{1}{N_1 P'_E} + \frac{1}{N_2 P''_E} + \ldots$$

Here, N_1 and N_2 – number of homogeneous atoms in subsystems.

The isomorphism degree and mutual solubility are evaluated in many (over one thousand) simple and complex systems (including nanosystems). The calculation results are in compliance with theoretical and experimental data.

The nomogram of dependence of structural interaction degree (ρ) upon coefficient α, the same for a wide range of structures, was constructed based on all the data obtained.

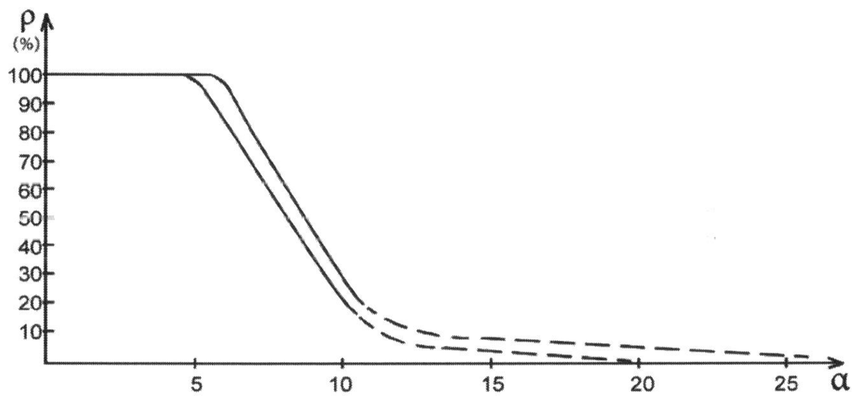

FIGURE 21-1 Nomogram of structural interaction degree dependence (ρ) on coefficient **α**.

This approach gives the possibility to evaluate the degree and direction of the structural interactions of phase formation, isomorphism and solubility processes in multiple systems, including molecular ones.

Such nomogram can be demonstrated[7] as a logarithmic dependence:

$$\alpha = \beta\,(\ln\rho)^{-1}, \qquad (21\text{-}21)$$

where coefficient β–the constant value for the given class of structures. β can structurally change mainly within ± 5% from the average value. Thus coefficient α is reversely proportional to the logarithm of the degree of structural interactions and therefore can be characterized as the entropy of spatial-energy interactions of atomic-molecular structures.

Actually the more is ρ, the more probable is the formation of stable-ordered structures (e.g., the formation of solid solutions), that is, the less is the process entropy. But also the less is coefficient α.

The equation (21-21) does not have the complete analogy with Boltzmann's equation (21-1) as in this case not absolute but only relative values of the corresponding characteristics of the interacting structures are compared, which can be expressed in percent. This refers not only to coefficient α but also to the comparative evaluation of structural interaction degree (ρ), for example – the percent of atom content of the given element in the solid solution relatively to the total number of atoms. Therefore, in equation (21-21) coefficient k = 1.

Thus, the relative difference of spatial-energy parameters of the interacting structures can be a quantitative characteristic of the interaction entropy: $\alpha \equiv S$.

21.6 ENTROPIC NOMOGRAM OF SURFACE-DIFFUSIVE PROCESSES

As an example, let us consider the process of carbonization and formation of nanostructures during the interactions in polyvinyl alcohol gels and metal phase in the form of copper oxides or chlorides. At the first stage, small clusters of inorganic phase are formed surrounded by carbon containing phase. In this period, the main character of atomic-molecular interactions needs to be assessed via the relative difference of P-parameters calculated through the radii of copper ions and covalent radii of carbon atoms.

In the next main carbonization period, the metal phase is being formed on the surface of the polymeric structures.

From this point, the binary matrix of the nanosystem C→Cu is being formed.

The values of the degree of structural interactions from coefficient α are calculated, that is, $\rho_2 = f\left(1/\alpha_2\right)$–curve 2 given in Figure 21-2. Here, the graphical dependence of the degree of nanofilm formation (ω) on the process time is presented by the data from ref. [7]–curve 1 and previously obtained nomogram in the form $\rho_1 = f\left(\frac{1}{\alpha_1}\right)$–curve 3.

The analysis of all the graphical dependencies obtained demonstrates the practically complete graphical coincidence of all three graphs: $\omega = f(t)$, $\rho_1 = f\left(\frac{1}{\alpha_1}\right)$, $\rho_2 = f\left(\frac{1}{\alpha_2}\right)$ with slight deviations in the beginning and end of the process. Thus, the carbonization rate, as well as the functions of many other physical-chemical structural interactions, can be assessed via the values of the calculated coefficient α and entropic nomogram.

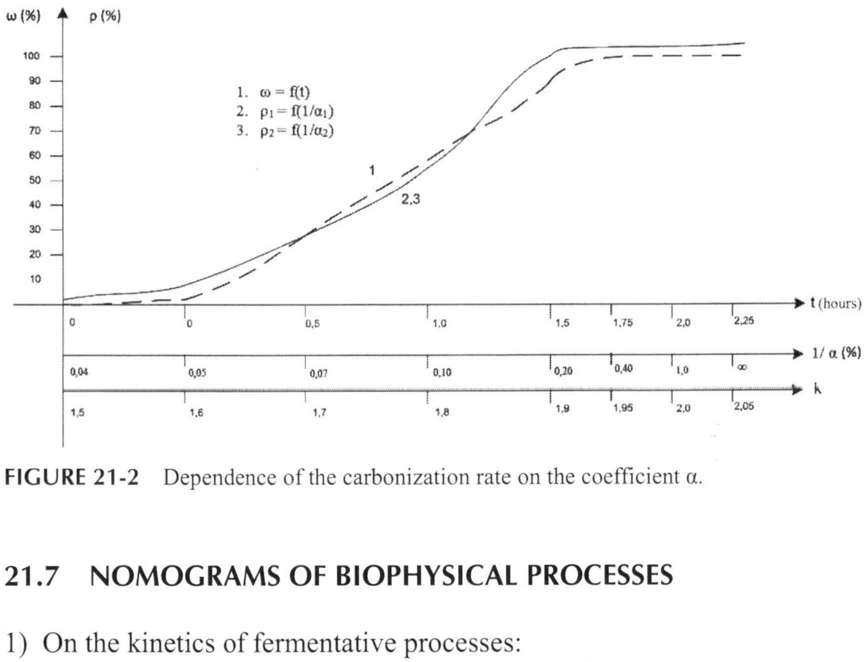

FIGURE 21-2 Dependence of the carbonization rate on the coefficient α.

21.7 NOMOGRAMS OF BIOPHYSICAL PROCESSES

1) On the kinetics of fermentative processes:

"The formation of ferment-substrate complex is the necessary stage of fermentative catalysis ... At the same time, n substrate molecules can join the ferment molecule"[3] [p. 58].

For ferments with stoichiometric coefficient n not equal one, the type of graphical dependence of the reaction product performance rate (μ) depending on the substrate concentration (c) has[3] a sigmoid character with the specific bending point (Figure 21-3).

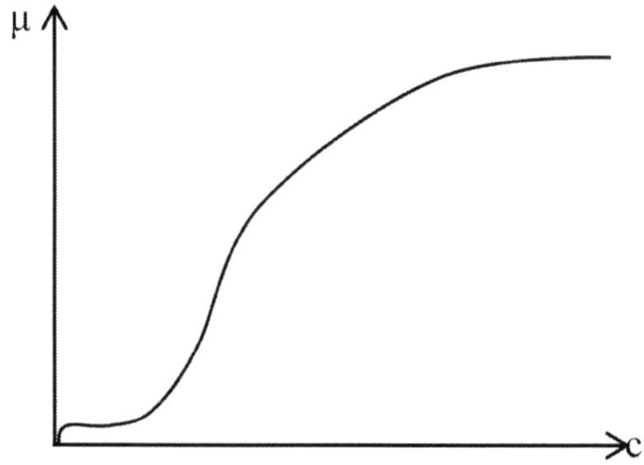

FIGURE 21-3 Dependence of the fermentative reaction rate (μ) on the substrate concentration (c).

In Figure 21-3 it is seen that this curve generally repeats the character of the entropic nomogram in Figure 21-2.

The graph of the dependence of electron transport rate in biostructures on the diffusion time period of ions is similar.[3][p. 278].

In the procedure of assessing fermentative interactions (similarly to the previously applied for surface-diffusive processes), the effective number of interacting molecules over 1 is applied.

In the methodology of *P*-parameter, a ferment has a limited isomorphic similarity with substrate molecules and does not form a stable compound with them, but, at the same time, such limited reconstruction of chemical bonds which "is tuned" to obtain the final product is possible.

2) Dependence of biophysical criteria on their frequency characteristics:

   a) The passing of alternating current through live tissues is characterized by the dispersive curve of electrical conductivity – this is the graphical dependence of the tissue total resistance (z-impedance) on the alternating current frequency logarithm (log ω). Normally, such curve, on which the impedance is plotted on the coordinate axis, and log ω– on the abscissa axis, formally, completely corresponds to the entropic nomogram (Figure 21-1).
   b) The fluctuations of biomembrane conductivity (conditioned by random processes) "have the form of Lorentz curve"[18] [p. 99]. In this graph, the fluctuation spectral density (ρ) is plotted on the coordinate axis, and the frequency logarithm function (log ω) on the abscissa axis.

The type of such curve also corresponds to the entropic nomogram in Fig. 1.

21.8 LORENTZ CURVE OF SPATIAL-TIMEDEPENDENCE

In Lorentz curve,[19] the space-time graphic dependence (Figure 21-4) of the velocity parameter (θ) on the velocity itself (β) is given, which completely corresponds to the entropic nomogram in Figure 21-2.

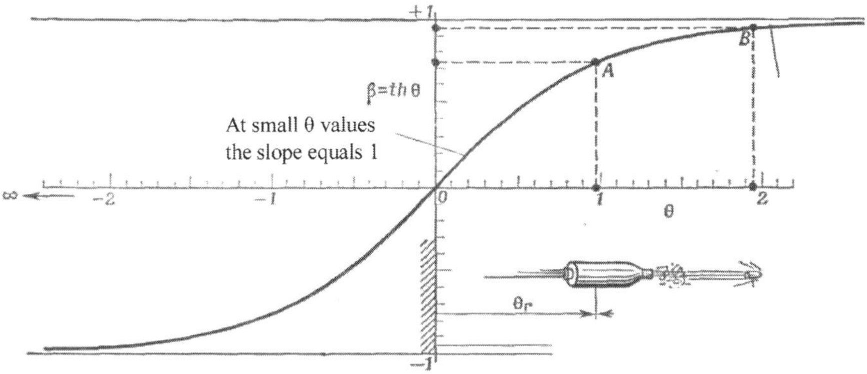

FIGURE 21-4 Connection between the velocity parameter θ and velocity itself β = thθ.

21.9 ENTROPIC CRITERIA IN BUSINESS AND NATURE

The main properties of free market providing its economic advantages are 1) effective competition, and 2) maximal personal interest of each worker.

But on different economy concentration levels these ab initio features function and demonstrate themselves differently. Their greatest efficiency corresponds to small business – when the number of company staff is minimal, the personal interest is stronger and competitive struggle for survival is more active. With increase in the number of companies and productions, the number of staff goes up, the role of each person gradually decreases, the competition slackens as new opportunities for coordinated actions of various business structures appear. The quality of economic relations in business goes down, that is, the entropy increases. Such process is mostly vivid in monostructures at the largest enterprises of large business (syndicates and cartels).

The concept of thermodynamic probability as a number of microstates corresponding to the given macrostate can be modified as applicable to the processes of economic interactions that directly depend on the parameters of business structures.

A separate business structure can be taken as the system macrostate, and as the number of microstates–number of its workers (N), which is the number of the available most probable states of the given business structure. Thus, it is supposed that such number of workers of the business structure is the analog of thermodynamic probability as applicable to the processes of economic interactions in business.

Therefore, it can be accepted that the total entropy of business quality consists of two entropies characterizing: 1) decrease in the competition efficiency (S_1) and 2) decrease in the personal interest of each worker (S_2), that is: $S = S_1+S_2$. S_1 is proportional to the number of workers in the company: $S \sim N$, and S_2 has a complex dependence not only on the number of workers in the company but also on the efficiency of its management. It is inversely proportional to the personal interest of each worker. Therefore, it can be accepted that $S_2 = {}^1\!/_\gamma$, where γ –coefficient of personal interest of each worker.

By analogy with Boltzmann's equation (1) we have:

$$S = (S_1 + S_2) \sim \left[\ln N + \ln\left(\frac{1}{\gamma}\right)\right] \sim \ln\left(\frac{N}{\gamma}\right)$$

or $S = k\ln\left(\frac{N}{\gamma}\right)$,

where k–proportionality coefficient.

Here, N shows how many times the given business structure is larger than the reference small business structure, at which N = 1, that is, this value does not have the name.

For nonthermodynamic systems when we consider not absolute but relative values, we take k = 1. Therefore,

$$S = \ln\left(\frac{N}{\gamma}\right) \tag{21-22}$$

In Table 21-2 we can see the approximate calculations of business entropy by the equation (21-22) for the three main levels of business: small, medium, and large. At the same time, it is supposed that number N corresponds to some average value from the most probable values.

When calculating the coefficient of personal interest, it is considered that it can change from 1 (one self-employed worker) to zero (0), if such worker is a deprived slave, and for larger companies it is accepted as $\gamma = 0.1–0.01$.

Despite of the rather approximate accuracy of such averaged calculations, we can make quite a reliable conclusion on the fact that business entropy, with the aggregation of its structures, sharply increases during the transition from the medium to large business as the quality of business processes decreases.

TABLE 21-2 Entropy growth with the business increase

Structure parameters	Business		
	Small	**Average**	**Large**
$N_1 - N_2$	10 – 50	100 – 1000	10000 – 100000
γ	0.9 – 0.8	0.6 – 0.4	0.1 – 0.01
S	2.408 – 4.135	5.116 – 7.824	11.513 – 16.118
(S)	3.271	6.470	13.816

In thermodynamics, it is considered that the uncontrollable entropy growth results in the stop of any macrochanges in the systems, that is, to their death. Therefore, the search of methods of increasing the uncontrollable growth of the entropy in large business is topical. At the same time, the entropy critical figures mainly refer to large business. A simple cut-down of the number of its employees cannot give an actual result of entropy decrease. Thus, the decrease in the number of workers by 10% results in diminishing their entropy only by 0.6% and this is inevitably followed by the common negative unemployment phenomena.

Therefore, for such supermonostructures controlled neither by the state nor by the society, the demonopolization without optimization (i.e., without decreasing the total number of employees) is more actual to diminish the business entropy.

Comparing the nomogram (Figure 21-1) with the data from the Table 21-2, we can see the additivity of business entropy values (S) with the values of the coefficient of spatial-energy interactions (α), that is, S = □.

It is known that the number of atoms in polymeric chain maximally acceptable for a stable system is about 100 units, which is 10^6 in the cubic volume. Then, we again have lg $10^6 = 6$.

21.10 S-CURVES ("LIFE LINES")

Already in the past century, some general regularities in the development of some biological systems depending on time (growth in the number of bacteria colonies, population of insects, weight of the developing fetus, etc.) were found,[20] The curves reflecting this growth were similar, first of all, by the fact that three successive stages could be rather vividly emphasized on each of them: slow increase, fast burst-type growth, and stabilization (sometimes decrease) of number (or another characteristic). Later, it was demonstrated that engineering systems go through similar stages during their development. The curves drawn up in coordinate system, where the numerical values of one of the most important operational characteristics (for example, aircraft speed, electric generator power, etc.) were indicated along the vertical and the "age" of the engineering system or costs of its development along the horizontal, were called S-curves (by the curve appearance), and they are sometimes also called "life lines".

As an example, the graph of the changes in steel specific strength with time (by years) is demonstrated[20] (Figure 21-5).

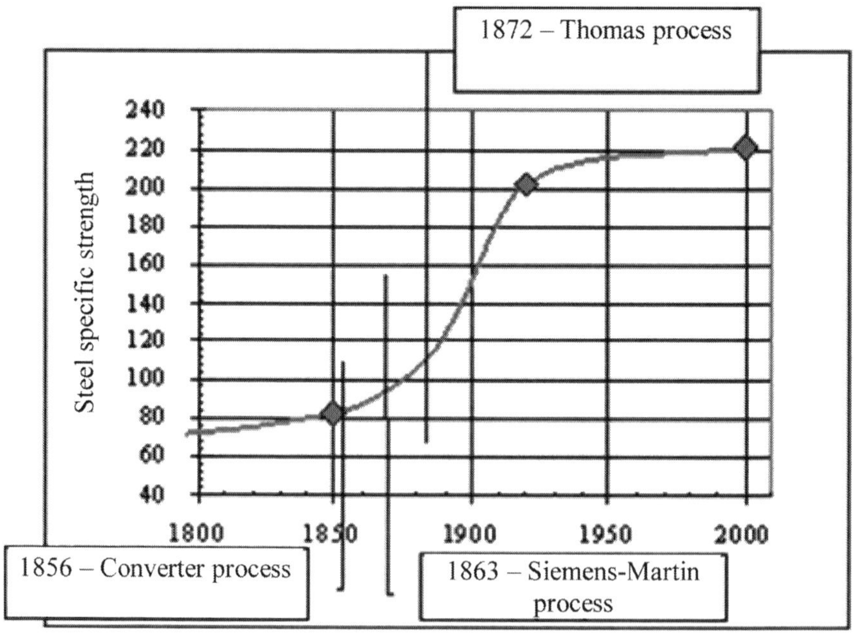

FIGURE 21-5 Dependence of steel specific strength on time.

Thus, the similarity between S-curves and entropic nomogram in Figure 21-2 is observed.

And in this case, the same as before, the time dependence (t) is proportional to the entropy reverse value ($1/\alpha$). As applicable to business, such curves characterize the process intensity, for example, sale of the given products.

At the same time, entropic nomograms in accordance with Figure 21-1 assess the business quality (ordinate in such graphs).

It is known that the entropy of isolated systems decreases. The entropy growth in open systems is compensated by the negative entropy due to the interaction with the environment.

All the aforementioned systems can be considered as open ones. This also refers to spatial-energy processes, when any changes in quantitative energy characteristics are conditioned by the interaction with external systems.

It is obviously observed in engineering and technological systems, the development of which is followed by additional innovations, modifications, and financial investments.

The entropy in thermodynamics is considered as the measure of nonreversible energy dissipation. From the point of technological and economic principles, the entropy is mainly the measure of irrational energy resource utilization. With the increase in time dependence, such processes stabilize in accordance with the nomogram to more optimal values—together with the growth of anti-entropy, that is, the value $1/\alpha=1/\rho$.

The similar growth with the time of rationality of technological, economic, and physical and chemical parameters proves that such nomograms are universal for the majority of main processes in nature, technology and economy.

21.11 GENERAL CONCLUSION

The idea of entropy is diversified in physical and chemical, economic, engineering, and other natural processes that is confirmed by their nomograms.

KEYWORDS

- **Biophysical Processes**
- **Business**
- **Engineering Systems Entropy**
- **Nomogram**
- **Spatial-Energy Parameter.**

REFERENCES

1. Reif, F. Statistic physics. M.: Nauka, 1972, 352 p.
2. Gribov, L. A.; Prokofyeva, N. I. Basics of physics. M.: Vysshaya shkola, 1992, 430 p.
3. Rubin, A. B. Biophysics. Book 1. Theoretical biophysics. M.: Vysshaya shkola, 1987, 319 p.
4. Blokhintsev, D. I. Basics of quantum mechanics. M.: Vysshaya shkola, 1961, 512 p.
5. Yavorsky, B. M.; Detlaf, A. A. Reference-book in physics. M.: Nauka, 1968, 939 p.
6. Christy, R. W.; Pytte, A. The structure of matter: an introduction to modern physics. Translated from English. M.: Nauka, 1969, 596 p.
7. Korablev, G. A. Spatial-energy principles of complex structures formation. Netherlands, Brill Academic Publishers and VSP, 2005, 426p. (Monograph).
8. Eyring, G.; Walter, J.; Kimball, G. Quantum chemistry. M., F. L., 1948, 528 p.
9. Fischer, C. F. Atomic Data, 1972, V. 9, № 4, 301–399.
10. Waber, J. T.; Cromer, D. T. J.Chem. Phys, 1965, V. 42, № 12, 4116–4123.
11. Clementi, E.; Raimondi, D. L. Atomic screening constants from S.C.F. functions, 1. J.Chem. Phys., 1963, V.38, №11, 2686–2689.
12. Clementi, E.; Raimondi, D. L. J. Chem. Phys., 1967, V. 47, № 4, 1300–1307.
13. Gombash, P. Statistic theory of an atom and its applications. M.: I.L., 1951, 398 p.
14. Clementi, E. J.B.M. S. Res. Develop. Suppl., 1965, V. 9, № 2, 76.
15. Korablev, G. A.; Zaikov, G. E. J. of Applied Polymer Science, USA, 2006, V.101, №.3, 2101–2107.
16. Korablev, G. A.; Zaikov, G. E. Progress on chemistry and biochemistry. Nova Science Publishers, Inc. New York, 2009, 355–376.
17. Kodolov, V. I.; Khokhriakov, N. V.; Trineeva, V. V.; Blagodatskikh, I. I. Activity of nanostructures and its manifestation in nanoreactors of polymeric matrixes and active media. Chemical physics and mesoscopy, 2008. V. 10. №4. p. 448–460.
18. Rubin, A. B. Biophysics. Book 2. Biophysics of cell processes. M.: Vysshaya shkola, 1987, 303 p.
19. Taylor, E.; Wheeler, J. Spacetime physics. Mir Publishers. M., 1987, 320 p.
20. Kynin, A. T.; Lenyashin, V. A. Assessment of the parameters of engineering systems using the growth curves. http://www.metodolog.ru/01428/01428.html.

Information about the authors

1. Grigory Andreevich Korablev – Professor, Doctor of Science in Engineering, Head of the Physics Department of Izhevsk State Agricultural Academy, e-mail: korablevga@mail.ru
2. Vladimir Ivanovich Kodolov, Doctor of Science in Chemistry, Professor, Head of the Department of Chemistry and Chemical Engineering of Kalashnikov Izhevsk State Technical University, 426000, Izhevsk, Studencheskaya St., 7, tel.: (3412) 582438, e-mail: kodol@istu.ru.
3. Gennady Efremovich Zaikov – Doctor of Science in Chemistry, Professor of N.M. Emmanuel Institute of Biochemical Physics, RAS, e-mail: chembio@sky.chph.ras.ru

INDEX

G

H

I

K

L

M

N

V

W

X

Y

Z